CW01639498

About BPEC

BPEC has become an industry leading provider of qualifications, assessments and learning materials through working closely with a UK wide network of colleges, private training centres and industry stakeholders within the Building Services Engineering (BSE) sector.

BPEC'S aim is to be trusted as the natural choice to support learning and development. Whether people are embarking on a career or currently working in the sector, we want to be a key part of their 'Learning for Life' experience.

Equal Opportunities

BPEC fully supports the principle of equal opportunities and we are committed to satisfying this principle in all BPEC activities and published material.

Copyright

First Edition: October 2019 Editor: BPEC Ltd

Author: Michael Maskrey Design: BPEC Ltd

Publications

Every effort has been made to ensure that the information contained in this publication is true and correct at the time of going to press. However, BPEC products and services are subject to continuous development and improvement and the right is reserved to change products and services from time to time.

E: services@bpec.org.uk

T: 01332 376000

BPEC, 1-2 Mallard Way, Pride Park, Derby, DE24 8GX

BPEC Plumbing series

This series has been developed alongside the BPEC Level 3 Diploma in Plumbing and Domestic Heating. This qualification sits within the new Plumbing and Domestic Heating Technician (Level 3) Apprenticeship Standard.

Disclaimer

BPEC do not accept any responsibility for any acts or omissions on the part of any person installing, servicing or commissioning plumbing and heating installations, which may arise from the use of this learning material. The material and information contained in this learning manual is for the purpose of study and reference only which may lead to gaining a recognised plumbing qualification, it is not a licence to practise.

The material contained within this manual was correct at the time of publication.

BPEC do not accept any liability if this learning material is used for any alternative purpose from that which it is intended.

Plumbing Core Units

Phase 2

The Level 3 Diploma in Plumbing and Domestic Heating is divided into a series of phases. Within each phase, there are a series of units that must be completed successfully before progressing to the next phase. At the end of each phase you must successfully complete a series of practical and theory based assessments.

In phase 2, you will complete selected Assessment Criteria from a range of Learning Outcomes. These Learning Outcomes and the relevant Assessment Criteria are listed below:

Unit 03

Scientific Principles

Scientific Principles

Science, and physics and chemistry in particular, play a very significant role within the plumbing industry. Every task we perform, every type of material we use, and every calculation we make has its roots in a scientific principle.

Understanding key scientific principles is crucial if we are to design and install safe cold water, hot water, central heating and sanitation systems that work effectively and efficiently, contributing to a safer and healthier way of life for every man, woman and child in the United Kingdom.

In this module, we will look at some key principles of chemistry that enables us to assess the correct materials for the systems we install and the principles of physics that make our systems work. In addition, we will investigate the SI system of measurement and its derivatives and briefly look at some simple principles of electrical theory. There are 6 Learning Outcomes for this module: However this book will cover:

LO2: Understand properties of materials

- AC2.7. Describe basic **properties of gases**

LO3: Understand the relationship between energy, heat and power

- AC3.1. Identify the relationship between the Celsius and Kelvin temperature scales
- AC3.2. Describe the principles associated with a **change of state**
- AC3.3. Identify the terms latent and sensible heat as they apply to liquids and gases
- AC3.4. Define methods of **heat transfer**
- AC3.5. Distinguish how **units of energy and heat** are related and derived
- AC3.6. Carry out heat, energy and power **calculations**

LO4: Understand principles of force and pressure and their application in the plumbing and heating industry

- AC4.1. State **how units of force and pressure** are derived from SI units
- AC4.2. Identify **pressure and flow** rate units of measurements
- AC4.3. Describe the application of pressure and flow rate measurements
- AC4.4. Carry out simple **force and pressure** calculations
- AC4.5. Explain the relationship between **velocity, pressure and flow rate** in systems
- AC4.6. Identify how **restrictions** in the pipework affects the flow of liquids and gases
- AC4.7. Describe the principles of a siphon

LO6: Understand principles of electricity in the plumbing and heating industry

- AC6.1. Identify **basic principles** of electron flow theory
- AC6.2. Describe the purpose and application of simple **units of electrical measurement**
- AC6.3. Carry out simple **electrical calculations**
- AC6.4. Identify the requirements for earthing of electrical circuits

LOs 1 and 5 you will find in the BPEC phase 1 textbook

Learning Outcome 2

Polybutylene
Low Carbon Steel
Ceramics
Polyethylene
Copper
Brass
Stainless Steel
Vitreous China
Gun Metal
Polypropylene

Understand Properties of Materials

AC2.7 Describe the basic properties of gases

All gases (with exception of steam) are liquids that have boiled and evaporated at ultra-low temperatures. In normal atmospheric temperature and pressures, these liquids exist in their gaseous state. Here we will look at gases from two perspectives:

1. The gases and their properties
2. Gas laws

1. The gases and their properties

Air

The use of air within plumbing is limited:

a) It is used for heating in warm air systems. The air is warmed by a warm air heater, usually gas-fired, and then distributed throughout the property by ductwork.
b) It is used as a pressure charge in expansion vessels installed on sealed heating systems and unvented hot water systems.
c) It is often used to clear blocked drains when administered at very high pressures.

Steam

The use of steam as a heating medium has declined over recent years. However, steam is now used in many modern applications:

a) Combined Heat and Power (CHP). With these systems, the steam is used to generate electricity and the excess steam is used to heat a district heating system.
b) Steam is still used for heating, especially for large industrial systems.
c) Hot water production using large hot water calorifiers for industrial applications.
d) Electricity generation.

Liquid Petroleum Gas (LPG)

LPG is a flammable gas that is heavier than air. Most LPG is a by-product of oil production. In its clean state, LPG can be used in areas where access to the mains Natural Gas is difficult. It is used to heat homes via LPG boilers and gas fires, and for cooking appliances. Plumbers also use LPG with their blowtorches. There are two basic types:

- **Butane** – mainly used as a camping gas. It has a high calorific value and a high temperature flame. However, it suffers from the fact that its boiling point is rather higher at -5°C. This means that on very cold days, the liquid butane will not expand to a gas.

- **Propane** – this does not suffer the problems of butane, as its boiling point is -45°C. Its calorific value is slightly less than butane with a slightly cooler flame. However, its ability to boil at low temperatures makes this the most widely used LPG type.

Natural gas

Natural gas is the most widely used fuel for home and business heating and cooking in the UK. It is primarily a mixture of a number of gases:

- Methane CH_4 (92%)
- Ethane C_2H_6 (3%)
- Propane C_3H_8 (3%)
- Butane C_4H_{10} (1%)
- Nitrogen N_2 (1%)
- Carbon dioxide CO_2 (trace)
- Hydrogen Sulphide H_2S (trace)

Four of these gases (methane, ethane, propane and butane) are flammable. Nitrogen is inert. There are also traces of carbon dioxide (CO_2) and hydrogen (H) mixed in.

Natural gas is always found when drilling for oil. The gas is extracted from deep within the earth, pumped ashore and cleaned of a waxy oil deposit called Naptha. At the refinery, the distinct smell is added. This is a chemical called mercaptan and is necessary because NG has no natural smell.

NG is lighter than air having a specific gravity of 0.6 – 0.8 of that of air.

Carbon dioxide

Carbon dioxide (CO_2) is used as a freezing agent with pipe freezing kits. It is also a very effective method of firefighting when used in fire extinguishers.

Refrigerant gases

Please refer to AC2.5 - Describe principle applications and basic properties of liquids (2) for information regarding refrigerants in liquid and gaseous states.

2. Gas Laws

Gases behave very differently from the other two states of matter, liquids and solids. Gases have neither a fixed shape nor a fixed volume. Their shape is determined by the container in which they are held. We measure gases by three variables:

a) Pressure

b) Temperature

c) Volume

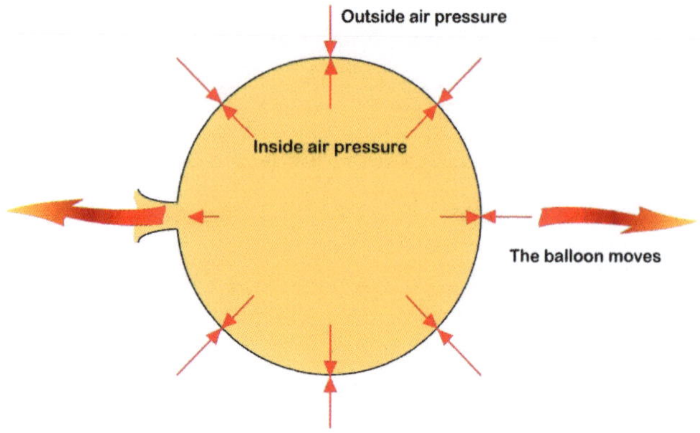

high pressure to low pressure movement

Pressure

The gas exerts this force on the walls of its container equally. An example of this is an inflated balloon. The balloon expands because the gas inside is at a higher pressure than the air that surrounds it. The balloon expands equally because the pressure inside is acting equally across the whole of the balloon's surface. If the balloon is released, the air, because of its higher pressure, will move quickly from the area of high pressure inside to the area of low pressure outside. This action will create a forward motion.

Temperature

The temperature of a gas is a measure of its mean kinetic energy. Gas molecules are in constant but random motion. The greater the motion, the higher the temperature. As the temperature decreases, the kinetic energy and therefore the motion decreases.

Volume

The volume of a gas is affected by both pressure and temperature. The pressure is constant and stable if the temperature is constant and stable. If the temperature either rises or falls, then both pressure and volume increase or decrease.

These three factors (pressure, temperature and volume) form the basis for two very important gas laws:

- Charles' Law
- Boyles' Law

Charles' Law

In 1802 Jacques Charles discovered that if a known quantity of gas is held at a constant pressure and temperature, then its volume will also remain constant. If the temperature is raised, then its pressure and volume also increase.

Gases expand when they are heated and contract when they are cooled. Consider the following analogy:

> ***A sealed copper tube, pressurised to 20mb at room temperature, is placed in direct sunlight where the pipe can warm up. As the gas inside warms, it expands and the pressure inside rises. When the pipe is taken away from the sunlight, the pipe cools to its original temperature and the gas inside contracts to its original volume and therefore its original pressure.***

This analogy shows that the rise in pressure is inversely proportional to the rise in temperature. The mathematical equation for this analogy is as follows:

$$V_1 \div T_1 = V_2 \div T_2$$

Where:

V = Volume

T = Temperature

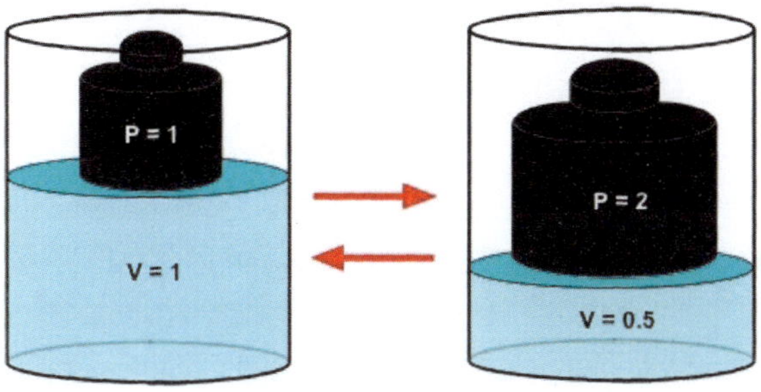

The Volume of a gas is inversely proportional to its pressure

Volume reduced = pressure increased

Boyles' Law

Boyles' Law, discovered by Robert Boyle in 1662, states that a volume of gas at a fixed temperature is inversely proportional to the pressure exerted on it. In other words, if $1m^3$ of gas at 1bar pressure is compressed into a space 0.5m^3 then its pressure will double to 2 bar. This happens because the molecules of gas have less room to move because the volume that the gas has to occupy has halved. This can be expressed as a mathematical equation:

$$P_1 \, x \, V_1 = P_2 \, x \, V_2$$

Where:

P_1 = starting pressure
V_1 = starting volume
P_2 = finishing pressure
V_2 = finishing volume

The table below illustrates the point

Table 1 - Sample of Gas at Constant Temperature and Varying Pressure				
Test	Pressure	Volume	Formula	Total
1	25 kPa	100 cm^3	P x V = K	25 x 100 = 2500
2	50 kPa	50 cm^3	P x V = K	50 x 50 = 2500
3	100 kPa	25 cm^3	P x V = K	100 x 25 = 2500
4	200 kPa	12.5 cm^3	P x V = K	200 x 12.5 = 2500
5	400 kPa	6.25 cm^3	P x V = K	400 x 6.25 = 2500

Learning Outcome 3

Understand the Relationship between Energy, Heat and Power

LO3: Understand the Relationship between Energy, Heat and Power

There are six assessment criteria within Learning Outcome 3.

AC3.1. Identify the relationship between the Celsius and Kelvin temperature scales

AC3.2. Describe the principles associated with a change of state

AC3.3. Identify the terms latent and sensible heat as they apply to liquids and gases

AC3.4. Define methods of heat transfer

AC3.5. Distinguish how units of energy and heat are related and derived

AC3.6. Carry out heat, energy and power calculations

Energy, heat and power are closely linked. One cannot exist without the other two. The first law of thermodynamics states that:

'Energy can neither be created nor destroyed. Energy is transferred from one form to another.'

In other words, energy, heat and power are all related. Where there is one, there will always be another. Consider the following:

- **The unit of power** - The watt is the SI unit for power. 1 Watt = 1 joule per second (1 J/s). In electrical units it is one volt ampere (1 V·A) (DC circuits).
- **The unit of heat** - The joule is unit of heat. 4.186 joule of heat energy = 1 calorie. 4.186j is needed to raise the temperature of 1 g of water from 0°C to 1°C. 4.186j is also known as the Specific Heat Capacity of Water.
- **The unit of energy** – see joule (above).
- **Specific Heat Capacity (SHC)** – The Specific Heat Capacity is the amount of heat needed to change the temperature of a mass of a substance by one degree in temperature. It is measured in kilojoules per kilogram per degree Celsius (kJ/kg/°C). The SHC will vary depending on the substance.

AC3.1 Identify the relationship between the Celsius and Kelvin temperature scales

Temperature can be defined as:

The degree or intensity of heat present in a substance or object, especially as expressed according to a comparative scale and shown by a thermometer or perceived by touch.

In other words, temperature is a measurement of how hot or cold an object is. It is measured using a thermometer, of which there are many different types. A thermometer measures how much energy is contained within the object. Temperature is usually expressed in terms of units or degrees on a standard scale, usually Celsius (°C) or Kelvin (K). In some parts of the world the Fahrenheit (°F) scale is used.

- Celsius (°C) – Increments of 1 degree (1°). Celsius is the most widely used unit of temperature in the plumbing and heating industry. It has a zero point (0°C) that relates to the temperature at which water freezes. The degree symbol (°) should be used against the temperature i.e. 18°C.
- Kelvin (K) – again, Kelvin has the same increments as the Celsius scale. This means that if the temperature rises or falls by 1°C, then the temperature also rises or falls by 1K. However, the Kelvin scale has a minimum temperature that corresponds to the point at which all molecular actvity stops. 0 K is known as absolute zero and is equal to -273°C.

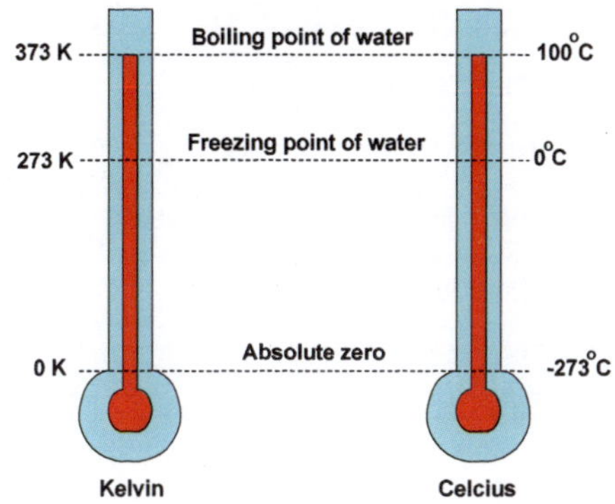

Kelvin/Celcius relationship

Therefore:

- Zero K (0K) = -273°C

  or,

- 0°C = 273K

AC3.2 Describe the principles associated with a change of state

Within this Assessment Criteria, the range will cover:

- Melting
- Freezing
- Boiling
- Evaporating
- Condensing

The States of Matter

Everything on Earth and in space is made of matter. Matter can exist in 3 states:

1. Liquid
2. Solid
3. Gas

Each of these states can be changed by the addition or removal of heat. For example, if heat is applied to ice (solid), it becomes water (liquid) and if more heat is applied, the water will evaporate to become steam (gas). These transitions will also work in reverse when heat is taken away – steam to water to ice. Each of these phases is given a name:

State	Given name
Ice (solid) to water (liquid)	Melting
Water (liquid) to steam (gas)	Boiling
Water (liquid) to steam (gas)	Evaporation/vaporisation (below boiling point)
Steam (gas) back to a water (liquid)	Condensation
Water (liquid) to ice (solid)	Freezing (solidification)
Under certain conditions, it is possible to miss out certain states. For instance, ice under controlled conditions will go straight from ice to a gas, completely missing out the liquid water stage:	
Solid to gas	Sublimation
Gas to solid	Deposition

Melting

When a solid is heated, the particles are given more energy and start to vibrate faster. At a certain temperature, the particles vibrate so much that their ordered structure breaks down. At this point the solid melts into liquid. The temperature at which this change from solid to liquid happens is called the melting point. Each solid has a set melting point at normal air pressure. At lower air pressure, such as up a mountain, the melting point lowers.

Boiling

When a liquid is heated, the particles are given more energy. They start to move faster and further apart. At a certain temperature, the particles break free of one another and the liquid turns to gas. This is the boiling point. The boiling point of a substance is always the same; it does not vary.

Evaporation

Even without boiling water in a kettle, some of the liquid water changes to gas. This is evaporation. It occurs when a liquid turns into a gas far below its boiling point. There are always some particles in a liquid that have enough energy to break free from the rest to become a gas.

Condensation

Dewdrops are often found on a spider's web early in the morning after a cold night. Water that is present as a gas in the air cools down and changes into tiny drops of liquid water on leaves and windows. This change from gas to liquid is called condensation.

Freezing

Lava is liquid rock which erupts through a volcano at temperatures as high as 1,500ºC (2,732ºF) through a volcano. However, the red-hot lava cools as it meets the Earth's surface, and turns back into solid rock again. This change from liquid to solid is called freezing or solidifying. It is the opposite process to melting.

Solids

The table shows some of the properties of solids:

Property	Why	
Solids have a fixed shape and cannot flow	The particles cannot move from place to place	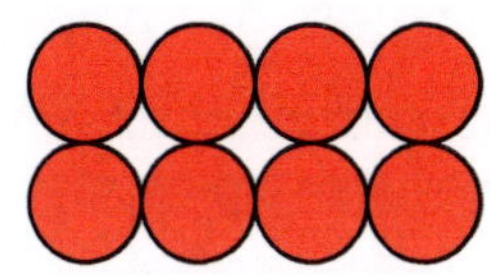
Solids cannot be compressed or squashed	The particles are close together and have no space to move into	

Liquids

The table shows some of the properties of liquids:

Property	Why	
Liquids flow and take the shape of their container	The particles can move around each other	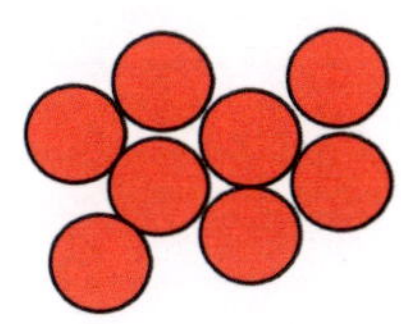
Liquids cannot be compressed or squashed	The particles are close together and have no space to move into	

Gases

The table shows some of the properties of gases:

Property	Why	
Gases flow and take the shape of their container	The particles can move quickly in any direction	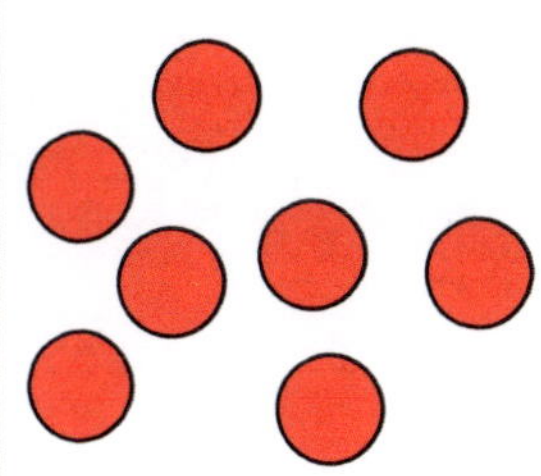
Gases can be compressed or squashed	The particles are far apart and can move into any space	

AC3.3 Identify the terms Latent and Sensible Heat as they apply to Liquids and Gases

What is the difference between latent heat and sensible heat?

Latent heat and sensible heat are types of energy released or absorbed in the atmosphere. Latent heat is related to changes in phase between liquids, gases, and solids. Sensible heat is related to changes in temperature of a gas or object with no change in phase.

Sensible heat	Latent heat
When heat is directly applied to water, its temperature will rise until it reaches its boiling point. The resulting increase in the heat of the water is known as sensible heat because the application of heat has led to hotter water: **Water @ 0°C to water @ 100°C = a change in temperature due to the application of heat.** <u>There is no change of state</u> The process is reversible	Changes of state are also a result of heat added, but here, the added heat leads to a change in state without an increase in temperature. Latent heat does not affect the water's temperature. For example: **Water @ 100°C to Steam at 100°C = a change in state due to the application of heat.** <u>There is no change in temperature</u> The process is reversible

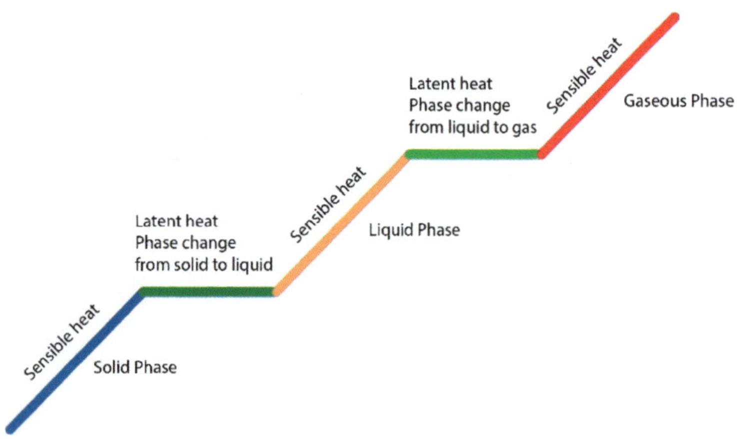

Sensible to Latent temperature

Heat energy is needed to increase the temperature of an object. The amount of energy needed depends on the mass of the object, the type of material it is made from, and the temperature increase.

Heat energy is also absorbed when substances melt or boil, but the temperature does not alter during a change of state. The amount of energy needed to melt or boil something depends upon the mass of the object and the type of material it is made from. Take a look at the table below. It shows how much heat in kilojoules per kilogram that are needed to change 0°C of water to 100°C of steam.

This table shows how much energy is required to raise 1kg of water from 0°C to 100°C and change its state into steam					
Pressure		Temp	Water *Sensible heat*	Evaporation to Steam *Latent heat*	Total Heat Required
bar	kPa	°C	kJ/kg	kJ/kg	kJ/kg
0	0	100.00	419.0	2257.0	2676.0

AC3.4 Define Methods of Heat Transfer

Within this Assessment Criteria, the range will cover:

- Conduction
- Convection
- Radiation

Heat can be transferred from one place to another by three methods: **conduction** in solids, **convection** of fluids (liquids or gases) and **radiation** through anything that will allow radiation to pass. The method used to transfer heat is usually the one that is the most efficient.

Conduction

Conduction occurs when heat travels ***through*** a substance, the heat passing directly from one molecule to another.

If a piece of copper tube is heated at one end, the heat will quickly be felt at the other end. This occurs because kinetic energy in the form of heat is directly transferred from one copper atom to another very quickly.

When the copper is cold, the molecules move slowly but, as heat is applied, the atoms gain energy and collide with other atoms. The heat (or energy) is quickly passed from atom to atom, the cooler atoms gaining energy to collide with other cooler atoms.

Thermal Conductivity of common materials	
Material	**Thermal Conductivity W/m/K**
Silver	406.0
Copper	385.0
Gold	310.0
Brass	109.0
Steel	50.2
Lead	34.7
High Density Polyethylene	0.5
Wood	0.04 – 0.12
Expanded Polystyrene	0.03

Not all materials conduct heat at the same rate and some do not conduct heat at all. Those materials that are not good conductors, such as wood and plastics, are known as insulators.

The same principle can be applied to the conduction of electricity, as good conductors of heat are generally also good conductors of electricity. The rate at which a material will conduct heat is known as the coefficient of thermal conductivity. It is measured in W/m^2K.

$$Thermal\ conductivity\ = \frac{heat\ x\ distance}{area\ x\ temperature\ difference\ (\Delta t)} = W$$

Convection

Convection occurs when heat is transferred through a fluid i.e. a gas or a liquid. Convection takes place because the warm fluid is less dense than the cool fluid. As the lighter, warm fluid rises, it releases heat into the surrounding environment. When this happens, the warm fluid loses its heat, becomes denser and falls due to gravity, to begin the process again.

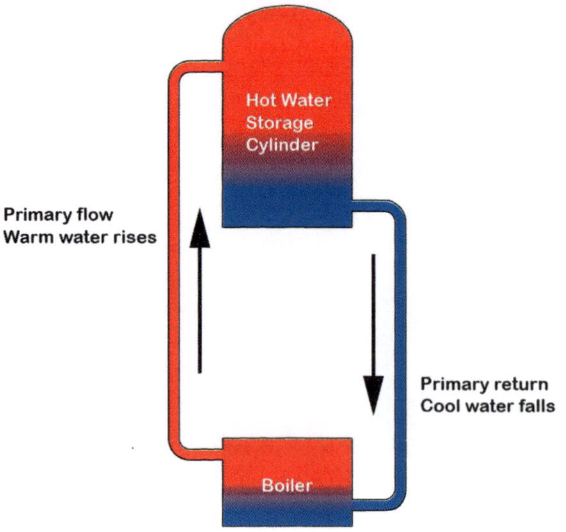

How convection works

In a hot water system, this is known as gravity circulation as hot water rises and cooler water moves down to replace it, causing a circular motion.

Modern convector/radiators use two methods of heat transfer to warm rooms and spaces. One of these methods is convection; the convector/radiator releases warm air out at the top and draws in cooler air at the bottom. The other method is radiation, which is described below.

Radiation

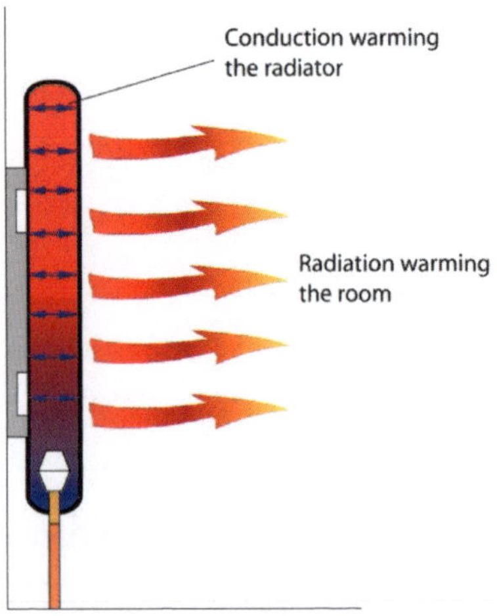

How radiation warms a room

All objects will transmit and receive heat in the form of thermal radiation (often called infrared radiation). The hotter an object is the more thermal radiation it will emit.

Thermal radiation is a type of electro-magnetic radiation and, unlike conduction and convection, the heat does not pass from atom to atom. Instead, it travels in waves.

On Earth, we get the heat to warm our planet from the Sun. The Sun's heat travels over 90,000,000 miles through the cold vacuum of space to warm the Earth's surface. The heat is felt because it travels in waves without heating the space between them.

We use the same principle with central heating radiators even though there is no visible flame. The heat from the hot water in the heating system heats the steel radiator through conduction and the radiator heats the room by radiation.

AC3.5 Distinguish how units of Energy and Heat are Related and Derived

Within this Assessment Criteria, the range will cover:

- Energy – Joules (J)
- Specific heat capacity (kJ/kg/°C)
- Power – Watts (W)
- Maximum density
- Coefficient of Linear Expansion

SI units of energy, heat and power

- **Energy – Joules (J)** - the SI unit of work or energy, equal to the work done by a force of one newton when its point of application moves one metre in the direction of the force. It is equivalent to one 3600th of a watt-hour.
- **Specific Heat Capacity (kJ/kg/°C)** – This is the amount of energy required (in joules or kilojoules) to increase the temperature of 1kg of a substance by 1°C. Water, for example, has a SHC of 4.19 kJ/kg/°C. This means that it takes 4190 joules of energy to raise the temperature of 1kg of water by 1°C.
- **Power – Watts (W)** - The watt (abbreviated W) is the International System of Units standard unit of power (or energy per unit time). It is the equivalent of one joule per second. The watt is used to specify the rate at which energy is dispersed, or the rate at which electro-magnetic energy is radiated, absorbed, or dissipated. The watt can be used to describe electrical power or heat power.
- **Maximum Density** – The maximum density of a substance is the highest attainable density of the substance under given conditions. Nearly all substances therefore reach a density maximum at very low temperatures and very high pressures. Water has its maximum density at 4 degrees Celsius. When the temperature changes from either greater or less than 4 degrees, the density will become less.
- **Coefficient of Linear Expansion** - This is how the size of an object changes with a change in temperature. Different materials expand by different amounts; just how much a particular material expands depends on the forces of cohesion between neighbouring atoms or molecules.

AC3.6 Carry out Heat, Energy and Power Calculations

Within this Assessment Criteria, the range will cover:

- Quantity of heat energy required to raise the temperature of a substance
- The amount of power required to heat a substance

To recap, the S.I. units of measurement of energy, heat and power are:

- Energy — The Joule (J)
- Heat — The Joule (J)
- Power — The Watt (W)
- Specific Heat Capacity — kilojoules per kilogram per degree Celsius (kJ/kg/°C)

REMEMBER! The Specific Heat Capacity of water is 4.19kJ/kg/°C

Calculations using the Specific Heat Capacity of Water

Example 1 - How many kilojoules would it take to heat 200 litres of water from 40°C to 80°C?

The formula for this is:

$$L \; x \; \Delta t \; x \; SHC \; of \; water$$

Where:

L = litres

Δt = temperature difference

SHC of water = 4.19

Therefore:

$$200 \; x \; (80 - 40) \; x \; 4.19 = \mathbf{33520kJ}.$$

Example 2 - This equation can be further developed to calculate how many kilowatts it would take to raise the temperature of the 200 litres water by 40°C. To do this, we need to include an amount of time (ie 1 hour, 2 hours, 3 hours etc). Assume that you require 200 litres of water at 80°C in 2 hours' time. The calculation would then become:

$$\frac{L \; x \; \Delta t \; x \; SHC \; of \; water}{Time \; (in \; seconds)}$$

Where:

L	=	litres
Δt	=	temperature difference
SHC of water	=	4.19
2 hrs in secs	=	2 x (60 x 60) = 7200

Therefore:

$$\frac{200\ x\ (80-40)\ x\ 4.19}{7200} = 4.65\ kW$$

This shows that to increase the temperature of 200 litres of water from 40°C to 80°C in 2 hours would take 4.65 kW of heat.

This can be even taken a step further.

Example 3 – How many hours would it take for 200 litres of water to be heated by 40°C using a 6kW of power?

REMEMBER! Water has a specific heat capacity of 4.19 kJ/kg/°C and that 1 W = 1 J/s.

The formula for this is:

$$\frac{l\ x\ t\ x\ SHC}{kW}$$

Where:

L	=	litres
t	=	Temperature
kW	=	kilowatts
SHC	=	Specific Heat Capacity

Therefore:

$$\frac{200\ x\ 40\ x\ 4.19}{6} = 5586.666667\ seconds$$

To convert seconds into hours:

$$\frac{seconds}{60\ x\ 60} = hours$$

Therefore:

$$\frac{5586.666667}{3600} = 1.55hrs\ or\ 1hr\ 33\ minutes$$

Learning Outcome 4

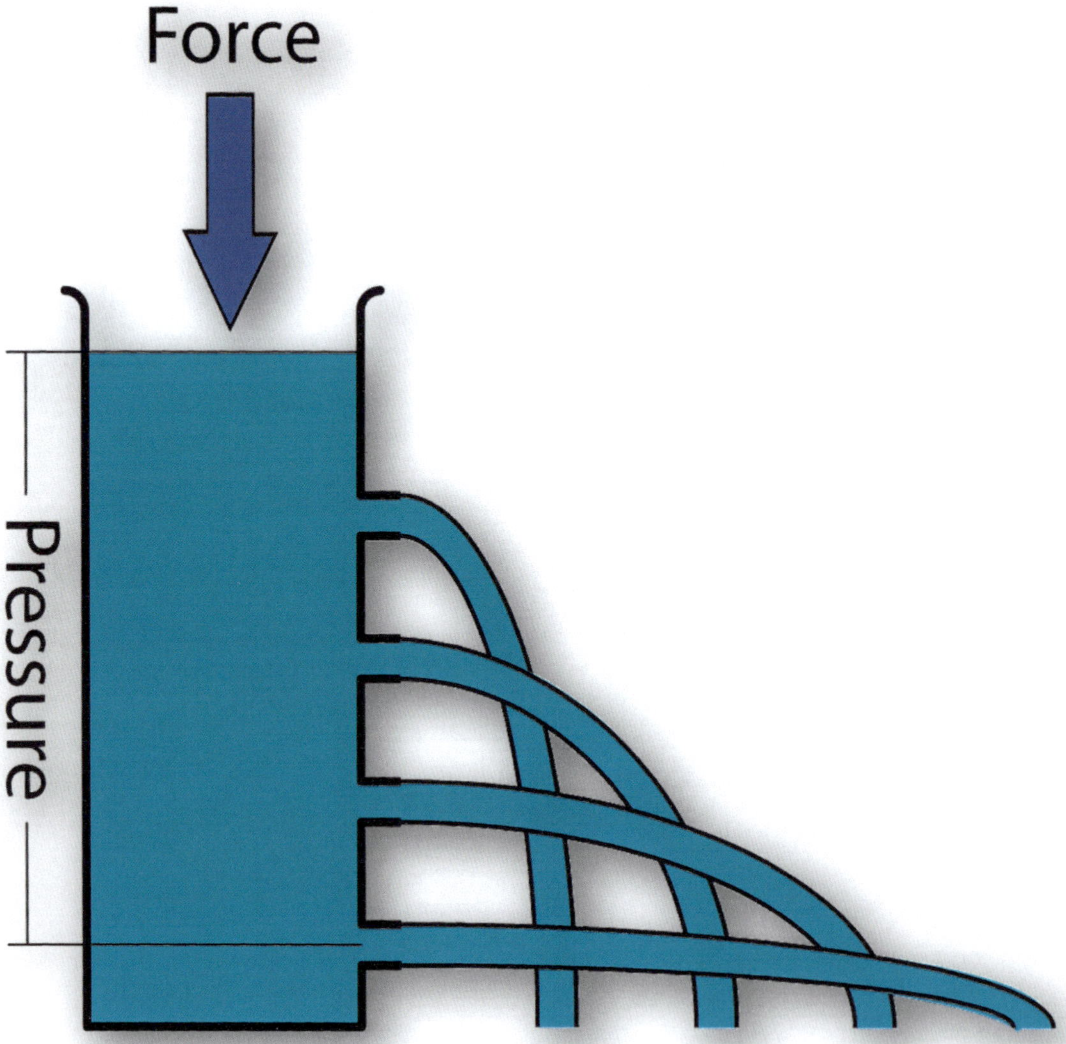

Understand the Principles of Force & Pressure and their Application in the Plumbing and Heating Industry

LO4: Understand the Principles of Force and Pressure and their Application in the Plumbing and Heating Industry

There are seven assessment criteria within Learning Outcome 4.

- AC4.1. State how units of force and pressure are derived from SI units
- AC4.2. Identify pressure and flow rate units of measurements
- AC4.3. Describe the application of pressure and flow rate measurements
- AC4.4. Carry out simple force and pressure calculations
- AC4.5. Explain the relationship between velocity, pressure and flow rate in systems
- AC4.6. Identify how restrictions in pipework restricts the flow of liquids and gases
- AC4.7. Describe principles of a siphon

Force, pressure, flow rate and velocity play a major part in plumbing systems. To fully understand how plumbing and heating systems work, we must first master the difference between flow rate and pressure. These subjects, although very different, are interlinked. Where there is one, you will always find the other.

In this part of the science unit, we will investigate those elements that are fundamental in plumbing theory – Force, Pressure, Velocity and Flow rate. We will also look at how these elements can be adversely affected and learn how to conduct simple pressure and flow rate calculations.

AC4.1 State how units of Force and Pressure are derived from SI units

Within this Assessment Criteria, the range will cover:

- Acceleration (m/s^2)
- Force due to gravity
- Force - Newton (N)
- Pressure (N/m^2)
- Atmospheric pressure
- Flow rate (m^3/s)

The table below shows the derived units of force and pressure:

Units of force and pressure			
Velocity	The rate and direction of the change in the position of an object	m/s	Metres per second
Acceleration	Is defined as the rate at which an object changes its velocity	m/s^2	Metres per second squared
Flow rate	The amount of fluid that flows in a given time	m^3/s	Metres cubed per second
Force	A force is a push or pull upon an object that causes a change in the motion of that object	N	Newton (equal to $kg/m/s^2$)
Force due to gravity	The force of gravity is the force exerted by the gravitational field of a massive object on anybody within the vicinity of its surface.	Fg	
Pressure	Pressure is an expression of force exerted on a surface per unit area.	pa	Pascal (equal to N/m^2)
Atmospheric Pressure	An atmosphere (atm) is a unit of pressure based on the average atmospheric pressure at sea level. The actual atmospheric pressure depends on many conditions, e.g. altitude and temperature.	atm	Bar/Pascal/Millibars

AC4.2 Identify Pressure and Flow Rate Units of Measurements

Within this Assessment Criteria, the range will cover:

Pressure

- Bar / millibar
- kPa
- Psi
- Metres head

Flow rate

- m^3/s
- l/s
- kg/s

Pressure

The standard unit of pressure is the pascal. However, other units, such as pounds per square inch (psi), bar and metres head, may also be used when describing pressure. The relationship between these four units is shown in the table below:

Unit of Pressure Conversion Chart			
Kilopascals (Kpa)	**Bars**	**Metres head of water**	**Pounds per square inch (p.s.i)**
10	0.1	1	1.42
20	0.2	2	2.84
30	0.3	3	4.27
40	0.4	4	5.68
50	0.5	5	7.11
100	1	10	14.22
150	1.5	15	21.33
200	2	20	28.44
250	2.5	25	35.55
300	3	30	42.66
350	3.5	35	49.77
400	4	40	56.88
450	4.5	45	63.99
500	5	50	71.10

Flow rate

In plumbing and heating, the volumetric flow rate, is the volume of fluid, which passes per unit time. The SI unit is m^3/s: in other words, the volume of water (in m^3) that flows every second (s).

However, since we know that $1m^3$ of water has a mass of 1000kg and that 1 kg of water is equal to 1 litre, then the flow rate can also be represented in various other ways too:

- m^3/s
- l/s
- kg/s

For example, if a tap discharges $0.3m^3/s$, then:

$$0.3m^3 = 300\ litres = 300kg$$

In this case, the flow rate could be quoted as $0.3m^3/s$ or 300 l/s or 300 kg/s.

AC4.3 Describe the Application of Pressure and Flow Rate Measurements

Pressure and flow rate are the key points that any plumbing system will be measured by. If the flow rate or the pressure is poor, then this can be the result of poor design, poor installation or both. Good plumbing design and installation will deliver the desired pressure and flow rate for a given specification. However, we must also be aware that too much pressure or too high a flow rate can be detrimental as they can cause noise and erosion problems in pipework and fittings and this may lead to breakdown or failure of the pipework or components. In heating systems, too much flow rate can be detrimental as this causes unwanted noise.

We will examine pressure and flow rate later in this section.

AC4.4 Carry out Simple Force and Pressure Calculations

Within this Assessment Criteria, the range will cover:

- Force calculations
- Pressure head
- Pressure calculations
- Static pressure
- Dynamic pressure
- Draught
- Forced draught

Force

In physics, force is the influence on a static object that causes the motion of the object to change. If the object is subjected to force, it will begin to move. Consider the following example:

A cold water cistern is connected to a pipe with a tap at the end. The cistern is situated in the roof space and the tap is in the bathroom. While the tap is closed, the water is at rest. It is not moving. When the tap is opened, the water moves from the cistern, down the pipe and out of the tap. The water is being acted upon by the force of gravity causing it to move. We can calculate the amount of force acting downwards on the cistern that causes the water to flow.

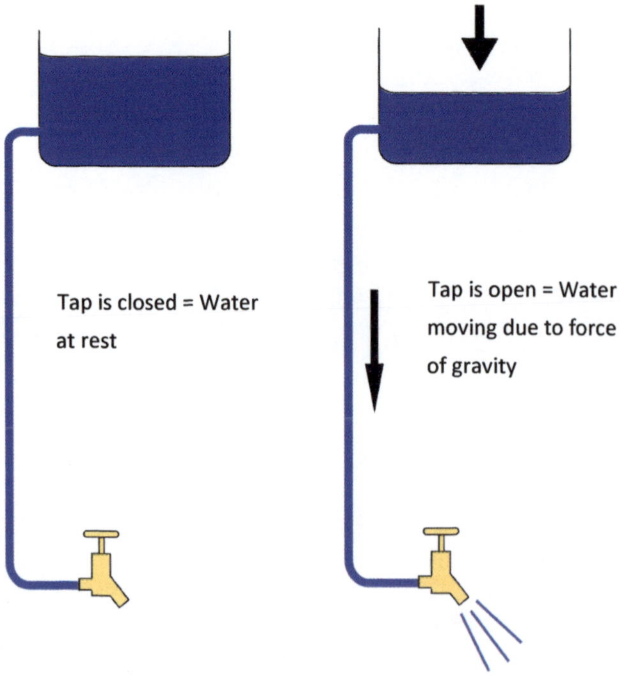

Gravity acting on water in a cistern

The cistern contains 100 litres of water. Since 1 litre of water = 1 kg, then the mass of the cistern is 100kg.

Therefore:

$$Kg \; x \; 9.81 = N$$

$$100 \; x \; 9.81 = 981N$$

So, the force acting on the water is 981 N.

Force, however, does not give the pressure of the water at the tap. Force only gives an indication of the gravitational pull downwards on the water. Water pressure is calculated slightly differently, and this will be discussed a little later in the unit.

Pressure

Physics defines pressure as force per unit area. The weight of, say, a cistern sitting on a platform in a roof space is measured in N/m2 or Newtons per square metre. If the cistern was of equal dimensions (length, width and height), then the weight being exerted onto the platform would be the same irrespective of whether the cistern was on its bottom, side or end.

If the cistern has unequal dimensions for the length, width and height, then different pressures will be exerted depending on which of the sides is face down. Take a look at the following calculation:

Example:

A cistern measuring 0.9m long x 0.8m wide x 0.7m high is to be installed on a platform in a roof space. What pressure would it exert on the platform if:

- The cistern was placed on its bottom (length x width)
- The cistern was placed on its side (length x height)
- The cistern was placed on its end (width x height)

First, we must first determine the mass of the cistern in kg. The formula for this is:

$$\text{Mass} = Length \; x \; width \; x \; height = volume \; in \; m^3$$

$$0.9 \; x \; 0.8 \; x \; 0.7 = 0.504m^3$$

Then, we must convert the volume into mass of water. To do this we must multiply the volume by 1000 to convert the mass into litres.

$$Volume \; x \; 1000 = litres$$

$$0.504 \; x \; 1000 = 504 \; litres$$

Since 1 litre of water has a mass of 1 kg, then:

$$Litres = kg$$

$$504 \; litres = 504 \; kg$$

From an earlier calculation we know that:

$$Kg\ x\ 9.81 = N$$

So:

$$504\ x\ 9.81 = 4944.24\ N$$

Therefore, the cistern has a force of 4494.24 N

The formula for finding the pressure exerted is:

$$\text{Pressure} = Force \div Area = Pressure$$

Area of the ***bottom of the cistern***:

$$0.9\ x\ 0.8 = 0.72m^2$$

$$4944.24 \div 0.72 = 6867N\ pressure$$

Area of the ***side of the cistern***:

$$0.9\ x\ 0.7 = 0.63m^2$$

$$4944.24 \div 0.63 = 7848N\ pressure$$

Area of the ***end of the cistern***:

$$0.8\ x\ 0.7 = 0.56m^2$$

$$4944.24 \div 0.56 = 8829N\ pressure$$

From these calculations we can see that the greater the surface area for a known mass of water, then the less force is exerted by the mass on to the platform. This is very important when we are deciding on where to place large cisterns that will contain a lot of water. The greater area we can present to the platform, the better the spread of the weight across the platform.

Water Pressure

The pressure that is exerted by water is due to its mass and is determined by the height of the source from the outlet. Water pressure can be measured when the water is flowing and when it is not flowing. These two methods are known as:

Static pressure (head) – Water that is not flowing

Dynamic pressure – Water that is flowing

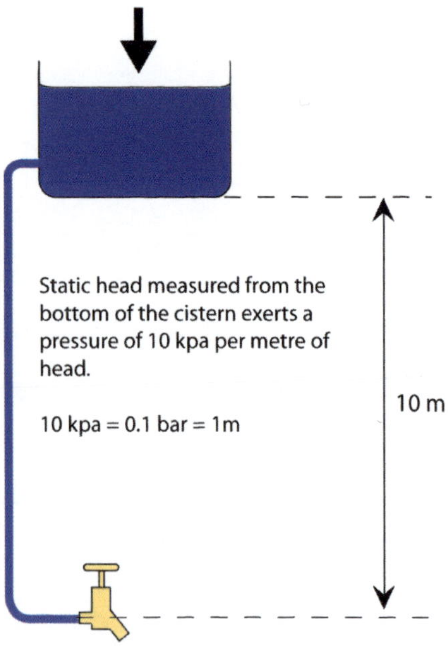

How static head acts on a cistern

Static pressure (head) – Water that is not flowing

The image shows a cistern full of water with a pipe connected to a tap. It is the distance between the cistern and the tap (measured from the bottom of the cistern to the outlet of the tap) that creates the water pressure. It is measured in Pascals (pa). This is known as the 'static head' or 'head of water'. Every metre of height (or head) between the two increases the water pressure by 10 Kpa (kilopascals):

As we have seen previously, water pressure is also measured in bar and psi.

$$1m\ of\ head = 10kpa = 0.1\ bar$$

So:

$$10m\ head = 100\ kpa = 1\ bar$$

Dynamic Pressure – Water that is Flowing

Dynamic pressure is also known as working pressure. It is the pressure of the water when it is flowing. This will generally be less than the static pressure because it is affected by other factors, such as changes of direction, length of the pipe and the type of pipe material. All these serve to lower pressure.

If the static pressure of the water is increased, then the effect would be that the flow rate of the water and its velocity would also increase. However, this can cause additional problems with unwanted noise and pipe erosion.

Remember: Flow rate + pressure = noise!

Draught and Forced Draught

Draught is one of the most essential requirements which will supply the correct amount of air for combustion and removes the products of combustion. To move the air and produce a flow of hot gases through the chimney or boiler, a difference of pressure is required. This difference of pressure maintains the constant flow of air and discharge the gases through the chimney or boiler to atmosphere. It is known as draught.

Draught can be obtained by the use of a chimney or fan or even a combination of both. When draught is produced naturally with the use of a chimney this is known as natural draught, when a fan is introduced this is known as forced draught.

AC4.5 Velocity, Pressure and Flow Rate

Within this Assessment Criteria, the range will cover:

- Effects of increasing/reducing pressure
- Effects of increasing/reducing pipe size

Velocity, Pressure and Flow Rate

These three subjects, velocity, pressure and flow rate, are linked.

- Increasing the static water pressure WILL increase the velocity of the water, and therefore the flow rate, because the water is flowing with more force through the pipe.
- Decreasing the static pressure will also decrease the water velocity and therefore reduce the flow rate.
- Decreasing the pipe size will reduce the flow rate and increase the velocity because of the Bernoulli Effect.

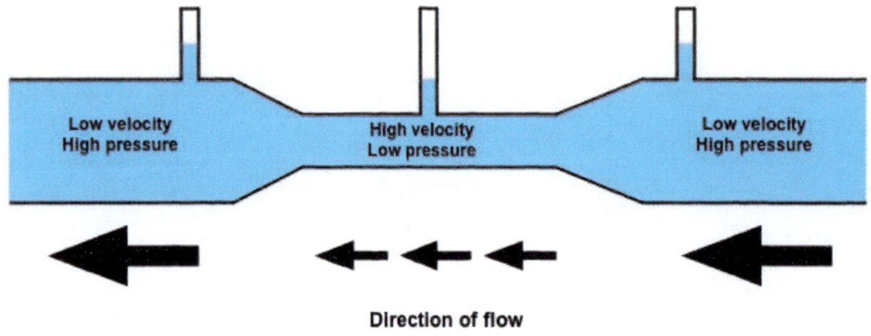

The Bernoulli effect - a venturi tube

The Bernoulli Effect describes the relationship between flow velocity and pressure of a fluid, i.e. water. It states that when the water passes through a reduction in pipe size, then the velocity of horizontal flow will increase and the pressure will decrease. This effect was discovered by Daniel Bernoulli in 1738.

If the water is moving uniformly through the pipe, then the only forces acting on the water are its own weight and the static pressure of the water itself. If the pipe reduces in size, the water must speed up, because the same amount of water is trying to flow through a smaller space. The only way the water will move faster is if the pressure behind the fluid is greater than the pressure in front. Thus, the pressure must decrease as velocity increases.

Increasing the pipe size has a reverse Bernoulli Effect and this increases the pressure and reduces the velocity.

AC4.6 Identify how Restrictions in the Pipework Restricts the flow of Liquids and Gases

Within this Assessment Criteria, the range will cover:

- Changes of direction, bends and tees
- Pipe size
- Pipe reductions
- Roughness of material surface
- Constrictions such as valves

The material of the pipe – different materials offer different resistance to the flow of water. Copper tubes and plastics are smooth internally and give little resistance to the flow of water. Low Carbon Steel, in contrast, is very rough internally and offers a greater resistance. This in turn decreases the pressure and flow rate through the pipe.

Changes of direction – elbows, bends and the junction of tee fittings also offer resistance to flow and this too will reduce the pressure and flow rate. An elbow, for example, has the same resistance as around 0.5m of pipe. So if the pipe run has 6 elbows (or 6 changes of direction), this is equivalent to an extra 3m of pipe. Machine made bends offer less resistance to flow, generally around 0.3m per bend.

Length of pipe run – again, the greater the length of pipe, the more resistance is encountered. This will effectively reduce the pressure and the flow rate of both water and gas.

Pipe size – the bigger the pipe, the greater the volume of water or gas.

Restrictions such as valves and stop taps – these reduce both pressure and flow rate because they offer resistance to smooth water flow.

Reductions in pipe size – a pipe size reduction reduces the pressure and the flow rate BUT increases the velocity. It is known as the Bernoulli Effect.

AC4.7 Describe Principles of a Siphon

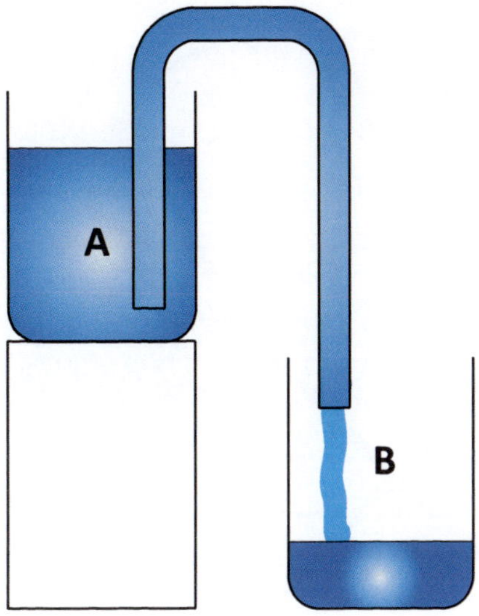

The principle of a syphon

The principle of a siphon is to move water from a high place to a low place using only atmospheric pressure and the cohesive properties of water.

In the diagram, the two containers are both at atmospheric pressure but container A is higher than container B. This means that the pressure at container A will be slightly lower than container B. For the siphon to work, the hose inlet at container A, must be higher than the hose outlet at B. When suction is applied to the hose outlet (B), the water will flow upwards, over the top of container A where the pressure is lower. At this point, gravity and water's cohesive qualities take over and the contents of container A simply flow by gravity into container B.

Learning Outcome 6

Understand Principles of Electricity in The Plumbing and Heating Industry

Learning Outcome 6: Understand Principles of Electricity in The Plumbing and Heating Industry

There are four assessment criteria within Learning Outcome 6. These are:

AC6.1. Identify basic principles of electron flow theory

AC6.2. Describe the purpose and application of simple units of electrical measurement

AC6.3. Carry out simple electrical calculations

AC6.4. Identify the requirements for earthing of electrical circuits

Electricity is crucial to modern living. It powers the appliances we use every day, both at home and at work and enables us to communicate with people all over the world instantly. It is something we all take for granted. But what is electricity?

In this, the last Learning Outcome of the Unit, we will find out about electricity, how it is generated, its laws of physics and basic circuitry.

AC6.1 Identify basic principles of electron flow theory

Basic principles

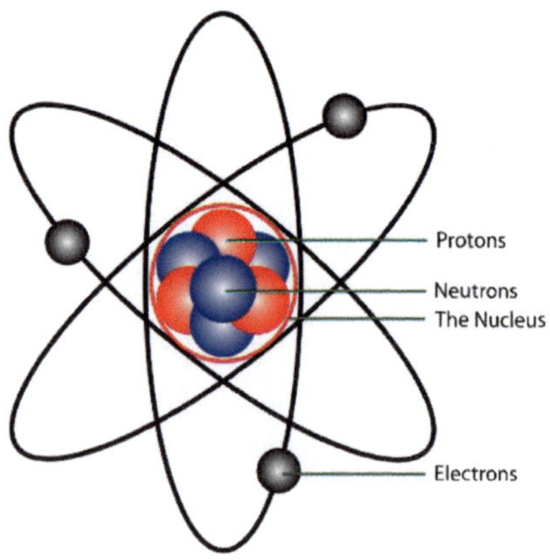

The atom

Everything on Earth is made up of atoms. Atoms are not solid. At the centre of every atom is a nucleus, which consists of an equal number of positively (+) charged protons and negatively (-) charged neutrons, making the atom electrically neutral. The neutrons play no role in the electrical properties of atoms. Their sole purpose is to hold the nucleus together. Without them, the nucleus would fly apart. Revolving around the nucleus is the electron. These too have a negative (-) charge.

As can be seen from the diagram, the atom is very similar to a mini solar system, with electrons orbiting the nucleus like the planets around a star, and, just like a solar system, those electrons nearer the nucleus orbit more quickly than those further away. The electrons that are furthest away are less attracted to the atom and are easily deflected from their weak orbits, to be attracted by other atoms. It is this constant movement of electrons from one atom to another that makes electricity possible. Materials that allow the free movement of electrons are called conductors. Those that restrict free movement are called insulators.

Measurements of electrical flow

Electricity is measured in two ways:

- By the number of electrons flowing. This is called the current and it is measured in Amperes or Amps.
- By the amount of pressure, or the push, which causes the current to flow. This is measured in Volts.

Voltage is created by the electrons which, being negatively charged, repel each other. When electrons are concentrated in one place, they will flow freely away provided the path is clear for them to do so. It is the Voltage (or push) that makes them move. If there are lots of electrons in one place, the Voltage is high, and many electrons will flow through the conductor. The more electrons that flow, the better the conductor.

The direction of movement of the negatively charged electrons is random unless a force acts upon them to move them in the same direction. This is Electro-Motive Force (EMF), better known as electricity. How well a material allows electron movement is called conductivity. How well a material resists electron flow is called resistivity.

Material conductivity and resistance

Conductivity – conductivity is the ability of the material to allow the movement of electricity. A good conductor allows free movement of electrons, whereas a poor conductor resists the flow of electrons. Silver is the best conductor of electricity with copper a close second. However, copper is much cheaper to manufacture into electrical cables and has less mass (silver = 10490kg/m3: copper = 8960kg/m^3).

Conductivity of common metals	
Material	Electrical Conductivity - σ - (1/Ω m, S/m, mho/m)
Silver	63 106
Copper	59.6 106
Gold	45.2 106
Aluminium	37.7 106
Tungsten	18.9 106
Zinc	16.6 106
Iron	9.93 106
Tin	9.17 106
Chromium	7.74 106

Resistivity – resistivity is the opposition to electron flow. It is the resistance of the material against the flow of electricity. When a metal resists the flow of electricity, the energy can be transformed into other forms of energy, such as heat and light. For example, when electricity passes through the resistance of a heating element in an electric kettle, the water will eventually boil. Similarly, when electricity passes through the filament of a light lamp, light is generated.

The amount of resistance will depend on the material.

Direct and alternating current

There are two different types of electricity:

- Direct Current (DC)
- Alternating Current (AC)

Direct Current (DC)

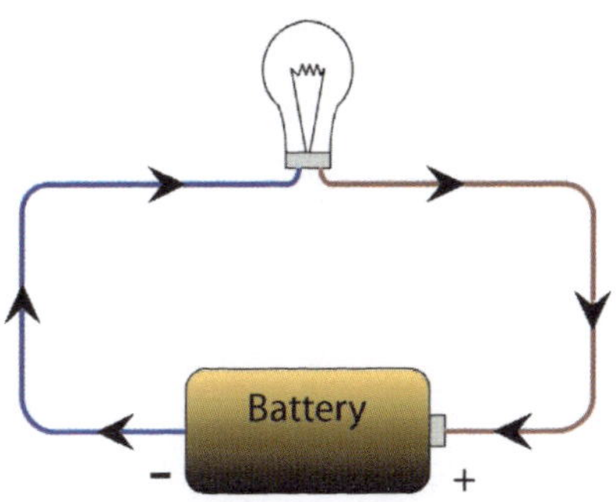

A simple DC circuit showing the direction of electron flow

A simple DC Circuit

Direct Current is produced by a number of sources. It can be generated by photo-voltaic (PV) panels or as a result of an electrochemical reaction. DC current can be stored in cells and batteries.

The Direct Current symbol

In a DC circuit, the electrons will always flow from the negative (-) pole to the positive (+) pole. The direction of the electrons, known as the polarity, never changes. The negative electrons will ALWAYS seek the positive. It never reverses.

Alternating Current (AC)

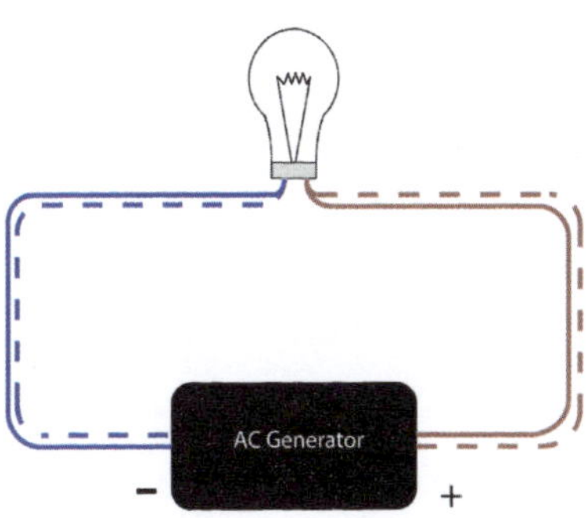

A simple AC circuit showing the alternating direction of electron flow

A simple AC Circuit

Alternating Current does not travel in one direction. It reverses its direction of travel from positive to negative and from negative to positive, constantly. In other words, it alternates. In the UK, AC current alternates 50 times every second. This is known as the frequency and is measured in Hertz. In the UK, the electricity supply is 230 V 50 Htz. Below is a table of the advantages and disadvantages of AC current:

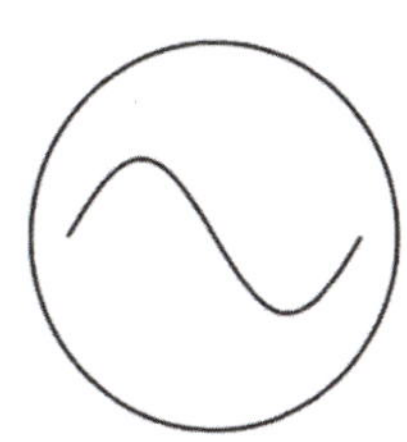

The AC current symbol

The advantages and disadvantages of AC current	
Advantages	Disadvantages
• Voltages can be transformed easily from one voltage to another. DC voltage is hard to transform. • AC current can be transported over long distances without excessive voltage loss making it more efficient than DC voltage.	• AC current cannot be stored. It is generated at power stations and bulky portable generators. • AC power tools need long cables, which can be a trip hazard. DC power tools use battery power and are much more portable than AC power tools.

AC6.2 Describe the purpose and application of simple units of electrical measurement

The terminology of electricity covers subjects such as current, voltage, resistance and power, but what are these measurements and what do they mean?

Voltage (volts)

In a circuit where there are more electrons in one part of the circuit than the other, the electrons will flow from the area of concentration to the area where they are absent. This difference in electron concentration is known as the 'potential difference' or voltage. The higher the voltage, then it can be assumed that there is a greater imbalance and the harder the electrons will repel each other. This leads to greater current or flow (in amps) in the circuit.
The voltage in a circuit can be calculated:

$$\boldsymbol{Amps\ (I)\ x\ Resistance\ (\Omega\ or\ R)}$$

Resistance

Resistance is how electrons move through a conductor. All circuits will have some resistance, and some will have more than others. Some circuits require as little resistance as possible to ensure high current flow, whilst others require high resistance to prevent electron flow. High resistance can create heat. The interaction between current (or electron flow), voltage (current flow) and resistance is shown in Ohm's Law. As the voltage increases, more electrons flow. Increasing the voltage leads to an increase current. However, if the resistance is increased, this decreases the current flow reducing the amps in the circuit.
The resistance in a circuit can be calculated:

$$\boldsymbol{Voltage\ (V)\ \div\ Amps\ (I) = Ohms\ (Resistance)(\Omega)}$$

Current

The rate at which electricity flows through an appliance is known as the current. It is measured in amps. The easiest way to understand current is through a water analogy.

A 15mm pipe at 2 bar pressure will deliver a set flow rate. If we want to increase that flow rate, we would need a bigger pipe. If we increased the pipe size to 22mm, then the pressure would be the same but the flow rate would be greater.

In electrical terms, the voltage is the pressure and the current is the flow rate. If we increase the cable size, the voltage (PRESSURE) remains at 230V but the current (FLOW RATE) is greater.

The SI unit of measurement of electrical current is the Ampere or amp (i). The current in a circuit can be calculated:

$$Voltage\ (V)\ \div\ Ohms\ (Resistance)(\Omega\ or\ R) = Amps\ (I)$$

Power

When electricity is converted into other forms of energy, such as light, heat or mechanical work, it is called power. It is equal to the sum of the current and the voltage.

An immersion heater rated at 3kW simply means that the electrical energy is converted into heat to warm the water. Electrical power, therefore, is the rate at which electrical energy is consumed and is defined as 1 joule per second.

The SI unit of measurement of electrical power is the Watt (W). The power in a circuit can be calculated:

$$Amps\ (I)\ x\ Volts\ (V) = Watts$$

AC6.3 Carry out simple electrical calculations

Previously, we have investigated current, voltage, resistance and power. To understand the interactions between these units, we must look at Ohm's Law, which states:

> **'The current through a conductor between two points is equal to the voltage across the two points, and inversely proportional to the resistance between them.'**

It defines the relationship current, voltage, resistance and power. According to Ohm's law:

$$1\ Ohm\ is\ the\ resistance\ through\ which\ 1\ volt\ will\ maintain\ a\ current\ of\ 1\ amp$$

In theory, if we have 2 electrical values, then we can calculate a third. In the following calculations:

$I = current$

$V = voltage$

$R = resistance$

$P = power$

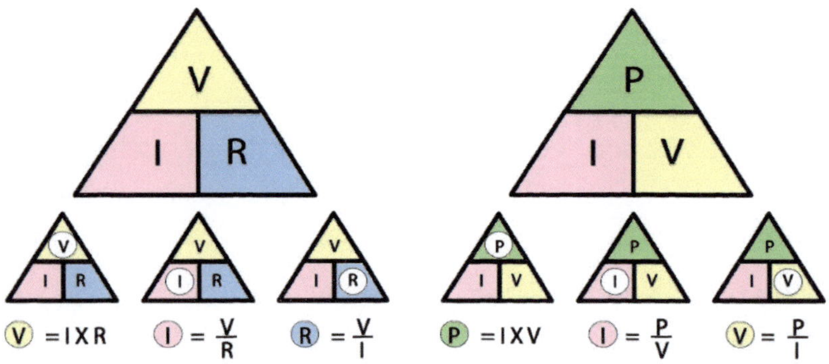

The Ohm's and Power triangles.

The diagrams left show Ohm's triangle and the power triangle. By covering the Unit, you are calculating with your finger, the remaining units are needed in the calculation. For example, if we are calculating voltage, cover the 'V' and we are left with $I \times R$. If we are calculating resistance, cover the 'R' and we are left with $\frac{V}{I}$

Simple calculations using Ohm's Law

Example 1

What size of over current protection device will be needed to protect a circuit that has a 3kW immersion heater installed on a 230V supply?

The formula for this is:

$$I = P \div V$$

Therefore:

First, convert the kilowatts to watts by multiplying by 1000:

$$3kw = 3000$$

Now complete the calculation:

$$3000 \div 230 = 13\ amps$$

Example 2

Using the formula below, Calculate the current flowing to a 10 kW shower fed by a 230 Volt AC supply.

Look at the Ohm's triangle. Since we only know the power and the volts, we will need to transpose the equation to find the amps (I).

$$I = P \div V$$

$$6\ kw = 6000\ watts$$

Therefore:

$$6000 \div 230 = 26\ amps\ (i)$$

Example 3 – Using the Ohm's Law formula below, calculate the voltage in a circuit, which has a resistance of 115 Ohms and a current of 2 amps.

The formula for this is:

$$V = I \times R$$

Therefore:

$$115 \ x \ 2 \ = \ 230 \ volts$$

Example 4 – Using the Ohm's Law formula Resistance = Volts ÷ Current, calculate the resistance in a circuit when the voltage is 110V and the current is 45 amps. The formula for this is:

$$\Omega = V \ \div I$$

Therefore:

$$110 \ \div 45 = 2.44 \ Ohms$$

Voltage, current and resistance in series and parallel circuits

There are two types of circuit:

- Series
- Parallel

Series circuits

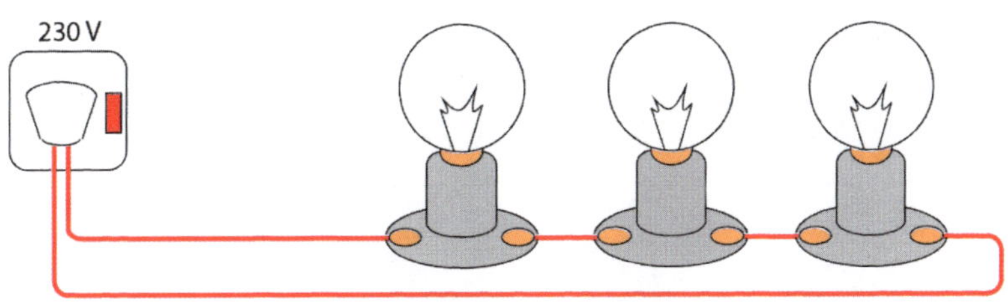

A series circuit

A series circuit has only one path from the power source through the circuit loads and back to the power source. In this instance, the current MUST flow through ALL loads.

This circuit depends upon all of the lamps working perfectly. If one of the lamps fails, then the whole circuit will fail and none of the lamps will light up. This is because the current flows from one lamp to the next. It is the ONLY path from the power source and back again.

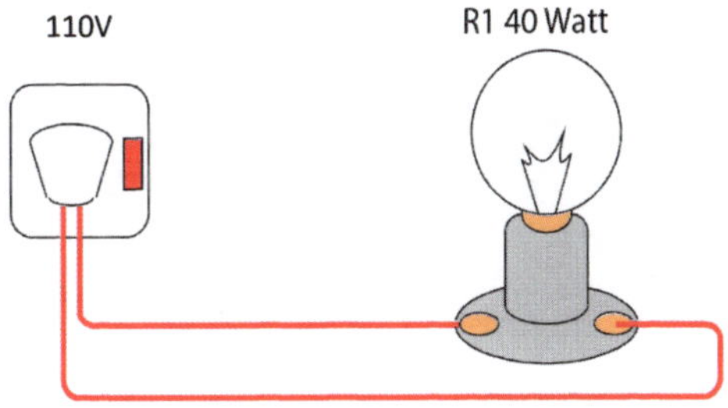

One 40-Watt lamp

Consider the circuit left. The circuit has one lamp connected to a 110V electricity supply. The lamp will glow at full brightness simply because it is receiving all of the current. To find the resistance in the circuit, we must first find the current (I). Since we know the POWER (40Watt) then:

$$I = \frac{W}{V}$$

Therefore, the current (amps) is:

$$\frac{40}{110} = 0.3636 \ amps$$

Now, we can apply Ohm's law to find the resistance:

$$R = \frac{V}{I}$$

Therefore:

$$\frac{110}{0.3636} = 302.53\ Ohms$$

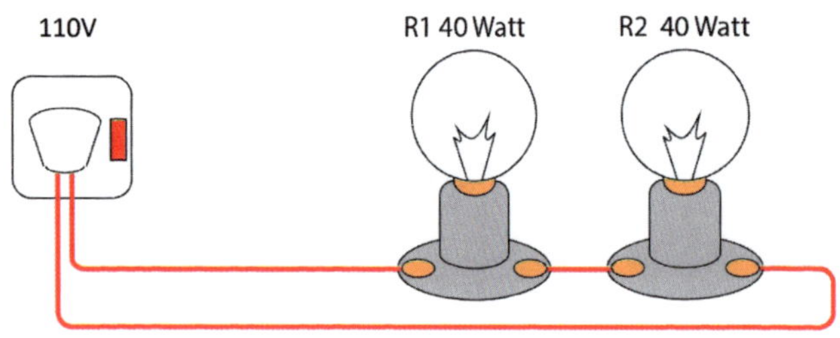

Two 40-Watt lamps

When a second lamp with the same 40 Watts is added to the circuit, then the resistance in the circuit will double and the current flow is half that when there was only one lamp. The voltage is now only 55V to each lamp and this halves the brightness emitting from each lamp. Both lamps have the same voltage drop.

Again, firstly, we must find the current:

$$I = \frac{W}{V}$$

Therefore, the current (amps) is:

$$Lamp\ 1 = \frac{40}{110} = 0.3636\ amps$$

$$Lamp\ 2 = \frac{40}{110} = 0.3636\ amps$$

Now, we can apply Ohm's law to find the resistance of each light lamp:

$$R = \frac{V}{I}$$

Therefore:

$$R1 = \frac{110}{0.3636} = 302.53\ Ohms$$

$$R2 = \frac{110}{0.3636} = 302.53\ Ohms$$

Now find the total resistance of the circuit:

$$RT = R1 + R2$$

$$RT = 302.53 + 302.53 = 605.06\ Ohms$$

To find the current across the circuit using the total resistance:

$$I = \frac{V}{R}$$

Therefore:

$$\frac{110}{605.06} = 0.1818\ amps$$

To find the voltage. Since each lamp is 40 Watts with a resistance of 302.53 Ohms and the total circuit current is 0.1818 amps, to find the voltage to each lamp:

$$V = I\ x\ R$$

$$0.1818\ x\ 302.53 = 54.99V$$

Because each lamp is only drawing 55V, each lamp will only glow with half its potential brightness. To find out how many Watts is being generated:

$$P = I\ x\ V$$

Therefore:

$$0.1818\ x\ 110 = 19.998\ Watts$$

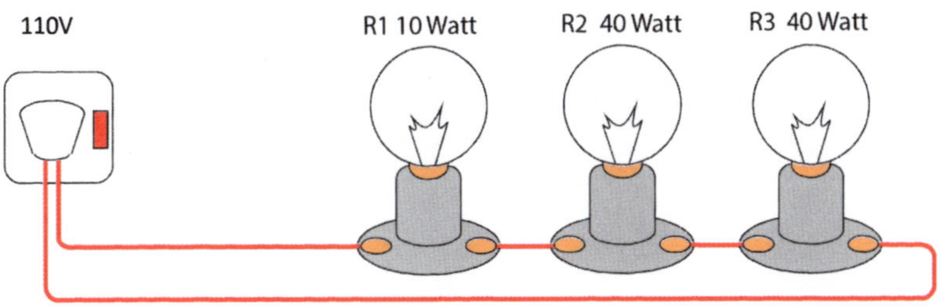

Now, let us add a third lamp but this time with less power.

Here, there are 3 lamps, 2 x 40Watt and 1 x 10Watt. We must find the current to each light lamp:

$$I = \frac{W}{V}$$

$$Lamp\ 1 = \frac{10}{110} = 0.0909\ amps$$

$$Lamp\ 2 = \frac{40}{110} = 0.3636\ amps$$

$$Lamp\ 3 = \frac{40}{110} = 0.3636\ amps$$

Now, we can apply Ohm's law to find the resistance of each light lamp:

$$R = \frac{V}{I}$$

Therefore:

$$R1 = \frac{110}{0.0909} = 1210.121\ Ohms$$

$$R2 = \frac{110}{0.3636} = 302.53\ Ohms$$

$$R3 = \frac{110}{0.3636} = 302.53\ Ohms$$

Now find the total resistance of the circuit:

$$RT = R1 + R2 + R3$$

$$RT = 1210.121 + 302.53 + 302.53 = 1815.181\ Ohms$$

To find the current across the circuit using the total resistance:

$$I = \frac{V}{R}$$

Therefore:

$$\frac{110}{1815.181} = 0.06060\ amps$$

To find the voltage to each lamp:

$$V = I\ x\ R$$

$$Volts\ @\ Lamp\ 1 = 0.06060\ x\ 1210.121 = 73.33\ Volts$$

$$Volts\ @\ Lamp\ 2 = 0.06060\ x\ 302.53 = 18.33\ Volts$$

$$Volts\ @\ Lamp\ 3 = 0.06060\ x\ 302.53 = 18.33\ Volts$$

Because each lamp is drawing less voltage, each lamp will only glow with a fraction of its potential brightness.

To find out how many Watts is being generated:

$$P = I\ x\ V$$

Therefore:

$$Watts\ @\ Lamp\ 1 = 0.0909\ x\ 73.33 = 6.66\ Watts$$

$$Watts\ @\ Lamp\ 2 = 0.3636\ x\ 18.33 = 6.66\ Watts$$

$$Watts\ @\ Lamp\ 3 = 0.3636\ x\ 18.33 = 6.66\ Watts$$

Parallel circuits

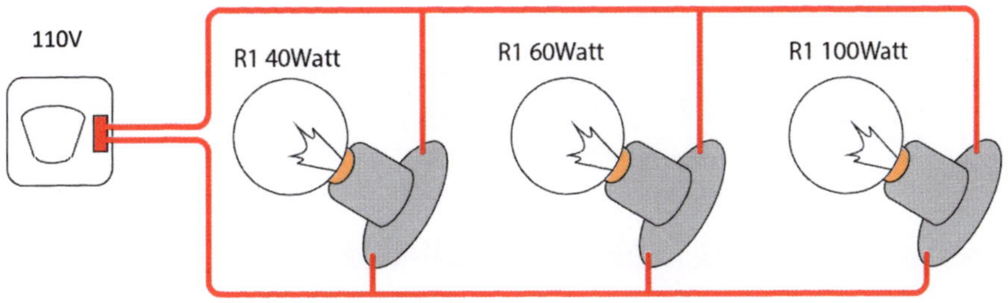

A parallel circuit

A parallel circuit has at least 2 individual paths in the circuit:

In a parallel circuit, each light lamp has its own independent path to the power source. This means that the lamps are unaffected by each other. If one lamp fails, the other two will continue to work. Voltage is equal across all components. However, the current increases when more light lamps are added. This means that if too many lamps are connected, then the circuit will overload and the fuse, Miniature Circuit Breaker or Residual Current Device will activate to cut off the electricity supply to the circuit.

To find the current at each lamp:

$$I = \frac{P}{V}$$

Therefore:

$$R1 = \frac{40}{110} = 0.3636\ amps$$

$$R2 = \frac{60}{110} = 0.5454\ amps$$

$$R3 = \frac{100}{110} = 0.9090\ amps$$

$$Total\ current = 0.3636 + 0.5454 + 0.9090 = 1.818\ amps$$

To find the resistance of each lamp:

$$R = \frac{V}{I}$$

$$R1 = \frac{110}{0.3636} = 302.53\ Ohms$$

$$R2 = \frac{110}{0.5454} = 201.68\ Ohms$$

$$R3 = \frac{110}{0.9090} = 121.01\ Ohms$$

To find total resistance of the circuit:

$$\frac{1}{RT} = \frac{1}{R1} + \frac{1}{R2} + \frac{1}{R3}$$

$$\frac{1}{RT} = \frac{1}{302.53} + \frac{1}{201.68} + \frac{1}{121.01}$$

$$\frac{1}{RT} = 0.003305 + 0.00495 + 0.00826$$

$$\frac{1}{RT} = 0.01651$$

$$RT = \frac{1}{0.01651} = 60.57\ Ohms$$

The chart below shows that the voltage stays the same and that the total resistance in the circuit decreases whilst the Watts and amps increase

	R1	R2	R3	Total
Volts	110	110	110	110
Amps	0.3636	0.5454	0.9090	1.818
Ohms	302.53	201.68	121.01	60.50
Watts	40	60	100	200

Ohm's law applies:

$$\frac{V}{I} = \frac{110\ V}{1.818\ A} = 60.50\ Ohms$$

AC6.4 Identify the requirements for earthing of electrical circuits

Electricity will always take the least line of resistance to earth, even if that line is us! If we accidentally touch an electricity source, the current will always flow through our bodies to get to the earth, causing a dangerous, sometimes fatal, electric shock. The reason for this is simple; electricity always wants to complete the circuit. The need, then, for a good earthing system is of utmost importance.

In this final part of the Unit, we will look at the need for good earthing and equipotential bonding of electrical circuits.

Earthing

Faulty electrical circuits can cause damage, fire and death. To prevent these from occurring, all electrical installations must be earthed. Earthing, in an electrical circuit, is provided by a dedicated wire, known as the

Earth wire, that permanently connects every socket and light fitting to a specific metal earth block at the consumer unit.

When an electrical fault occurs, the current is carried safely away from the fault to earth. This triggers a change in electrical flow, that will cause the fuse to blow or the Residual Current Device (RCD) to trip, disconnecting the electrical supply to the circuit where the fault is located.

The earth cable is coloured green and yellow and must be installed in $10mm^2$ cable.

Some electrical appliances also have a green/yellow earth flex. The earth wire must always be connected to the earth pin on the plug. Those electrical appliances that do not have an earth flex are known as double insulated appliances.

The ring main and lighting installations are usually completed in twin/earth cable. In this cable type, the earth cable is an unsheathed copper wire. A green/yellow sheathing has to be placed over it at installation stage.

Equipotential bonding

Metal fixtures in a property, such as central heating pipework, gas pipework, cold water pipework, stainless steel sinks and washbasins and cast iron baths must be 'equipotentially bonded' using a bonding clamp connected by a green/yellow cable. This is to ensure that all metal fittings have the same zero electrical potential.

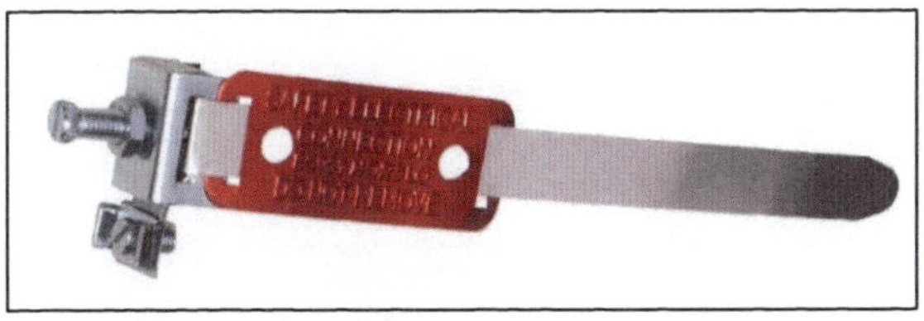

An earth bonding clamp

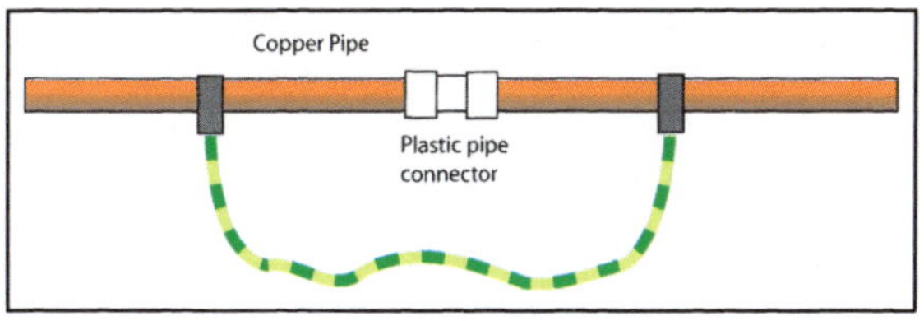

Cross bonding plastic pipe connections

When attaching a bonding clamp, it is essential that a good connection to the metal be made. This is done by cleaning the pipework of any paint or corrosion before attaching the clamp. If plastic connections have been made to copper pipework, this MUST be permanently bridged to ensure a good cross bonding of the pipework.

Bonding metallic sinks, basins, baths and shower trays can be completed using $4mm^2$ single-core earth wire.

Unit 05

Cold Water – Phase 2

Unit 05 – Cold Water – Phase 2

In Phase 2, there 3 Learning Outcomes:

LO 2: Install cold water systems:

- AC2.2. State advantages and disadvantages of **cold water systems**
- AC2.3. Identify types and typical pipe sizes used in **cold water systems** within dwellings
- AC2.4. Describe working principles of **cold water systems,** positioning, fixing, connection and operation of **components**
- AC2.5. Describe layout and installation requirements for protected **plastic storage cisterns**
- AC2.6. Describe insulation requirements, system frost protection and prevention of undue warming of **cold water systems**
- AC2.7. Identify the positioning and fixing of pipework within the **building fabric**
- AC2.8. Identify **sources of information** required when undertaking work on cold water systems
- AC2.9. Explain **backflow** risk and required **methods** of prevention
- AC2.10. **Install** cold water systems

LO3: Decommission Cold Water Systems

- AC3.1. Explain **procedures** for **decommissioning** systems
- AC3.2. Carry out decommissioning **procedures**

LO5: Perform a soundness test and commission cold water systems and components (AC5.1 TO AC5.6 only)

- AC5.1. Identify information sources required to complete testing and commissioning
- AC5.2. Describe how to fill and vent cold water systems
- AC5.3. Describe a visual inspection of a cold water system to confirm that it is ready to be soundness tested
- AC5.4. Describe a **soundness test** to industry requirements on cold water systems **pipework** and components
- AC5.5. State the **flushing requirements** including the use of **system additives** for new and existing cold water systems
- AC5.6. Carry out a soundness test

Learning Outcome 2

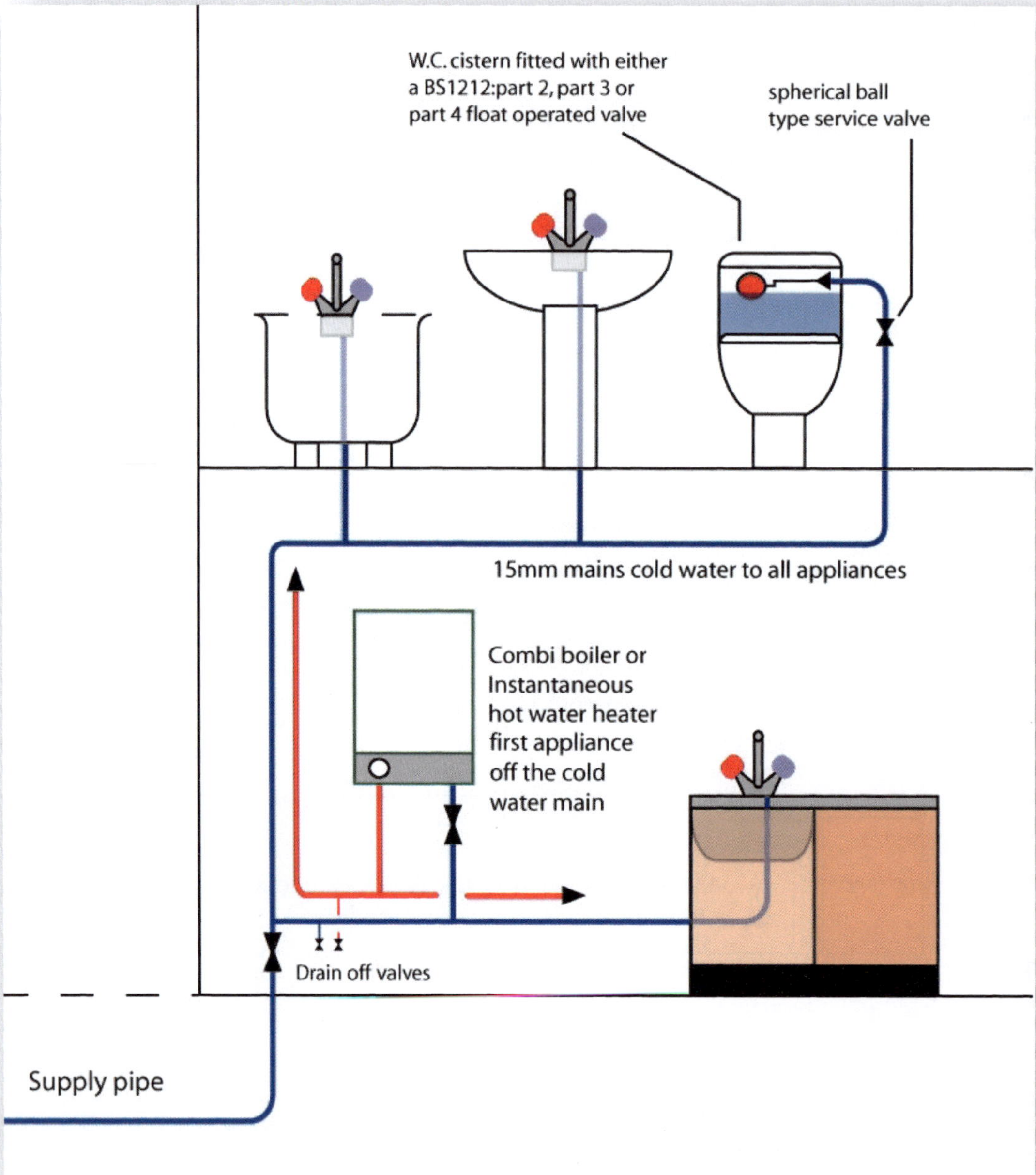

Install Cold Water Systems

AC2.2 State the advantages and disadvantages of cold water systems

In book 1, we investigated the different cold water systems available for both domestic and industrial/commercial installations. In book 2, we will look at the advantages and disadvantages of those systems used in domestic properties.

There are eight cold water systems within four sub headings:

1. Direct cold water system
2. Indirect cold water system
3. Borehole:
    a. Pumped supply with pressure control
    b. Pumped supply with level control.
4. Boosted:
    a. Direct boosted
    b. Direct boosted to a water header or cistern
    c. Indirect boosted to a cistern
    d. Indirect boosted with pressure vessel

For the purposes of this AC, we will concentrate on systems 1 to 3.

1. Direct Cold Water System

The Direct System of cold water supply is the most common system in use in the UK. With this system, all of the cold water outlets are fed with water directly off the water undertaker's mains cold water supply. Storage of cold water is required if the hot water supply is taken from a low pressure hot water storage vessel. Storage is not required when the hot water supply is supplied from an unvented hot water storage vessel or a combination (combi) boiler/instantaneous gas hot water heater.

Table 1: Direct Cold Water installations

Advantages	Disadvantages
It is the most cost effective cold water system installation	At times of peak demand, the water pressure may drop
Mains cold water is available at all cold water taps and outlets	There is no cold water storage if the cold water main is under repair
It requires less pipework than any other installation	Can be noisy due to the high flow rates and pressures of the water supply
Little or no structural support is required for cisterns in the roof space	Greater risk of contamination to the cold water main due to the fact that all connections, taps, outlets and appliances are connected direct to the mains supply
The most suited installation where mains cold water is required by appliances and components,	Greater wear on taps and valves, which can increase the noise within the system

such as electric showers and unvented hot water systems.	
Smaller pipe sizes can be used in the majority of installations	Water hammer can be a problem where poor installation practice has been used or where taps and valves are worn or faulty
Good pressure at ALL cold water outlets	Greater condensation build-up on pipework that is often mistaken for a leak
	Can cause a lot of water damage due to the high pressure and flow rate, should there be a leak

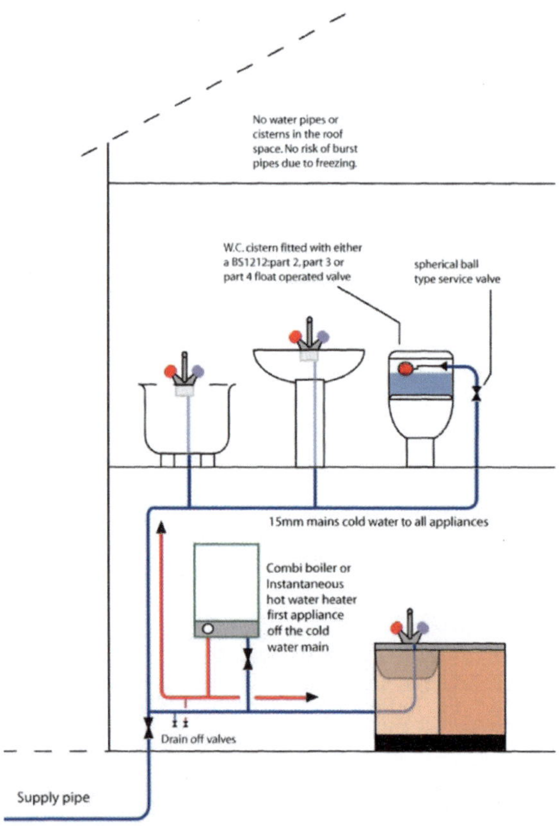

Direct Cold Water installation with instantaneous hot water supply

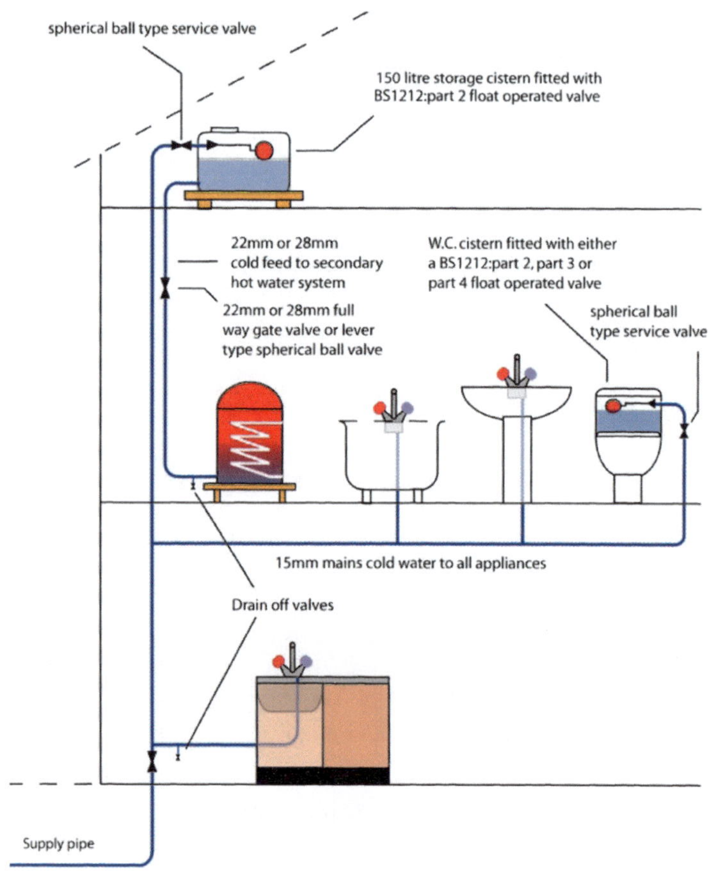

Direct Cold Water installation with stored water in the roof space

2. Indirect Cold Water System

With the indirect system of cold water supply, only the kitchen sink and the Cold Water Storage Cistern are supplied with water direct from the water undertaker's water main. All other appliances, bath, washbasin and WC, are supplied with water from a cold water storage cistern in the roof space. Because of the increased need for stored cold water, the size of the cistern will also increase.

Table 2: Indirect Cold Water installations	
Advantages	**Disadvantages**
Reduced risk of contamination	The rising main must be protected against backflow from the cistern
Less risk of water hammer and reverberation noise	Greater risk of frost damage of pipework and cisterns in the roof space
Low pressure supply reduces the impact of leakage damage	A large amount of water storage is needed
Reduces the demand on the cold water main at times of peak demand	Cisterns use a lot of storage space in the roof
Pipework can be sized to give very good flow rates despite the lack of pressure	More structural support required to carry the weight of the cisterns when full of water
Shower valves on low pressure supplies may be supplied with equal pressures and flow rates	Greater cost of installation
Reserve water in the cistern should the cold water mains fail	More pipework needed
	Reduced pressure at outlets and taps

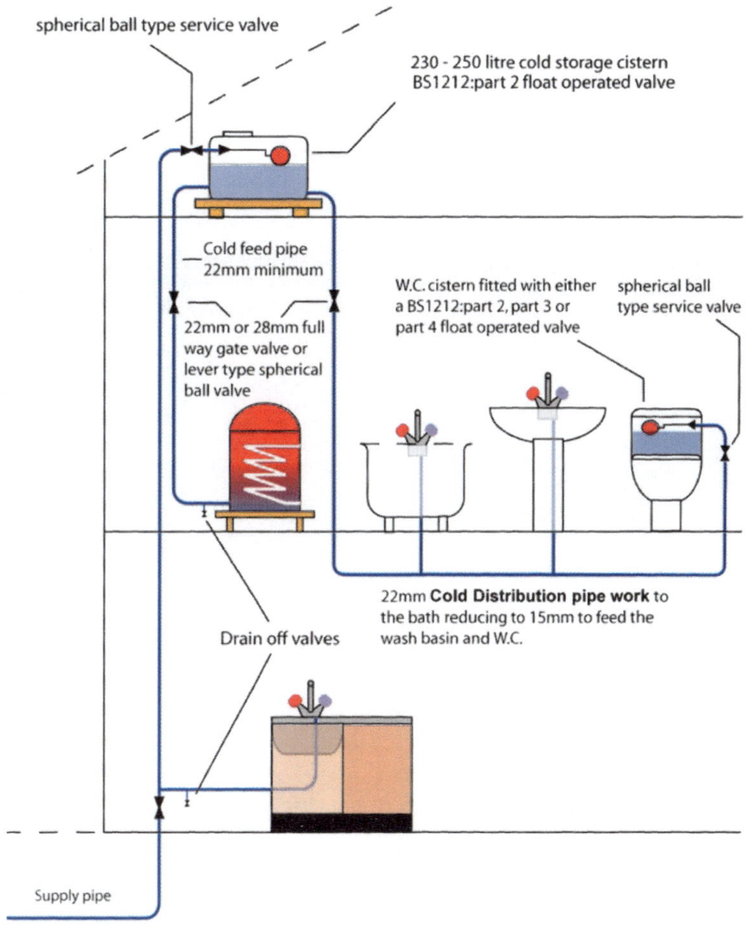

The indirect method of cold water supply

3. Systems supplied from a Borehole

There are two methods of supplying cold water to a property from a borehole or well:

a. Pumped supply with pressure control

b. Pumped supply with level control

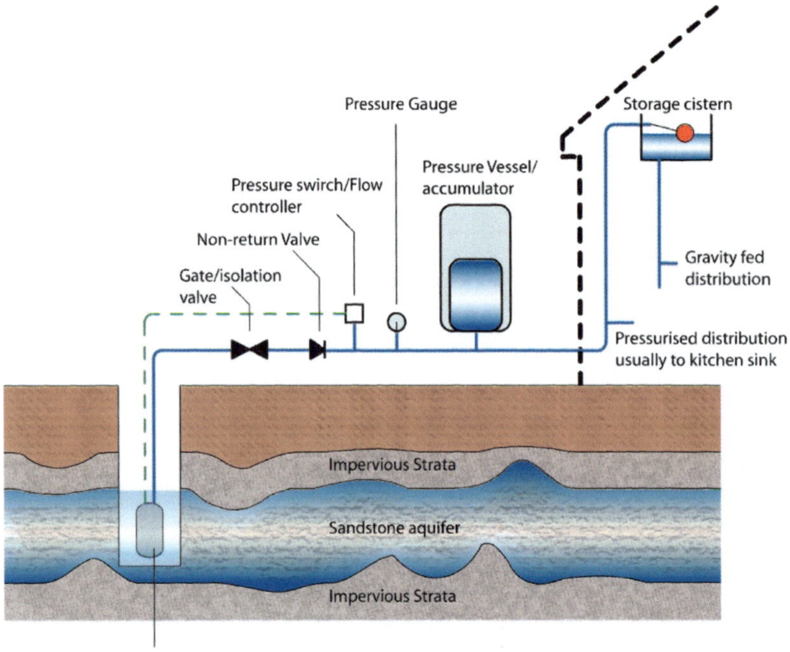

Pumped supply with pressure control cold water system.

a. Pumped supply with pressure control

Here, water is drawn by a pump directly from the borehole. The system contains a pump (either submersible or surface mounted) and a steel pressure vessel, known as an accumulator to maintain the system pressure. The accumulator also stores water for use when the pump is not running. These systems usually operate at pressures between 1.5 and 3 bar.

Table 3: Pumped supply with pressure control

Advantages	Disadvantages
Good pressure and flowrate to the kitchen sink outlet	Requires a large cistern in the roof space to supply all other outlets
Using an accumulator and a storage cistern means the pump is not running all the time	Greater risk of frost damage of pipework and cisterns in the roof space
It requires less pipework than any other installation	A large amount of water storage is needed
Water is available in the event of pump failure	Cisterns use a lot of storage space in the roof
	More structural support required to carry the weight of the cisterns when full of water
	Greater cost of installation
	Reduced pressure at some outlets and fittings
	Pumps can be noisy
	Accumulator needs to be sized carefully to ensure sufficient high pressure water supply
	Greater risk of contamination to the water source
	Water source need careful monitoring for water quality
	Cisterns must be protected against contamination

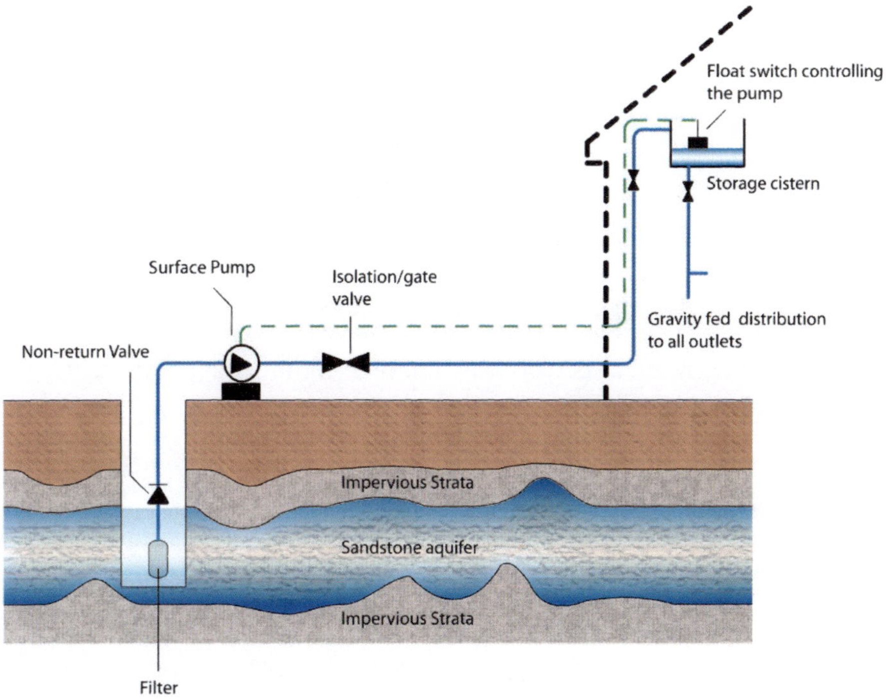

Pumped supply with level control

b. Pumped supply with level control

This system uses a electrically operated float switch to control the level of water in the cold water storage cistern. When the float switch is down, this means that the cistern requires water and the pump is switched on. The cistern continues to fill, until the water level is high enough to close the float switch and switch off electricity to the pump.

Table 4: Pumped supply with level control	
Advantages	**Disadvantages**
Water is available in the event of pump failure	Greater risk of frost damage of pipework and cisterns in the roof space
Little or no risk of water source contamination as all the dwelling's water is supplied from a protected cistern	A large amount of water storage is needed
Using a storage cistern means the pump is only required to fill the cistern	Cisterns use a lot of storage space in the roof
An expensive accumulator is not required	More structural support required to carry the weight of the cisterns when full of water
	Greater cost of installation
	More pipework required
	Reduced pressure at some outlets and fittings
	Pumps can be noisy
	Accumulator needs to be sized carefully to ensure sufficient high pressure water supply
	Greater risk of contamination to the water source
	Water source need careful monitoring for water quality
	Cisterns must be protected against contamination

AC2.3 Identify types and typical pipe sizes used in cold water systems within dwellings

For most domestic installations, a 25mm MDPE water supply will enter the building, terminating with a screw down stop tap. From here, the appliances and components installed will determine the pipe size.

In all cases, pipe sizing should be carried out to determine the exact size required.

Direct systems

Generally speaking, most direct systems utilise a 15mm rising main with 15mm pipework to all cold water outlets. This is sufficient for small installations where a combination boiler supplying hot water is installed, without en-suite bathrooms and shower rooms.

Where the hot water is supplied through an open vented hot water storage system, the cold water rising main will terminate at a cold feed cistern normally positioned in the roof space. The water storage capacity of the cistern must be at least the storage capacity of the hot water storage vessel.

For those installations where an unvented hot water storage system is to be connected, a 22mm minimum cold water rising main should be installed. This is because unvented hot water storage systems require a far greater flow rate which can only be supplied through a larger size pipe. Installations where the hot water is supplied direct from the mains cold water supply through a mains pressure hot water heater or unvented hot water storage system, the number of outlets is effectively doubled and this will increase the size of the mains cold water supply pipework.

Occasionally, a 28mm rising main may be required depending on the size of the property and the number of outlets being supplied.

Indirect systems

The indirect system of cold water supply is designed for areas where the cold water supply pressure is low and not capable for supplying the full water requirements of the system design. It utilizes a 15mm cold water rising main, feeding cold water storage cisterns in the roof space. Only the kitchen sink and possibly any downstairs cloakroom or utility room are taken directly off the cold water main. All other outlets are fed by a low pressure supply, minimum 22mm pipework and often 28mm pipework depending on the number of outlets, from the storage cisterns. This is known as the cold distribution pipework. 22mm pipework should be supplied to the bath to give a good flow rate.

To give an adequate pressure, the cold water storage cisterns should be positioned as high as possible. They must also be protected against undue warming and frost.

The amount of cold water stored in the roof space will need to be carefully calculated as, in most cases, the cisterns will also supply cold water to an open vented hot water installation and so will need to be large enough to supply both cold water and hot water household needs.

AC2.4 Describe working principles of cold water systems, positioning, fixing, connection and operation of components

The working principles of cold water systems

The working principle of any cold water system for domestic purposes is to deliver a supply of cold water to a tap or outlet that is clean, cold, wholesome, potable and fit for human consumption. The different cold water system designs vary on their approach of how this is achieved:

Domestic systems

Direct cold water system

Direct systems deliver cold water to all taps and outlets directly from the Cold Water Mains Supply. With this system, water quality does not pose any contamination issues because the water undertaker's mains cold water is regulated under section 67 of the Water Act.

Indirect cold water system

Often used in areas of low pressure and flow rate, the indirect system uses a protected cold water storage cistern to deliver water to the majority of the dwellings cold water outlets and taps. Only the kitchen sink and possibly any utility room and/or cloakroom are directly connected to the cold water mains supply.

Domestic Borehole systems

Domestic borehole systems are usually installed where access to the water undertaker's cold water supply is difficult. These are not covered by section 67 of the Water Act, nor or they regulated by the Water Supply (water fittings) Regulations. They are regulated by the Private Water Supply Regulations, which are enforced by the Local Authority. Careful monitoring of the water source must be maintained to ensure water quality. Water filters and disinfection equipment must be fitted in all installations. There are two common systems:

Pumped supply with pressure control

This system is very similar in design to the indirect system insomuch that the cold water is supplied to the kitchen tap straight from the borehole. The supply then terminates at a protected cold water storage cistern. Cold water for all other appliances is taken from the protected cistern.

The system utilises an electric submersible or surface pump to extract the water from the bore hole and a cold water accumulator or pressure vessel. The accumulator has two jobs:

1. To store water for delivery to the kitchen tap and the cistern,
2. To maintain system pressure to the stored water.

Pumped supply with level control.

Again, with this system, bore hole water is extracted by the submersible or surface pump and delivered directly to a large cold water storage cistern. All of the property's water is then delivered from the cistern, including the water required for the kitchen sink. The cistern must be protected against undue warming and frost. A float switch in the cistern operates the pump. When water in the cistern is low, the float switch activates the pump and water is then pumped to the cistern. When the cistern is full, the float switch switches the pump off.

Boosted systems for domestic installations in high rise buildings

Where inadequate supplies are noted or where the building is too high to be supplied from the cold water main, the water supply within the building must be boosted. However, it should be noted that the Water Supply (water fittings) Regulations forbid pumping direct from a mains supply when the flow rate is greater than 12 litres per minute, without prior consent of the water undertaker. Where the flow rate is less than 0.2 litres per second, drinking water may be directly boosted from the cold water main if the water undertaker agrees. In all cases, the water undertaker must be consulted.

Boosting pumps create aeration of the water. This can lead to the water having a milky appearance and should not be regarded as a deterioration in water quality.

Direct boosted

This system pumps water directly from the cold water main to a cistern at high level. From here, the water flows by gravity to all properties and outlets. Where boosting from a cold water main has been allowed by the water authority, pumps can be fitted directly to the incoming supply to allow an increase in the water pressure.

Direct boosted to a water header or cistern

With this system, high-level cisterns supply non-drinking water only. For drinking water, a header with a capacity of 5 to 7 litres is fitted to the boosted supply to supply the dwellings with drinking water when the pumps are not in operation. Excessive pressure should be avoided as this can lead to problems of splashing and water wastage at the sink taps.

Indirect boosted to a cistern

The indirect boosted to a storage cistern system incorporates a cistern that is filled from the mains cold water supply. This is called a break cistern and its function is to store water before it is boosted to the system. By doing this, water is not boosted directly from the cold water main because the cistern provides a 'break' between the mains supply and the cold water system in the building.

Indirect boosted with pressure vessel

This cold water system is usually used where multiple storage cisterns are fed at various levels making control by float switches alone impractical. The system incorporates a pneumatically operated pressure vessel to maintain the system pressure to the higher levels of the building.

Some of the lower floors are supplied un-boosted directly from the mains cold water supply whilst the remaining levels, being beyond the range of the mains cold water pressure, are fed from the break cistern and booster pumps.

Diagrams of all four boosted cold water systems can be found in the Level 3 Plumbing and Domestic Heating Qualification Textbook 1.

Positioning, fixing, connection and operation of components – Appliances

The installation of sanitary appliances is covered in detail in Unit 9, LO1, AC1.8 and AC1.14.

Baths

Connections to baths vary depending on the position of the taps. For domestic installations, bath taps have a ¾ inch BSP male thread for connection with a ¾" x 22mm tap connector.

WCs

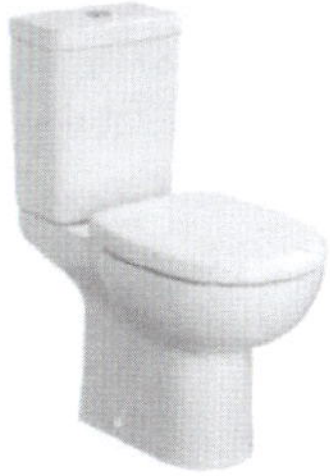

All WCs are provided with a float operated valve to the WC cistern. The float operated valve (FoV) for WC cisterns can either be bottom fed or side fed and all have ½ " BSP male threads to allow connection to the water supply via a ½" x 15mm tap connector. The FoV should be provided with a servicing valve for maintenance, repair and replacement

Over the rim bidets

Over-rim bidets installed in domestic premises can either have single or combination taps installed. All taps fitted to over-rim bidets must have an AUK2 air gap (air gaps and backflow prevention is covered in AC2.9) provided between the outlet of the tap and the spill-over level of the bidet. The Water Regulations Advisory Scheme (WRAS) Installation Requirement Note IRN R010 also stipulates that the water supply to combination taps shall have reasonably balanced pressures. Where unbalanced water pressures are used, then approved single check valves must be fitted immediately upstream of both hot and cold supplies.

Wash hand basins

Wash hand basins installed in domestic premises can either have single or combination taps installed. Single taps have ½ " male BSP threads while combination taps often have 10mm copper tails. All taps fitted to wash basins must have an AUK2 air gap (air gaps and backflow prevention is covered in AC2.9) provided between the outlet of the tap and the spill-over level of the basin.

Again, Installation Requirement Note IRN R010 also stipulates that the water supply to combination taps shall have reasonably balanced pressures. Where unbalanced water pressures are used, then approved single check valves must be fitted immediately upstream of both hot and cold supplies.

Sinks

All sinks for domestic purposes, whether manufactured from ceramics or stainless steel must have taps that give an AUK3 air gap. Sink taps can either be single high-neck pillar taps, combination taps or wall mounted bib taps, depending upon the installation requirements. The water in sinks should be regarded as fluid category 5 (see textbook 1). Single taps have ½ " male BSP threads while combination taps often have 15mm copper tails.

Urinals

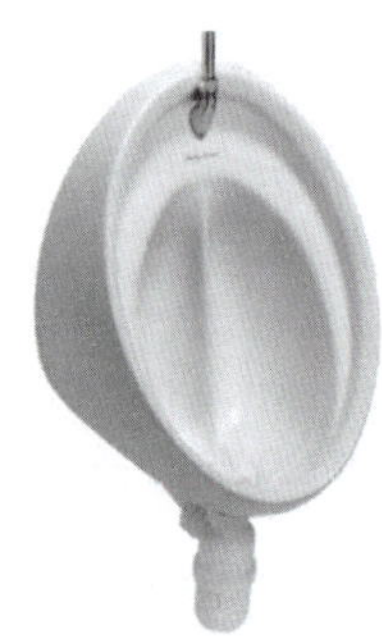

Urinals, although not a domestic appliance, require careful consideration. Connecting a urinal to the water supply can be completed in one of either two ways:

- Via an automatic flushing cistern, or;
- A manual or automatic flushing valve

Urinals using automatic flushing cisterns must be installed in such a way that does not constitute a wastage of water. WRAS note IRN R075 states that urinals using automatic flushing cisterns should be installed to give a flush rate not exceeding 10 litres per hour for a single urinal bowl and 7.5 litres per hour for each 700mm of a slab urinal or two or more urinal bowls. Urinals using flushing valves must not deliver a flush exceeding 1.5 litres per single flush per bowl each time the valve is operated.

In each case, the water supply must be protected by an air gap or some such no less effective mechanical device in accordance with fluid category 5.

Refrigerators

Many modern refrigerators contain ice making facilities that can be connected directly to a cold water supply. Where this is possible, the refrigerator connection should be installed in accordance with fluid category 2. The reason for this is that, although the water is fluid category 1, it undergoes both a change in temperature and state for it to become ice. Therefore, this kind of installation must be fitted with a single check valve or an approved EA/EB Backflow prevention device.

Washing machines /Dishwashers

Washing machines/Dishwashers are classified as fluid category 3 and must be protected against backflow. In most instances, an air gap to guard against backflow is already built into the appliance by the manufacturer. However, where multiple washing machines are installed on the same installation, then some form of mechanical backflow prevention is required. This is usually an approved type EC/ED backflow prevention device (or double check valve).

Modern washing and dishwashing machines only require a cold connection to the water supply. Internal heaters heat the water as required.

Taps, outlets and valves

Mixer taps

True mixer taps allow both hot and cold water to be mixed inside the tap body, and as such, constitute a cross connection between fluid category 1 and fluid category 2, which is not allowed under the Water Supply (water supply) Regulations 1999. Both hot and cold supplies must be protected by an approved EA/EB backflow prevention device (single check valve).

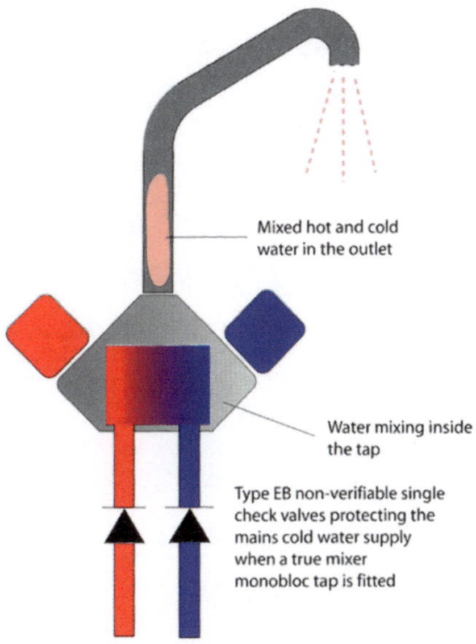

Bi-flow mixer taps

Bi-flow mixer taps are essentially two taps within a single tap body. The tap is divided through the body and into the spout so that the hot and cold water supplies do not mix until they emerge from the spout. They do not require backflow protection.

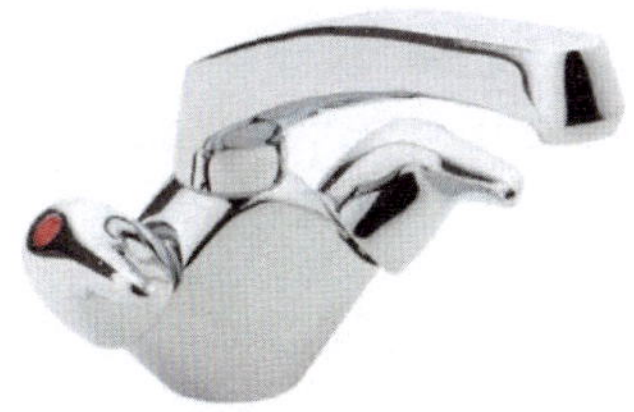

Pillar taps

These are available with both ½" and ¾" male BSP tails for use on washbasins and baths. They are designed to give and AUK2 airgap at the appliance, the outlet of the tap being higher than the spill-over level of the appliance.

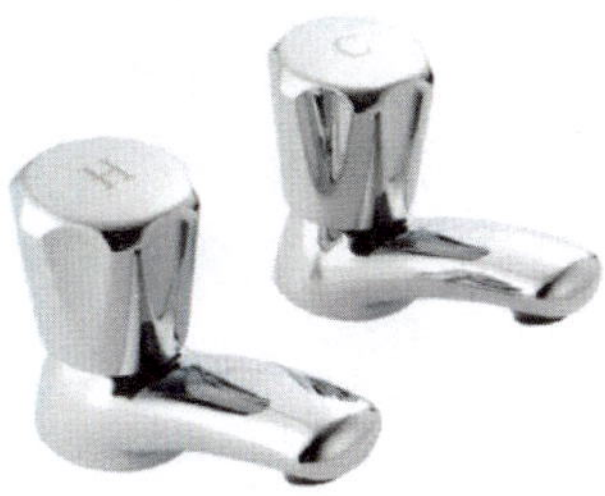

High Neck Pillar taps

Similar in design to pillar taps, high neck pillar taps are designed for use on kitchen sinks. The high neck provides an AUK3 air gap suitable for fluid category 5, which is a mandatory requirement for kitchen/utility sinks, cleaners sinks, Belfast sinks and any sink or washbasin installed in a health care facility.

Bib taps

Bib taps are predominantly used in conjunction with Belfast sinks, cleaners' sinks and situations where an AUK3 air gap is a requirement, such as hospitals, dental and doctors surgeries and health care facilities. In most cases, they are fixed to the wall using back-plate elbows, although as can be seen in the photograph, they can also be installed with concealed pipework.

Hose Union Bib taps (Outside taps)

Similar in design to bib taps, hose union bib taps have threaded hose connection at the spout for easy connection of a garden hose. These must be regarded as a fluid category 3 risk and be protected by an approved type EC/ED backflow prevention device (or double check valve).

Stop valves (BS5433 and BS1010)

Stop valves are designed for the isolation of high/mains pressure cold water supplies. Their internal design restricts the smooth flow of water and so they should not be used on low-pressure cold water supplies. Stop taps are not suitable for hot water systems.

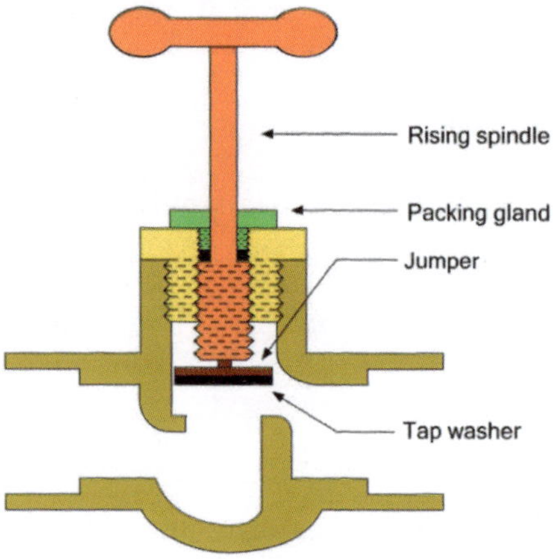

Stop valves are available with compression and capillary connections for use on copper, polyethylene and polybutylene pipe.

Stop valves have an arrow moulded into the body of the valve to show the direction of the flow of water.

Taps

Taps fall into three categories:

- Taps with rising spindles (BS1010)
- Taps with non-rising spindles (BS5412)
- Ceramic disc type

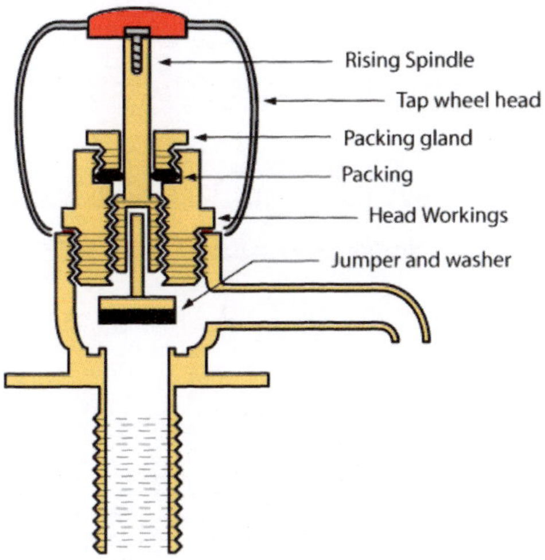

Taps with rising spindles (BS1010)

BS1010 taps contain a tap washer that is attached to a jumper plate. The jumper plate is then inserted into the rising spindle, which rises when the tap is rotated counter-clockwise, turning the tap on allowing water out of the tap spout. The jumper plate can either be fixed (usually for hot taps or low pressure cold water) or loose (usually used on high pressure water supplies).

Although BS1010 has been withdrawn for many years, BS1010 taps and head workings continue to be manufactured, because they are very reliable and can be repaired and refurbished easily and quickly.

BS1010 tap head workings are interchangeable, insomuch that tap head workings from one manufacturer will fit other taps.

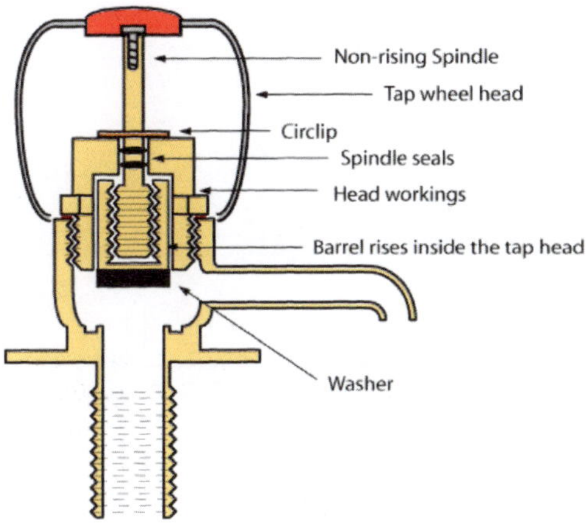

Taps with non-rising spindles (BS5412)

BS5412 taps do not have a rising spindle. The spindle remains in one position, retained by a circlip. At the end of the spindle is a thread. When the tap is rotated counter-clockwise, the thread lifts a hexagonal-shaped barrel with a washer attached to it.

BS5412 taps are for use on washbasins, baths, kitchen sink and bidets, in ½" and ¾" sizes.

Unlike BS1010 taps, the head workings are not interchangeable from manufacturer to manufacturer, with each company having their own design.

Ceramic disc taps

Ceramic disc taps do not contain a washer. Instead they have two thin ceramic plates, one of which is fixed while the other rotates through 90°. Each plate has a pair of corresponding slots cut into them where the water passes through.

Ceramic discs tap heads correspond to the type of water. Hot tap heads rotate clockwise while cold rotate anti-clockwise. They are colour coded for easy identification red and blue. Ceramic disc replacement needs to be carefully considered as the tap workings are not universal and the correct make and model must be available to be able to obtain the correct type.

Full-way gate valves

Full-way gate valves are used predominantly to isolate low pressure water supplies from storage cisterns. They use a 'gate' to stop the flow of water without the need for a rubber washer. When the valve head is rotated counter-clockwise, the gate rises allowing water to flow. The design of gate valves allows for water to flow at full bore with virtually no restriction in flow rate, hence the term 'full-way'.

They are not suitable for high pressure water supplies.

Spherical plug valves/ Servicing valves

These isolation/servicing valves contain a brass ball or sphere with a hole through the middle. When the hole is in line with the pipe, the valve is on and water will flow. When the valve is turned through 90°, the hole is across the flow of water and the water flow stops.

There are many versions of spherical ball valves available. Some have handles or levers, such as the types shown in the photographs, while others have a screwdriver slot to enable the valve to be turned on and off.

Drain valves (BS2878-2)

Drain valves, also known as drain-off valve and M/T valves, are used to allow systems to be completely drained of water for repair, maintenance or replacement. They should be placed at low points in the system, close to places where the drained water will not pose a nuisance.

There are many types of drain valve available, including types with male threaded ends and those with or without packing glands.

Drain valves must be fitted in accordance with the Water Supply (water fittings) Regulations.

Float operated valves (to BS1212 parts 1-4)

Float Operated Valves or FoVs are used to control the flow of water into cisterns. Manufactured to BS1212, they are purposely designed to shut off the inflow of water once it has reached a predetermined level. They can be used on high pressure water or low pressure, depending on the type of orifice that the FoV is fitted with. A white orifice denotes high pressure supply (marked HP) and a red orifice denotes a low pressure supply (marked LP).

There are four specific types:

B.S.1212 Part 1: Portsmouth pattern and Croydon types

B.S.1212 Part 2: Diaphragm type (brass)

B.S.1212 Part 3: Diaphragm type (plastic)

B.S.1212 Part 4: Torbeck equilibrium type for WC cisterns only

B.S.1212 Part 1: Portsmouth pattern and Croydon types

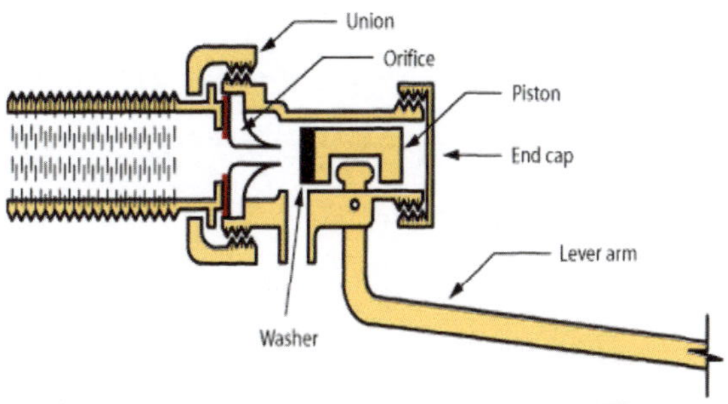

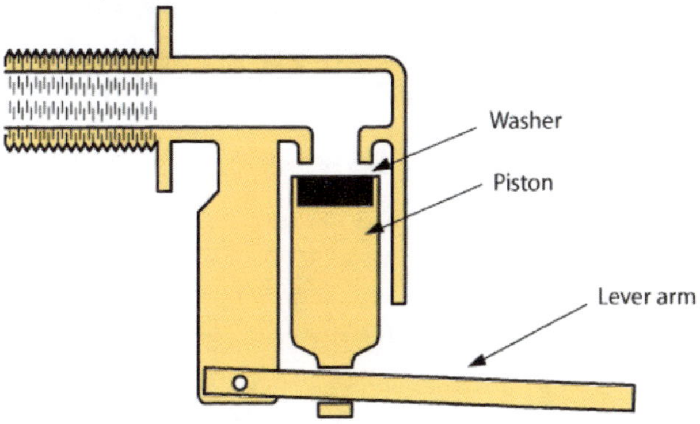

Portsmouth pattern

This type of FoV releases water from underneath the valve. Because of this, they are prohibited from new installations without a back-flow prevention device as the outlet may become submerged in water if the valve fails, creating a possible back-flow contamination issue. However, they may be repaired if they are part of an existing installation.

The Portsmouth-type valve are vulnerable to noise and water hammer.

Croydon pattern

The Croydon-type FoV again releases water from below the valve making it susceptible to back-flow issues. It is instantly recognisable due to the piston being in the vertical position.

Croydons tend to be very noisy in operation.

They are no longer manufactured but may still be found in older, high level WC cisterns.

B.S.1212 Part 2: Diaphragm type (brass) and B.S.1212 Part 3: Diaphragm type (plastic)

Parts 2 and 3 FoVs discharge water from the top of the valve rather than underneath, making them less likely to encounter backflow problems. The main difference with this kind of FoV is the size of the rubber washer. It is large, flat and very flexible.

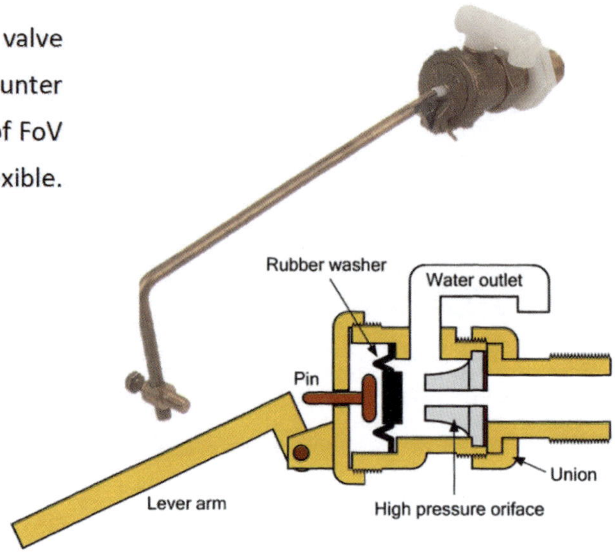

The valve has very few moving parts, making it quieter in operation than the part 1 and less likely to be associated with water hammer and reverberation of pipework.

Part 3 FoVs are identical to the part 2 except they are made of plastic making them ideal for WC cisterns but not recommended for storage cisterns in roof spaces due to the risk of freezing and splitting of the plastic body.

Parts 2 and 3 FoVs discharge water from the top of the valve rather than underneath, making them less likely to encounter backflow problems. The main difference with this kind of FoV is the size of the rubber washer. It is large, flat and very flexible.

The valve has very few moving parts, making it quieter in operation than the part 1 and less likely to be associated with water hammer and reverberation of pipework.

B.S.1212 Part 4: Diaphragm Equilibrium type (Torbeck)

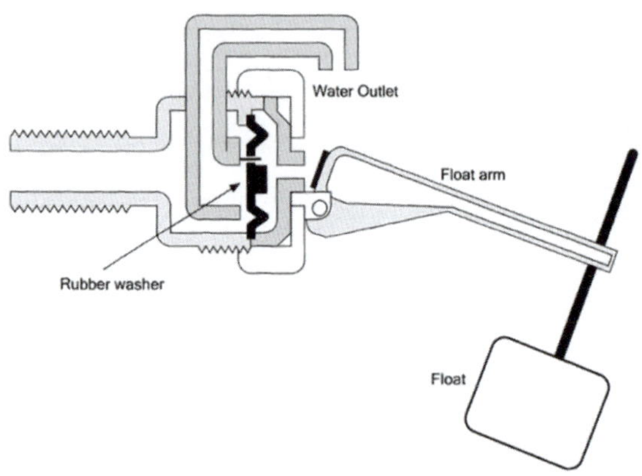

The Torbeck valve works by ensuring that there is equal pressure on both sides of the rubber washer when the valve is in the open position. When the valve is closed, the float arm closes the pressure relief orifice on the front of the valve creating a higher pressure in front of the rubber washer. This pushes the washer on to water outlet to stop the flow of water.

These valves are much quieter than other FoV types although the valve tends to close with a 'snap' action, which can create some pipe reverberation.

These are only for use inside WC cisterns.

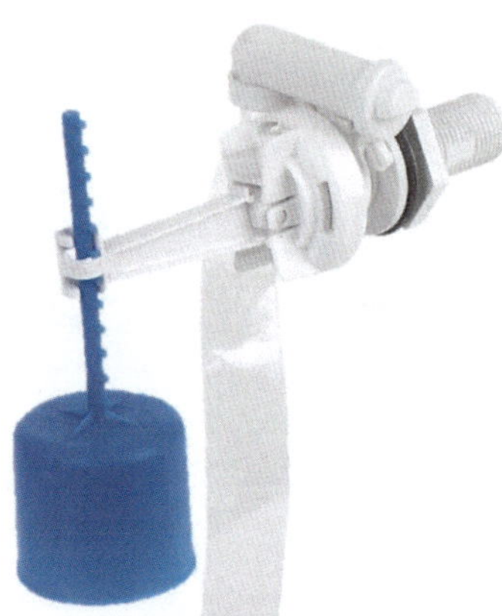

Water meters

External installations

A Groundbreaker© water meter housing

There are two types of external water meter installation:

- A water meter fitted in an underground housing at the boundary of the property, and;
- The 'Groundbreaker©' external water meter housing fitted on an external wall of the property (as shown in the photograph)

Fitting a water meter externally has several advantages

1. It prevents illegal tampering with the water meter.
2. The meter can be read without the need to access the building

Internal installations

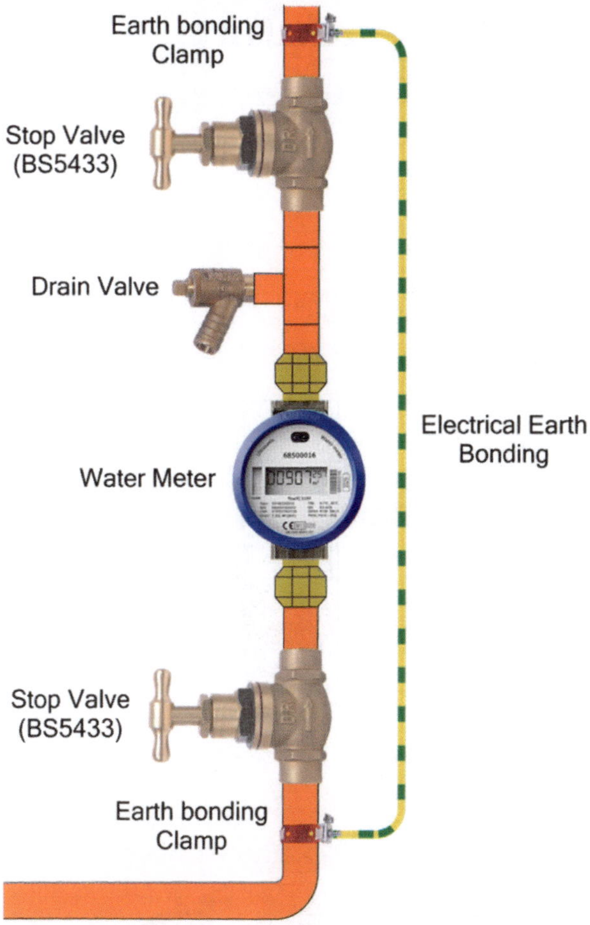

Water meters on existing cold water supplies are usually fitted internally close to the internal stop valve. It must be fitted in the manner shown in the image:

Stop Valve – Water Meter – Drain Valve – Stop Valve. This is to prevent a breach in the Water supply (water fittings) Regulations 1999 with regards to erroneous metering of water. Any water drawn through the drain valve on a live cold water supply would register on the meter as water consumed. If the drain valve was fitted below the meter, any water drawn off would not register.

Earth bonding should also be fitted from below the first stop valve to above the second stop valve so that earth protection is continued.

Showers

Gravity

Gravity showers use a mixing valve to blend cold water and hot water to the required temperature. There are several types:

- Manual-type mixing valves (non-thermostatic)
- Thermostatic mixing valves
- Bath/shower mixing taps

Manual-type mixing valves (non-thermostatic)

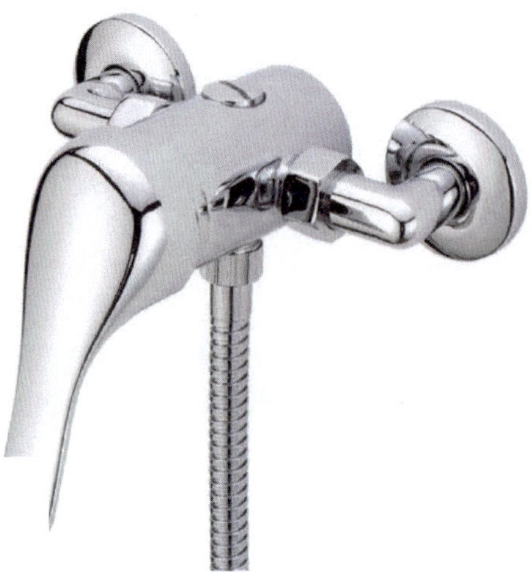

These are **not** thermostatic control showers. They rely on both hot and cold water supplies having the same pressure and flow rate. Manual non-thermostatic valves **do not** adjust the water temperature to compensate for flow and temperature fluctuations. Because of this, it is important that the temperature of the hot water supply is fairly stable. They are usually used on low-pressure water supplies, fed from a storage cistern.

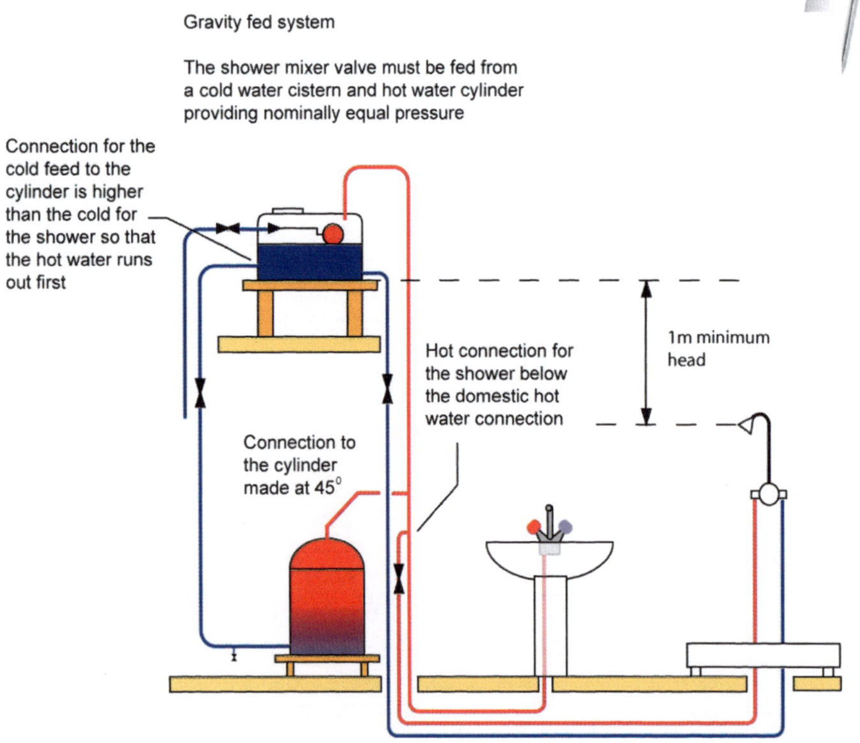

A gravity fed shower installation

Thermostatic mixing valves

Thermostatic mixing valves are fitted in exactly the same way as manual non-thermostatic mixing valves. The difference is that these shower valves maintain the showering temperature despite fluctuations in flow rates. The water temperatures generally are 15°C for cold water and 55°C for hot water will give a showering temperature of around 38°C to 42°C. There are two types of Thermostatic mixing valve:

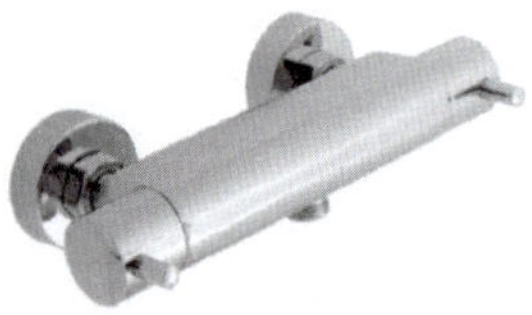

A thermostatic bar-type mixing valve

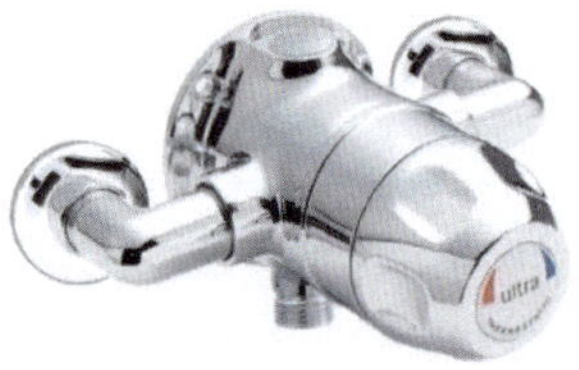

A thermostatic mixing valve

- Wax Capsule – this uses a copper capsule containing a mixture of heat sensitive wax and copper particles to control the water temperature. As the wax gets hot, it expands activating a shuttle, which effectively controls the flow of hot water. The hotter the water, the more expansion of wax takes place and the more the restriction is put on the flow of hot water.
- Bi-metallic coil – These use a coil of two different metals with differing expansion rates, usually brass and invar steel. The brass expands more quickly than the steel and this causes the coil to twist. This twisting motion activates a shuttle that either increases or decreases the flow of hot water to the valve.

Pressure compensating mixing valves

These are designed to give greater stability to the water temperature when the pressures and flowrates of the hot and cold supplies are different. Some types are designed for use on high/low pressure differences while others are designed for use on high/low flowrates where the pressures are equal. There are two specific types:

- Sequential control – these maintain temperature whilst adjusting to subtle pressure fluctuations.

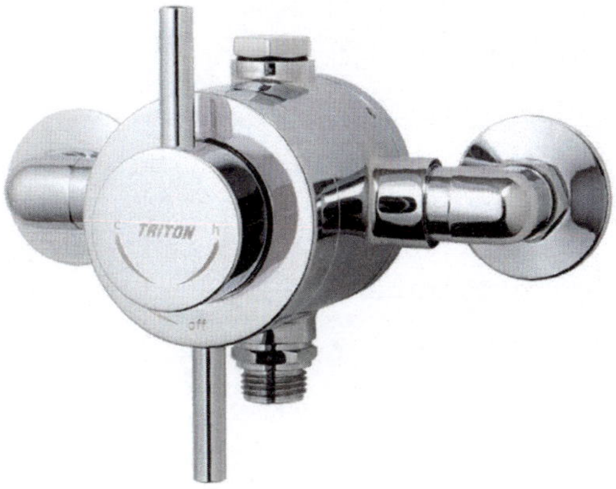

A pressure compensating shower valve

- Dual control – these have separate mechanisms for both temperature and flowrate control. They react to differences in pressure but not to differences in temperature. Ideal for combi boilers and instantaneous gas water heaters.

Digital shower valves

These use a digital panel that communicates with the shower mixing valve to maintain a strict showering temperature regime. The digital control panel can be located as far as 10m away from the valve, giving greater installation flexibility.

A digital shower valve

Bath/shower mixing taps

Bath/shower mixing valves rely on equal pressures of both hot and cold supplies to work effectively. They simply replace the normal pillar-type bath taps, using the same tap hole spacings, which are generic to almost all bath types.

The example shown right is thermostatic.

To prevent back-siphonage, bath/shower mixers have an inbuilt air gap, known as a HC Diverter. This

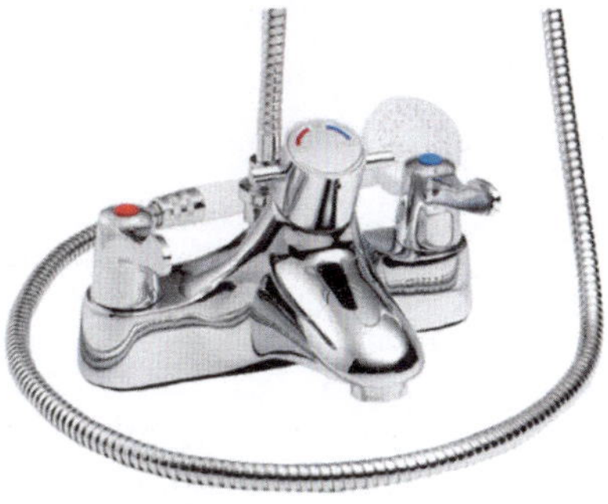
A bath/shower mixing valve

automatically drains the hose and shower head when the taps are turned off. Alternatively, single check valves (EA/EB Back flow prevention devices) can be fitted to both hot and cold supplies.

Shower pumps

Shower pumps are used to boost low pressure hot and cold supplies from storage cisterns, giving the 'power shower' feel. There are two different kinds:

- Single impeller outlet types, and;
- Twin impeller inlet types

Single impeller outlet shower pumps

A single impeller type shower pump is fitted AFTER the shower valve to boost the mixed water from the valve to the shower head.

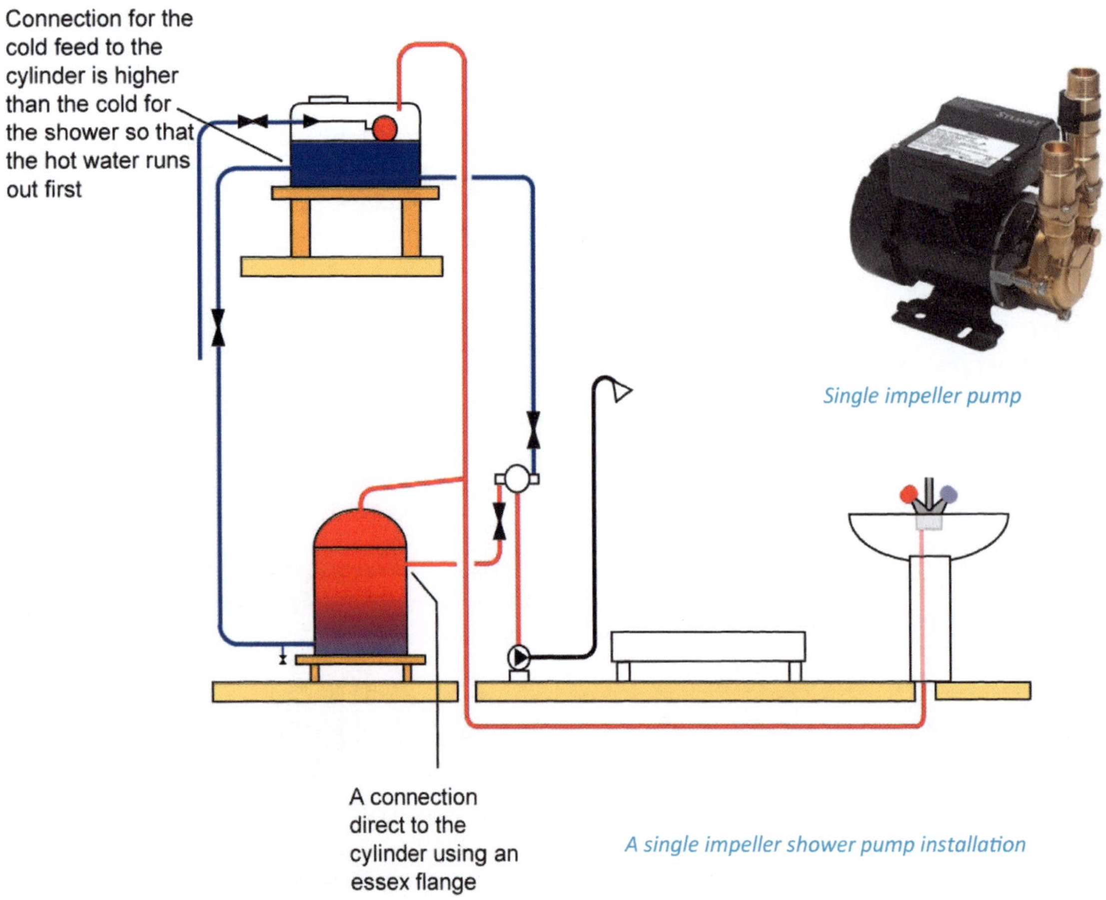

Single impeller pump

A single impeller shower pump installation

Twin impeller inlet shower pumps

Twin impeller shower pumps are installed before the shower valve and boost the individual low pressure hot and cold supplies to the shower mixing valve. They have two identical impellers driven by a single motor to ensure that both the pressure and the flow rate to both supplies is the same.

A twin impeller pump

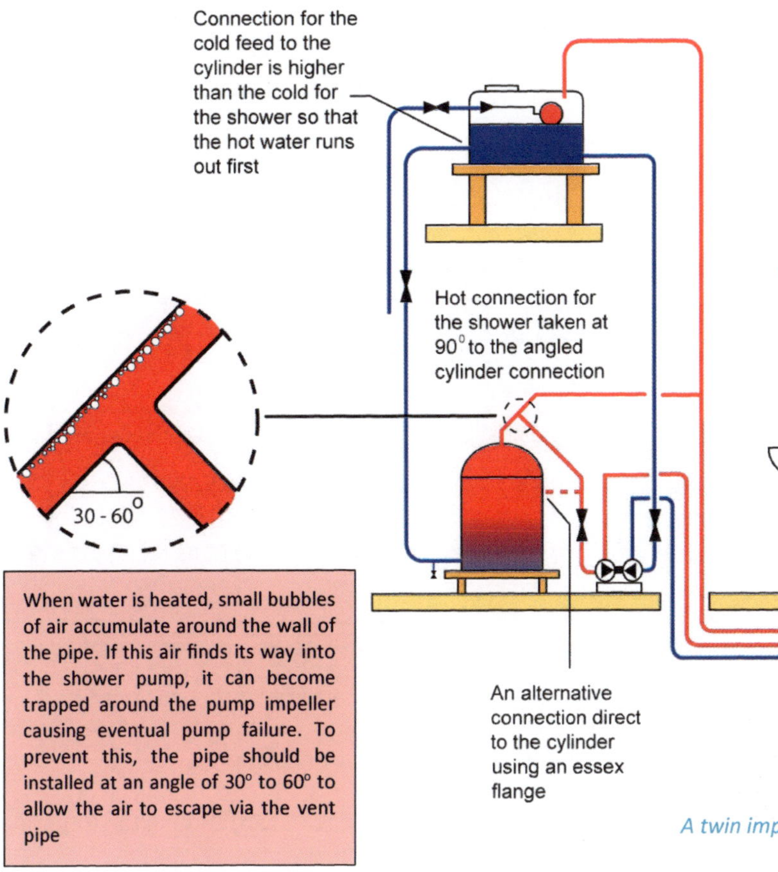

A twin impeller shower pump installation

Instantaneous electric showers

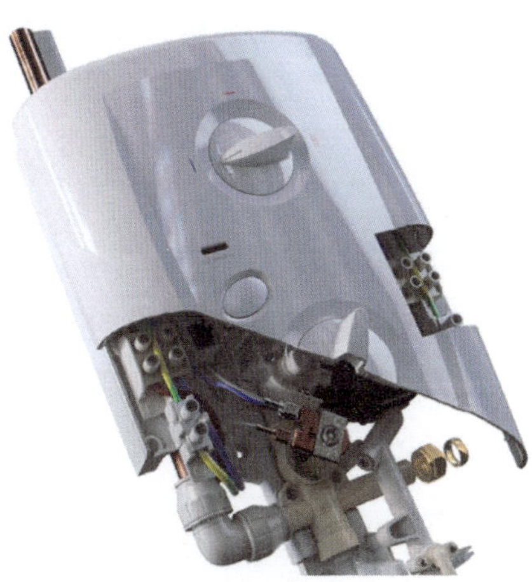

Inside an electric shower

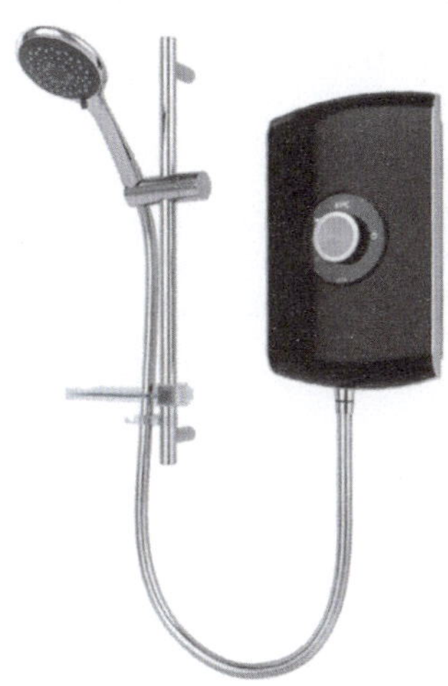

A typical electric shower

There are many different electric showers on the market with power ratings from 8kW to 12kW. Instantaneous electric showers are fed with water direct from the cold water mains supply. This is then heated by a small but powerful heater inside the shower itself.

Electric showers should be protected from back-siphonage by either an EC/ED backflow protection device (double check valve) or a hose retaining ring that is supplied with the shower. The retaining ring prevents the shower hose from entering bath or shower water.

Care should be taken when installing electric showers. The electrical installation must be completed by a qualified electrician so that the correct cable size, fuse and isolation switches are installed.

Water treatment

Water softeners

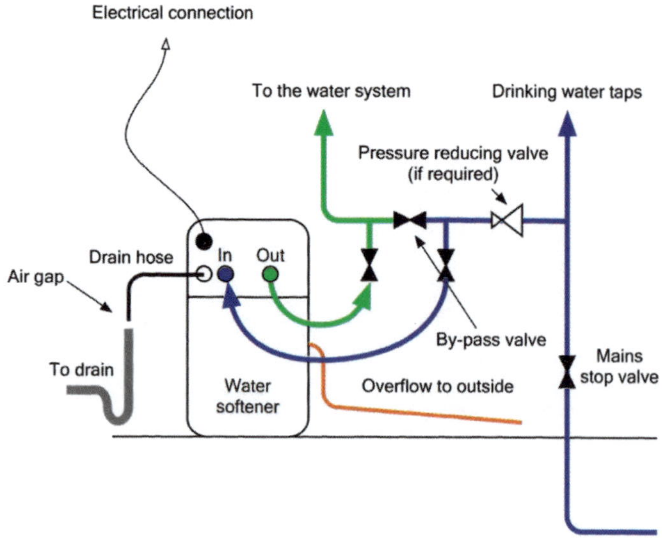

Installation of a water softener

A water softener removes water hardness in hard water areas. They can be installed in both domestic and commercial buildings. It is usual practice to install water softeners as close to the main internal stop valve as possible so that the entire cold water installation is served with softened water.

Most water softeners use a process called ion exchange and are known as 'base exchange water softeners'. In this process the calcium and magnesium salts are replaced with sodium. Water softeners are designed to automatically wash away the calcium and magnesium ions with salt water at least once a day. Every month, the softener the unit requires the replacement of the salt granules.

Water softeners reduce the water hardness from around 350mg/l to less than 10mg/l.

It should be remembered that water softener installation requires that at least 1 tap, usually the kitchen sink, remains unsoftened for drinking and culinary purposes.

Water filters.

Water filters are designed to improve the taste and appearance of water by removing impurities. There are two common types:

- Jug filters that are filled from a tap
- Plumbed in filters usually sited under kitchen sinks.

There are generally, six different forms of filter:

Activated carbon

Carbon in the form of powder, granules or block is used to remove chlorine, thereby improving the taste and odour of the water. The carbon has a large surface area and this attracts and absorbs organic substances in the water.

Ion exchange

Reduces lime scale formation and other metal-ion contaminants (including lead). They generally use tiny granules of sodium salts, which replace the mineral contaminant ions with sodium ions.

Sediment filters

These remove fine particles from the water by using a mesh through which the water passes. The finer the mesh, the more particles are removed.

Reverse osmosis These work under high-pressure by passing water through a semi-impermeable membrane. This process removes the dissolved mineral content.

Distillation The mineral content is removed by boiling the water to steam and condensing the steam back to water.

Disinfection These are usually used with water supplied from bore hole. They use ultra-violet light to kill bacteria and other micro-organisms.

Water conditioners

These work by supressing lime scale formation. This has the following benefits:

- A reduction of scale in pipework
- Easier cleaning of outlets, such as shower heads
- Less lime scale deposits on taps and outlets

There are three basic methods of conditioning water:

Electro-magnetic

These prevent scale build-up by influencing which type of calcium crystal precipitate. By doing this, it ensures that only aragonite crystals are formed, which have a needle-like structure, making it difficult for them to adhere to smooth surfaces.

They are generally used on individual appliances, such as combination boilers, to reduce internal scale build-up in hard water areas.

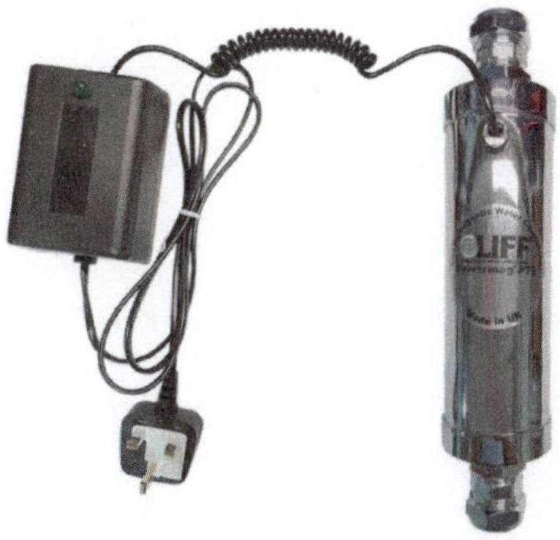

Electro-magnetic water conditioner

Electrolytic

These add a microscopic amount of zinc to the water, which prevents the formation of calcium crystals. Any crystals that are formed are washed through the system by the flow of the water.

An electrolytic water conditioner

Electro-chemical

These conditioners are quite large and require an electrical supply. They are filled with ceramic beads and work by causing the magnesium and calcium crystals to precipitate, or re-form.

Cisterns

A cistern is defined as a vessel containing water at atmospheric pressure. In plumbing, there are several different types:

Cold water storage cisterns (CWSC)

These store water for distribution in a low-pressure indirect cold water system. It supplies cold water only.

Cold water feed cisterns (CWFC)

Identical to cold water storage cisterns. Cold feed cisterns are designed to feed cold water, via the cold feed pipe, to an open vent hot water storage system. This must be of at least of the same capacity as the hot water storage vessel it serves.

Combined Cold water storage and feed cisterns (CWS&FC)

These are a combination of the first two cisterns and are primarily for use when the cistern is supplying cold water to both a cold water system and hot water system from the same cistern.

Feed and expansion cisterns (F&E)

These small cisterns that are designed for use with vented hot water heating systems and, as the name suggests, combine two very important roles:

1. It supplies cold feed water to a heating system, and;
2. Accommodates any expansion of the water due to the water being heated.

The cistern must be large enough to accommodate any expanded water and must be sized accordingly.

Break cisterns

Used in conjunction with boosting pumps, a break cistern is used to provide a 'break' in the cold water supply between the water undertaker's cold water main and the property's cold water installation. They are often used where the cold water installation requires more water than the cold water main can deliver or where a greater water pressure is required such as multi-storey installations. There are a number of advantages to using break cisterns:

- Break cisterns ensure that there is no surge on demand when boosting pumps are used,
- Break cisterns ensure that contamination between the building and the cold water main cannot occur,
- Using a break cistern maintains a sufficient supply of water to the property at times of peak demand,
- Break cisterns maintain a plentiful supply of water to other properties by not drawing excessive amounts of water from the main.

As with all cisterns, break cisterns must be fitted in accordance with the Water Supply (water fittings) Regulations 1999 and must be sized to ensure that stagnation of water does not occur.

Sectional (1000 litre+)

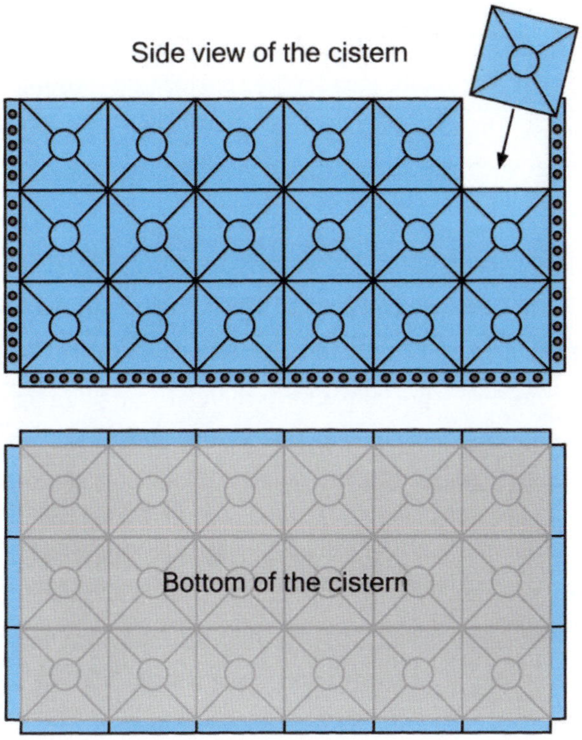

How sectional cisterns are constructed

Sectional cisterns are constructed from 1m^2 sections of steel, aluminium or Glass Reinforced Plastic (GRP) and can be sized to give any capacity. The sections are bolted together using stainless steel bolts and sealing mastic that conforms to the Water Supply (water fittings) Regulations 1999. They are usually used where water capacities required exceed 1000 litres.

WC/urinal flushing cisterns

These are installed with the sole purpose of clearing the contents of WC pans and Urinal bowls. Urinal cisterns are usually automatically flushed at set periods and should be fitted with a water saving device such as a cistermiser © valve. Urinals can be fitted with flushing valves that flush directly from the mains cold water. WCs must be flushed using a cistern fed flush arrangement using either a siphon or a dual flush valve (4 ltrs short flush, 6 ltrs long flush). They are covered in greater detail in a later chapter

Boosted cold water system components

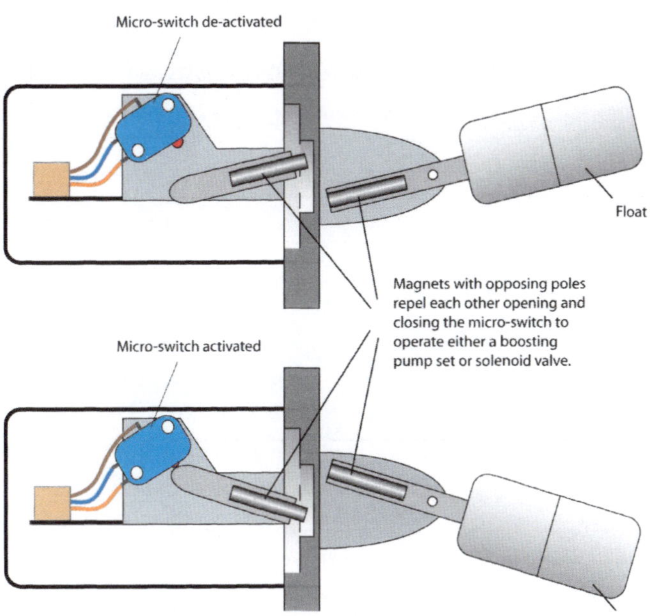

A magnetic toggle float switch

Boosted cold water systems contain components that are important for the correct operation of the system, performing functions such as start/stop, open/close and pressure indication and temperature sensing.

Float switch (also known as a 'Level Switch')

A float switch is a device that is installed inside cold water systems serving boosted cold water systems to detect water levels and activate or deactivate boosting pumps as required, open and close solenoid valves and activate water level alarms. There are many different types, including:

- Magnetic toggle switches
- Sealed float switches
- Electronic and ultrasonic float switches

Pressure switch/transducer

A transducer is an electronic sensor that senses system pressure in large multi-storey cold water systems. It is designed to sense pressure fluctuations and converts the pressure into an electrical signal. Transducers can operate boosting pumps or the compressor feeding the large pressure vessel when the pressure requires boosting to its normal operating pressure. In some instances, they may also be used to sense and control over-pressurisation.

Temperature sensors

The Water Supply (water fittings) Regulations 1999 advise that the temperature of stored water should not exceed 20°C. Above this temperature bacteria growth begins. Because of this, temperature sensors are often used to monitor the temperature of water stored in large cisterns, especially where the water may be used for drinking water purposes in large institutions, such as in hospitals or where there is a 'duty of care'.

Accumulator/pressure vessel

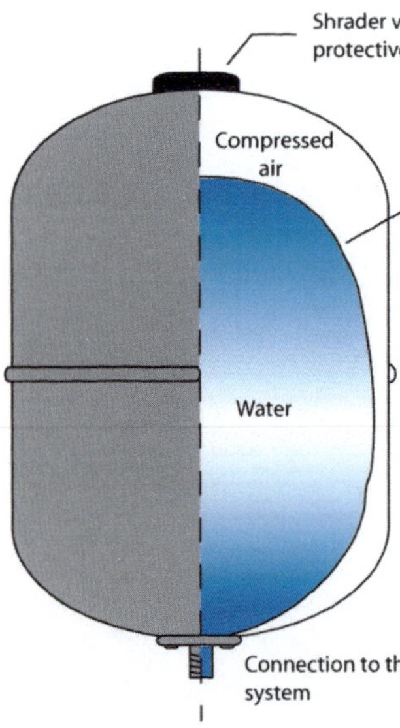

The accumulator

An accumulator is used to maintain the cold water system pressure. It also stores water for use when the system pump is not running. It is a steel vessel that contains a neoprene rubber bag or bladder. On the outside of the bladder is air that is usually at around 80-90% of the water system pressure. When water is pumped into the rubber bladder, the bladder expands and the air is compressed, which leads to an increase in the pressure inside the accumulator. When a tap is turned on, the pressure forces water out of the accumulator. At a predetermined pressure, the system pump starts and the system pressure is maintained. When the tap is turned off, the pump continues refilling the accumulator until it is at full pressure and capacity again. These systems usually operate at pressures between 1.5 and 3 bar.

Booster pump set with accumulator

Booster pump sets

Booster pump sets are used to boost the pressure of a cold water system to an acceptable operating pressure. There are many different types of booster pump, depending on the application and the system it is serving

Pressure relief valve

A pressure relief valve is designed to discharge water in the event of over-pressurisation. It uses a spring to open the valve when the pressure exceeds a set limit. They are usually sited near to the accumulator on boosted cold water installations.

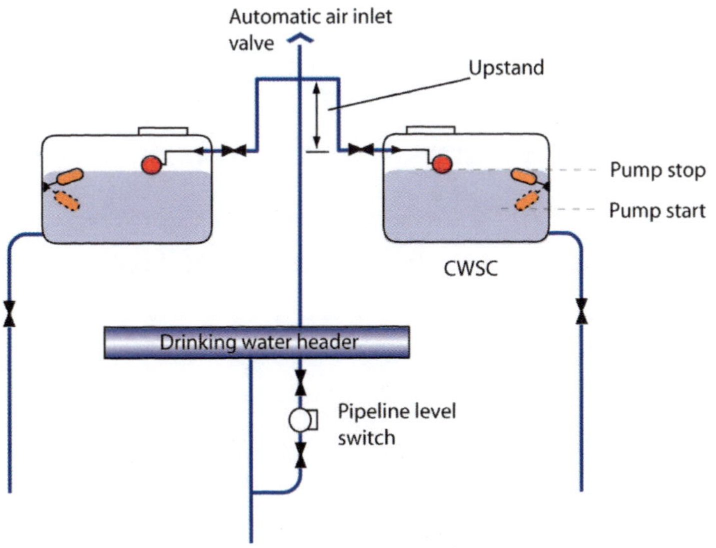

A drinking water header

Pressure gauge

Used to indicate the pressure of a cold water, hot water or central heating system. They are usually fitted at or near the cold water accumulator.

Drinking water header

A drinking water header is a large bore pipe that is designed to hold a quantity of drinking water on boosted cold water systems. It works via gravity for use when the booster pumps are not running.

AC2.5 Describe layout and installation requirements for protected plastic storage cisterns

General installation requirements for domestic cisterns

A cistern complying with Schedule 2 para 16 of the Water Supply (water fittings) Regulations 1999

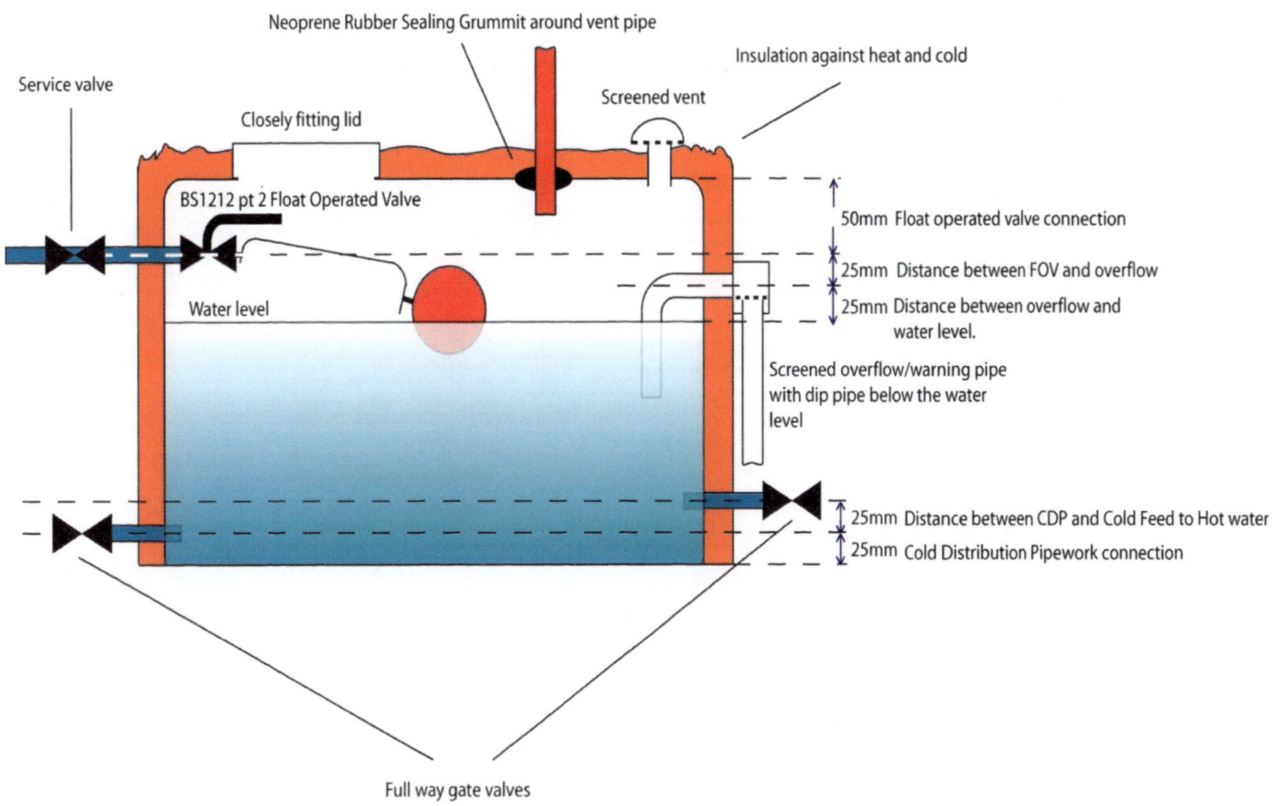

A cistern complying with schedule 2 para 16 of the Water Supply (water fittings) Regulations 1999

Schedule 2 paragraph 16 of the Water Supply (water fittings) Regulations 1999 states that a cistern supplying low pressure cold water to either a cold water or hot water system should be capable of supplying wholesome, potable water. Protection measures must be included during installation to ensure that the water cannot become contaminated in any way. Cisterns must therefore include:

Definition:

Potable – pronounced poe-table, is taken from the French word 'Potable' and simply means drinkable

a) An effective inlet control device to maintain the correct water level i.e. a float operated valve.
b) Service valves on inlet and outlet pipework connections to allow for maintenance and repair activities.
c) Screened warning/overflow pipes.
d) A rigid, close fitting lid, which is not air tight but excludes light and insects.
e) Insulation against freezing or undue warming.
f) Installation methods that eliminate the risk of contamination.
g) An arrangement of pipework so designed as to allow water to circulate, preventing stagnation.
h) Support to avoid distortion or damage which could lead to leaks
i) Access for maintenance and cleaning.

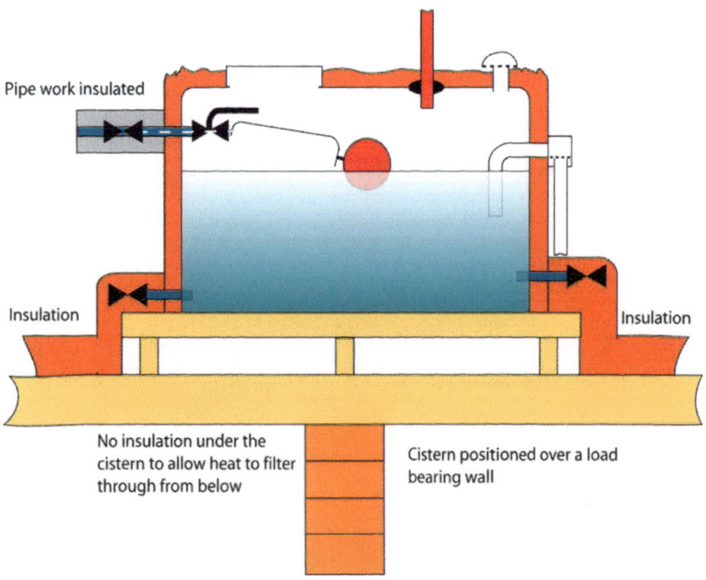

Cistern shown positioned over a load bearing supporting wall

Installation requirements for domestic cisterns

Cisterns that are installed in roof spaces and filled with water create stress on the roof timbers and joists. Water is heavy. At 4°C, 1 litre of water = 1kg. When this is multiplied by the average cistern water capacity of say 230litres, it has the mass of almost ¼ of a metric tonne. Good practice dictates that, wherever possible, cisterns should be positioned over a load bearing wall, as shown left. Many modern properties do not have internal supporting walls and if this is the case then a suitable platform (often called a stillage) must be constructed to spread the load across the roof trusses. The platform

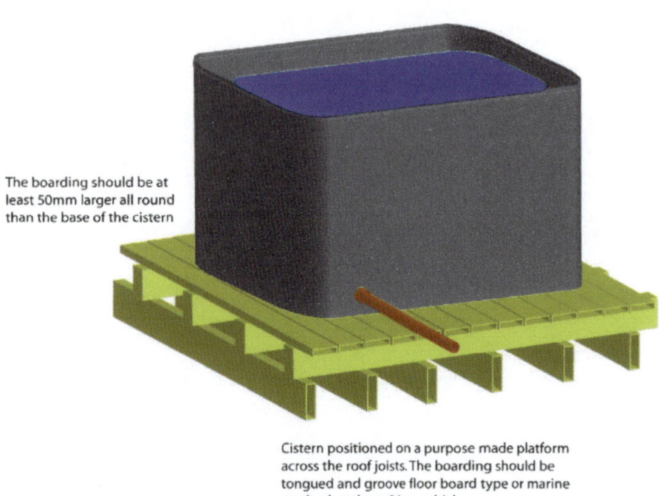

Construction of a suitable stillage. Positioning a cistern on a platform

should be covered in marine grade plywood or tongued and grooved boarding at least 22mm thickness. The platform should be at least 150mm larger than the base area of the cisterns on all sides

Clearance for access and maintenance of cisterns

The minimum access requirement for cisterns of less than 1000 litres for cleaning, maintenance and inspection is 350mm above the cistern. this allows the repair and replacement of float operated valves.

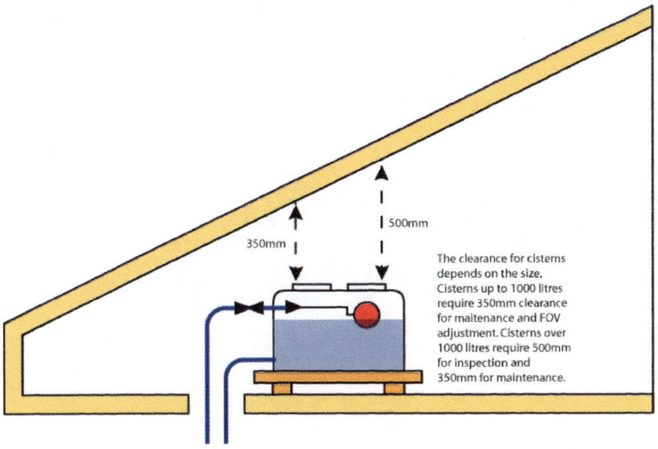

Clearance requirements for cisterns

Inlet requirements for domestic cisterns

The Water Supply (water fittings) Regulations 1999 state that a cistern should be fitted with an adjustable float operated valve (FOV) or other no less effective device for controlling the flow of water into the cistern. For domestic cisterns, this is a BS1212 FOV conforming to BS1212 Part 1 (fitted with suitable backflow protection) and BS1212 part 2 as these allow the adjustment of the water level within the cistern. BS1212 covers FOVs up to 54mm diameter. FOVs that exceed this size must be authorised by the local water undertaker. (see pages 64 and 65)

Outlet requirements for domestic cisterns

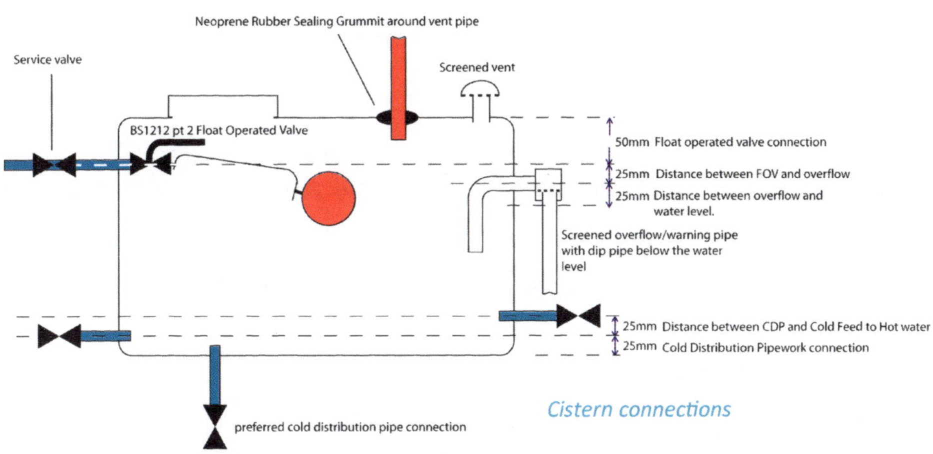

Cistern connections

The drawing left indicates the position of the outlets (cold feed pipe, cold distribution pipe) for domestic cisterns. The cold distribution pipe must be the lowest connection on the cistern to ensure that water to the hot water storage vessel (cylinder) via the cold feed pipe is the first to cease flow in the event of mains cold water failure. This prevents the possibility of scalding when showers and mixing valves are used. The cold distribution pipe should be taken from the bottom of the cistern to prevent sediment build up and possible legionella growth. An alternative position for the cold distribution pipe (25mm below the cold feed pipe and on the opposite side of the cistern) is shown in the drawing.

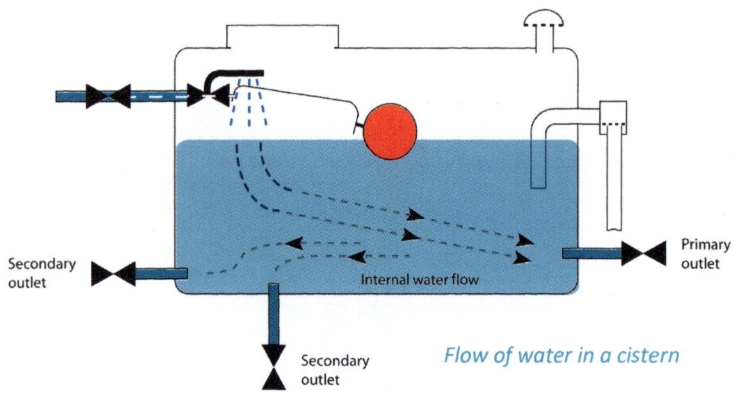

Flow of water in a cistern

Positioning the pipework correctly helps to prevent water stagnation by ensuring that water circulation takes place within the cistern when the outlets are used. If there is only one outlet connection, such as a cold feed to a hot water storage vessel, then this MUST be fitted on the opposite side to the FOV.

If two outlets are fitted, then the highest outlet, usually the cold feed to the hot water system, should be fitted on the opposite side to the inlet connection (FOV).

Connecting two or more cisterns

Providing large quantities of water can be problematic where there are space restrictions such as in roof spaces. In these circumstances, cisterns will need to be linked. However, there are four points that must be remembered if stagnation and possible legionella bacteria growth are to be avoided:

1. The cisterns must be small enough to encourage rapid turnover of water but large enough to prevent water running out.
2. The cisterns must have Inlet and outlet connections at opposite ends.
3. They should be regularly inspected and cleaned to prevent sediment build-up.
4. Cisterns should be dressed and installed in accordance with the Water Supply (water fittings) Regulations.

Cisterns can either be interconnected in two ways:

- In series
- In parallel

Cisterns connected in series

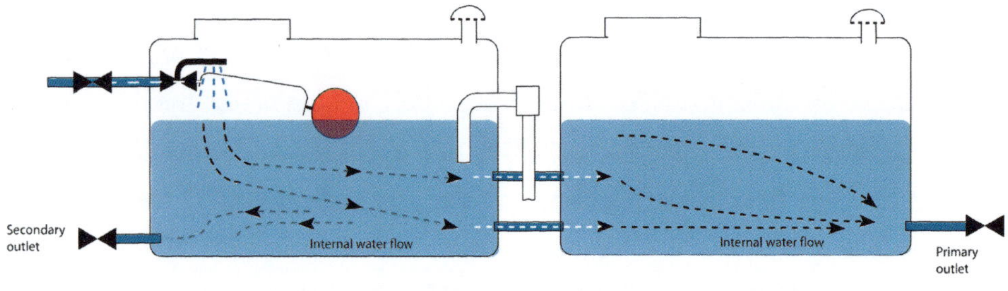

Cisterns connected in series

Take a look at the drawing left. This shows how cisterns should be connected in series. Here, the cisterns are connected at the ends in two places – the bottom of the cistern and the middle of the cistern. This allows water movement across the whole body of water and not just at the bottom. The primary connection i.e. the connection that gets the most usage usually the cold feed to the hot water storage vessel, should be connected on the opposite cistern to where the FOV is fitted, this encourages water movement and prevents stagnation of water. The secondary connection i.e. the cold distribution pipework, can be made on the cistern with the FOV if required. The overflow/warning pipe must be connected onto the cistern with the FOV.

Most importantly, both cisterns MUST be of the same size and capacity and no more than two cisterns should be connected in series.

Cisterns connected in parallel

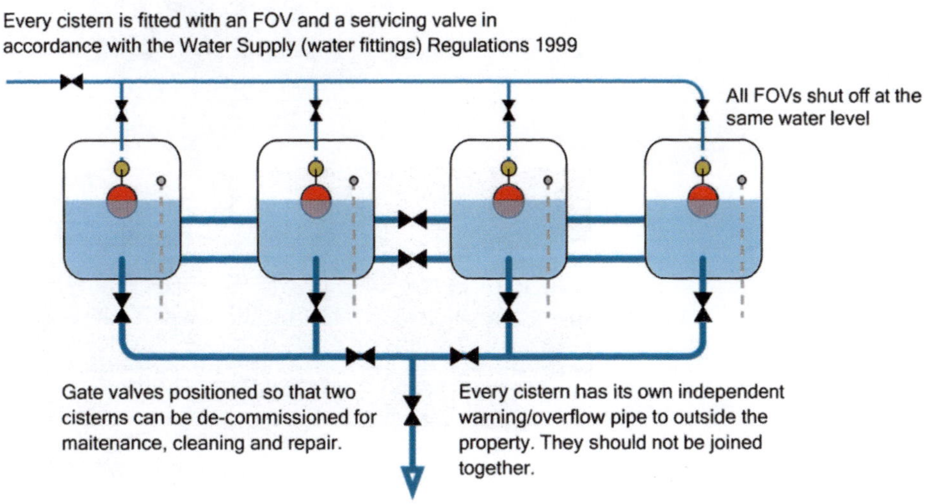

Cisterns in parallel

The drawing left shows four cisterns connected in parallel. In other words, they are connected end-on. This method is usually used where two or more cisterns are to be interconnected. With this method, every cistern has an FOV and a service valve and, since where a FOV is fitted, there should be an overflow, every cistern is fitted with its own independent overflow to outside the property. Overflows ***must not*** be connected together and must be placed in a visible and conspicuous position.

The cisterns must be arranged as follows:

a. All of the cisterns are connected in the middle and at the bottom to allow good flow-through of water.
b. Cistern capacities should be kept to a minimum to ensure a frequent and rapid turn-over of water to prevent stagnation.
c. Inlets and outlets are kept on opposite ends/sides of the cistern.
d. All cisterns should have independent gate valves on each outlet pipe.
e. Two cisterns can be isolated at any time for cleaning and maintenance. While two cisterns are out of commission, two cisterns will be working. This allows the cold water to remain in operation at all times.
f. Stratification is prevented by the use of delayed action float operated valves.

The materials used for fabricating domestic cisterns

For domestic properties, cisterns are usually manufactured from two types of plastic:

- Polyethylene
- Polypropylene

These two materials are chosen because they have a light weight construction, strength, good resistance to bacterial growth and corrosion and have the flexibility to be able to be passed through roof space openings. Cisterns must be properly supported over their entire base area.

A domestic one-piece cistern

A circular cistern

A rectangular cistern

Holes to allow the fitting of tank connectors must be cut using a hole saw. A 17mm hole saw will allow a ½" x 15mm tank connector to be fitted and a 25mm hole saw will suffice for a ¾" x 22mm tank connector. The following points should be noted when dressing a cistern:

- Heated pipework must not be used to make holes in plastic cisterns as this alters the structure of the plastic making it brittle and susceptible to cracking and leakage.
- No jointing linseed oil-type compound must be used when making connections to plastic cisterns as these tend to break down the plastic making it susceptible to cracking. The linseed oil base also provides a breeding culture for legionella pneumophila bacteria.
- Joints to the cistern can be made with plastic 'poly' washers.
- Do not overtighten the tank connectors as this may cause the cistern to split.

The materials used for fabricating large cisterns (1000litres+)

Large cisterns of 1000 litres and above can be supplied in $1m^2$ sections for fabrication on site or in one piece. Sectional cisterns can be made to suit any capacity requirements. The materials for large cisterns are:

- Sectional cisterns:
  - Glass Reinforced Plastic (GRP) BS EN 13280: 2001.
  - Steel to BS 1564: 1975 with protection against corrosion and subsequent water contamination (although this BS has been withdrawn, it is still the standard by which steel sectional cisterns are manufactured).
- One piece cisterns:
  - Glass Reinforced Plastic (GRP) BS EN 13280: 2001.
  - Plastic BS 4213:2004 and BS EN 12573 - 1:2000:
    - Polyproylene (PP).
    - Polyethylene (PE).
    - Polyvinyl Chloride (PVC).

A sectional Glass Reinforced Plastic (GRP) Cistern

A sectional galvanised steel Cistern

A one-piece Glass Reinforced Plastic (GRP) cistern

Warning and overflow pipes

Cisterns up to 1000 litres are fitted with overflow pipes, while cisterns 1000 litres+ are fitted with warning and overflow pipes:

- A warning pipe is a small overflow pipe that is intended to act as an early warning sign that the FOV has malfunctioned and that the overflow is about to start running. This is important because overflow pipes are generally one size larger than the inlet water pipe and FOV and, once they begin to run, they can very quickly waste vast amounts of water. Warning pipes are not required on cisterns of less than 1000 litres capacity.
- An overflow pipe, sited below the shut off level of the FOV, acts as a sign that the FOV has failed to shut off. It has two vital jobs:

- To show that a FOV has malfunctioned and requires repair.
- To divert water away from the building that would otherwise leak into the property and cause water damage due to FOV malfunction.

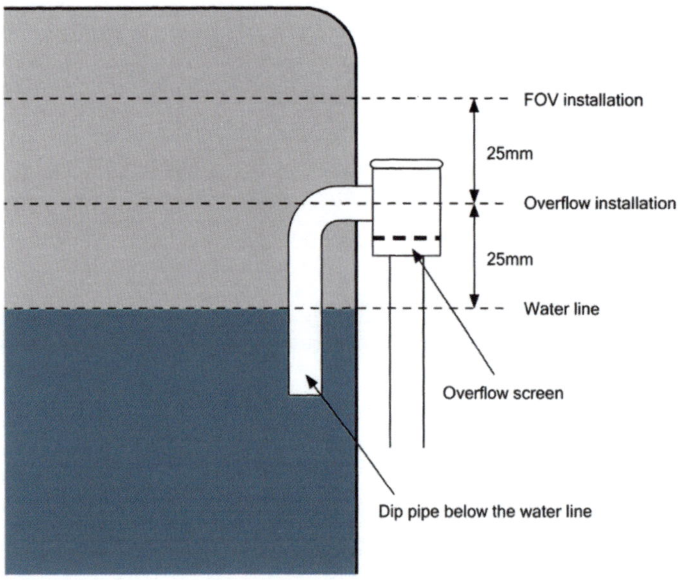

Installation of an overflow in a domestic cistern

Overflow pipes should be sized correctly so that in the event of total FOV failure and the water running at full bore into the cistern, the FOV can never become submerged in water as this could lead to a backflow situation and possible cold water contamination.

Domestic cisterns of less than 1000 litres capacity

Where domestic cisterns are concerned, it is general practice to make the overflow one size larger than the water inlet, so if a 15mm FOV is installed, the overflow would be 22mm etc. The overflow should have a continuous fall along its entire length and should terminate outside the dwelling in a visible and conspicuous position e.g. over an outside door. It should contain a mesh or screen to prevent the ingress of vermin and insects and should contain a dip pipe to prevent freezing air blowing across the surface of the water. The overflow should be positioned so that it is 25mm below the FOV and 25mm above the shut-off water level.

Overflow and warning pipe requirements of large – scale cisterns 1000 litres+

With large cisterns, the potential for water damage to buildings is far greater than domestic cisterns, due to the large size of the overflow pipe. Because of this, the warning and overflow pipe arrangement is different. It must:

- Contain a screen to prevent vermin and insect ingress
- Be capable of maintaining the airgap between the water and the FOV by draining the maximum flow.
- Contain an air break before connection to any drain
- Be of such length so as not to restrict the flow of water, causing the air gap to be compromised.
- Discharge in a safe and visible position
- The invert of the overflow must not be any less than 50mm above the cistern's working level

The general requirement is that the warning pipe invert should be located a minimum of 25mm above the maximum water level and the air gap not less than 20mm or twice the internal diameter of the inlet pipe, whichever is greater.

Cisterns with an actual capacity of 1000 litres to 5000 litres

- The discharge position of the FOV must be positioned at least x2 the diameter of the inlet pipe above the top of the overflow pipe.
- The invert of the overflow pipe must be positioned at least 25mm above the invert of the warning pipe.
- The warning pipe must be at least 25mm in diameter and positioned so that its invert is at least 25mm above the water level.

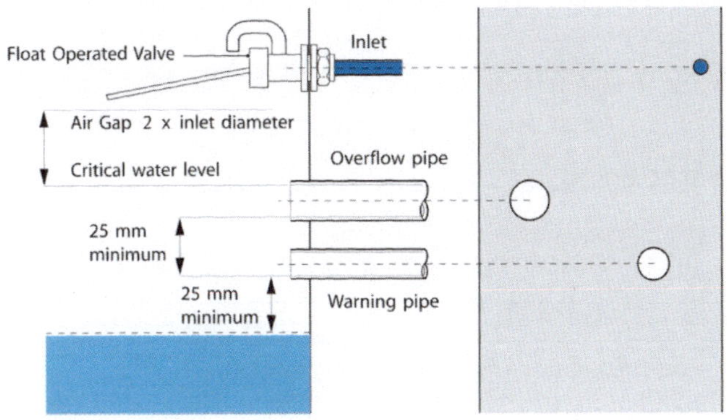

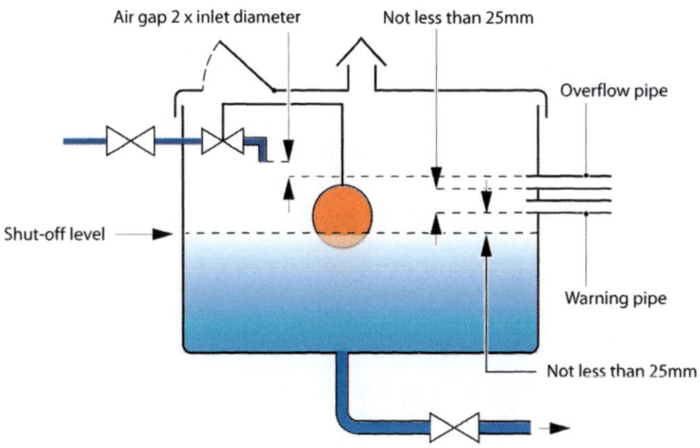

The position of warning and overflow pipes

Cisterns with an actual capacity greater than 5000 litres+

- The discharge level of the FOV must be x2 the diameter of the inlet pipe above the top of the overflow pipe.
- The invert of the overflow pipe must be positioned at least 25mm above the invert of the warning pipe.
- The warning pipe must be at least 25mm in diameter and positioned so that its invert is at least 25mm above the water level.
- The warning pipe may be discarded if a water level indicator is to be installed. However, the water level indicator must also incorporate a visual and audible alarm that sounds when the water level reaches a point not less than 25mm below the invert of the overflow pipe.

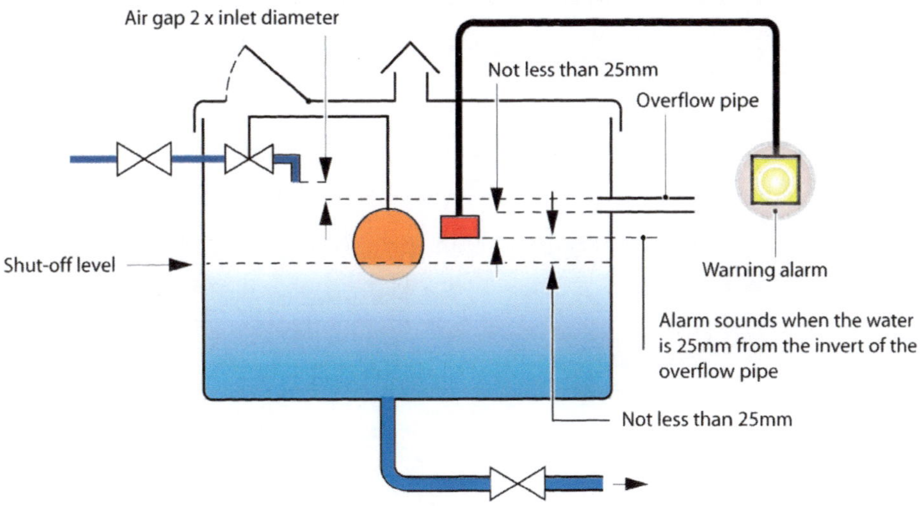

Cisterns greater than 5000 litres

General requirements for large cisterns

In both of the above cisterns, the overflow pipe size will depend on the type of air gap that is incorporated into the cistern. The type of air gap will depend on the fluid category of the contents of the cistern. For example:

- If the cistern contents is fluid category 3, then a Type AG air gap will be required and this will mean that the overflow pipesize must be a minimum of x2 the inlet pipe diameter.
- If the cistern contents is fluid category 4, then a Type AF air gap will be required and will mean that the cross sectional area of the overflow pipesize must be x4 the cross sectional area of the of the inlet pipe.

Cistern Capacities

The capacities of cold water cisterns will depend upon the system the cistern is supplying. As such, BSEN806 and BS8558 do not state a minimum or maximum capacity. These documents recommend that cold water storage requirements are calculated. Calculations are based upon:

a) The size of the dwelling
b) The number of outlets served
c) The number of occupants.

AC2.6 Describe insulation requirements, system frost protection and prevention of undue warming of cold water systems

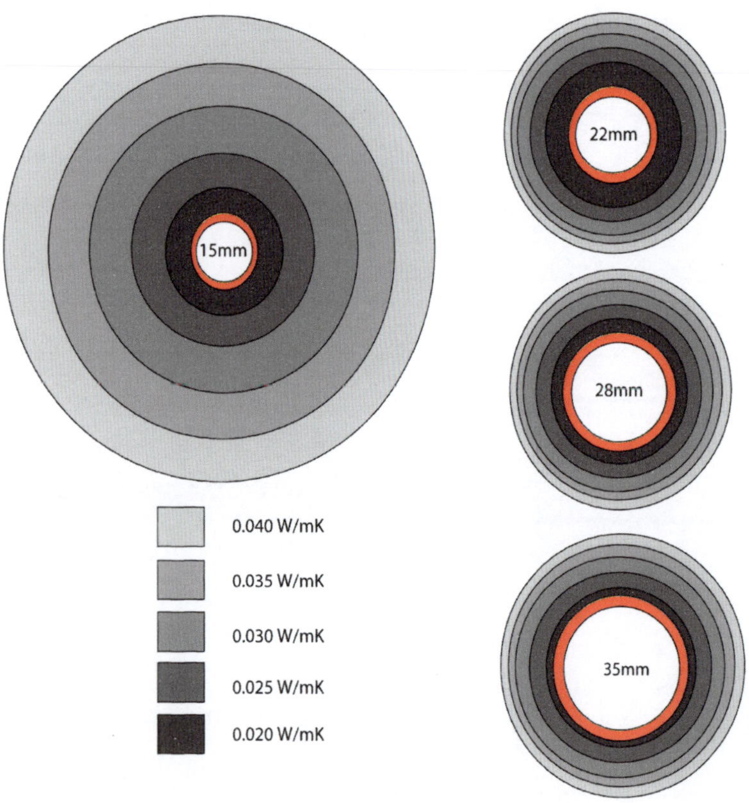

Insulation to pipe size

The Water Supply (water fittings) Regulations state that pipework, fittings and components need to be insulated. This is for two reasons:

a) To stop the water within the system from becoming too warm
b) To protect the system against freezing in cold weather.

Insulating pipework, fittings and components delays the effect of warming or freezing. No matter how much we insulate, if the temperature becomes cold enough, pipes will freeze. The act of insulating does not prevent the adverse effects of temperature, it merely delays them. Insulating helps to retain the heat already in the system. The thicker the insulation, the longer the system will retain the heat.

The Water Regulations Guide G4.2:

> ***'All cold water fittings located within a building but outside the thermal envelope, or those outside the building must be protected against damage by freezing.'***

Pipework that is installed in vulnerable positions within the dwelling must have insulation protection:

- Unheated cellars
- Roof spaces
- Under ventilated suspended floors
- Garages
- Outbuildings

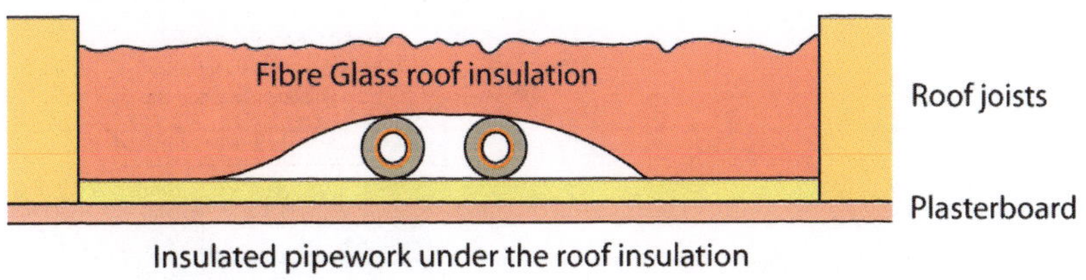

Pipework below the roof insulation blanket

Where pipework is installed in roof spaces, the pipework should be insulated even if it is placed underneath the fibreglass insulation blanket. This is so that any heat that passes through the ceiling from the room below does not unduly heat the water.

The thickness of the insulation will depend on the size of the pipework and its location. Pipework installed outside the dwelling should be insulated with external insulation that is waterproof.

Cisterns should be insulated with a PVC wrapped fibreglass jacket. Cisterns in roof spaces should have the fibreglass roof insulation below them removed. This will allow some heat from below to filter upwards to help prevent the cistern from freezing.

The effectiveness of insulation materials

Pipework insulation should conform to BS5422 and be fitted in accordance with BS5970. It should be water and moisture proof and resistant to mechanical damage and vermin attack. Recommended insulation materials are:

a. Rigid phenolic foam (less than 0.020 W/m^2K)
b. Polisocyanurate foam (0.020 – 0.025 W/m^2K)
c. PVC foam (0.025 – 0.030 W/m^2K)
d. Expanded polystyrene, extruded polystyrene, cross-linked polyethylene foam and expanded nitrile rubber (0.030 – 0.035 W/m^2K)
e. Expanded synthetic rubber, cellular glass and standard polyethylene foam (0.035 – 0.040 W/m^2K)

Thermal conductivity of insulation material at 0° (W/m²K)					
	0.020	**0.025**	**0.030**	**0.035**	**0.040**
mm	**mm**	**mm**	**mm**	**mm**	**mm**
15	**20 (20)**	**30 (30)**	**25* (45)**	**25* (70)**	**32* (91)**
22	**15 (9)**	**15 (12)**	**19 (15)**	**19 (19)**	**25 (24)**
28	**15 (6)**	**15 (8)**	**13 (10)**	**19 (12)**	**22 (14)**
35	**15 (4)**	**15 (6)**	**9 (7)**	**9 (8)**	**13 (10)**
42 and over	**15 (3)**	**15 (5)**	**9 (5)**	**9 (5)**	**9 (8)**

Insulation thicknesses marked * limit ice formation to 50% after 9,8 and 7 hours respectively. All other figures not in brackets limit ice formation to 50% after 12 hours. Insulation thicknesses marked in brackets () are those recommended for 12 hour protection from freezing based on an air temperature of -6 °C and a water temperature of +7°C

AC2.7 Identify the positioning and fixing of pipework within the building fabric

Pipework installed under suspended timber floors, in solid floors and embedded in walls

The Water Supply (water fittings) Regulations 1999, Schedule 2, Regulation 7 state that:

7 – (1) No water fitting shall be embedded in any wall or solid floor.

(2) No fitting which is designed to be operated or maintained, whether manually or electronically, or which consists of a joint, shall be a concealed water fitting.

(3) Any concealed water fitting or mechanical backflow prevention device, not being a terminal fitting, shall be made of gunmetal, or another material resistant to dezincification.

(4) Any water fitting laid below ground level shall have a depth of cover sufficient to prevent water freezing in the fitting.

(5) In this paragraph 'concealed water fitting' means a water fitting which –

- ***a) is installed below ground;***
- ***b) passes through or under any wall, footing or foundation;***
- ***c) is enclosed in any chase or duct; or***
- ***d) is in any other position which is inaccessible or renders access difficult.***

Obviously, when we install plumbing pipework in dwellings, there are occasions when we have to place pipes under timber floors and in studded timber walls. It is simply not possible to install all of the pipework on the surface. The Regulations allow the installation of pipes in walls and under floors provided that we follow the guidance given in the WRAS Water Regulations Guide.

<u>Guidance note G7.1</u>:

Unless they are located in an internal wall which is not a solid wall, a chase or a duct which may readily be exposed or under a suspended floor which may, if necessary be readily removed and replaced or to which there is access, water fittings should not be:

a) located in the cavity of a cavity wall

b) embedded in any wall or solid floor

c) installed below a suspended or solid floor at ground level.

The drawings below illustrate what this statement means:

Pipes laid under floors

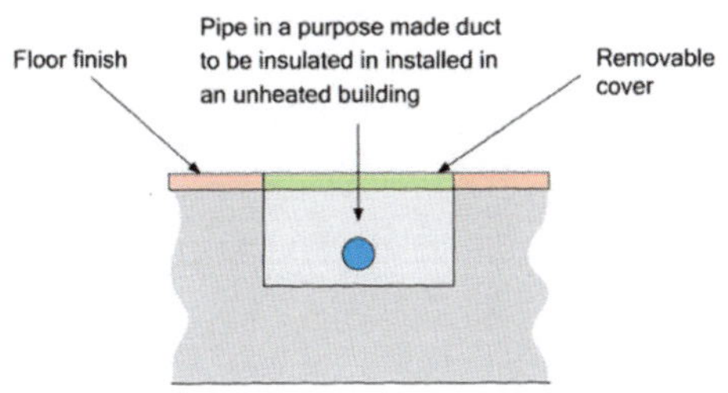

Pipe in a purpose made duct with a removable cover in a solid floor

Pipe in a purpose made duct with a removable cover in a solid floor

The enclosure within pipe ducts is acceptable providing that leaks would become apparent and the section of pipe can be exposed by the removal of superficial surface finishes such as tiles or screed.

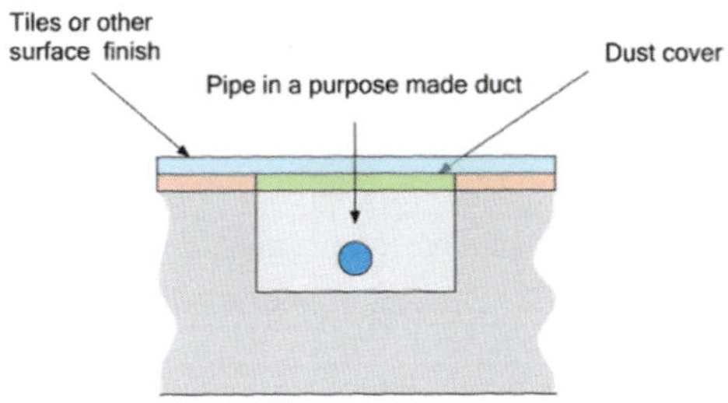

Pipe in a purpose made duct with no access

Pipe in a purpose made duct with no access

This method is only acceptable when there are few joints that are enclosed in the duct and the pipe can be withdrawn for inspection.

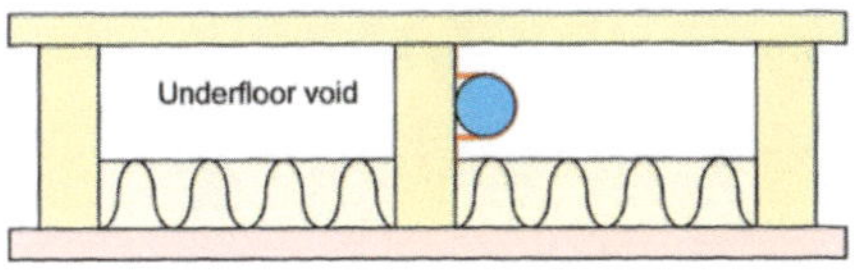

Pipe in an insulated ground floor

Pipe in an insulated ground floor

Acceptable without the need for thermal insulation. Regular clipping to the British Standards is required.

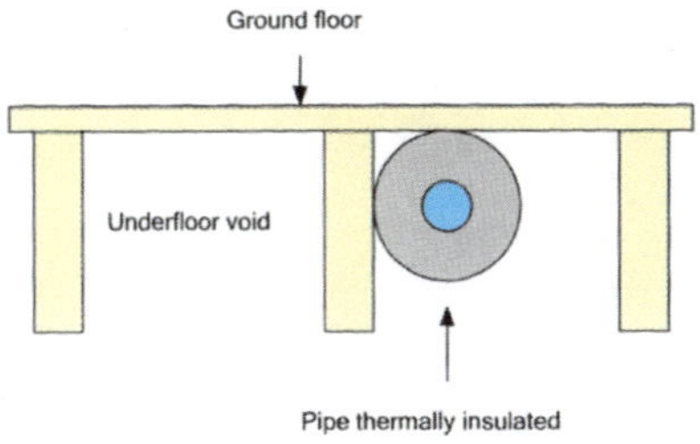

Pipe located under a non-insulated ground floor

Pipe located under a non-insulated ground floor

Access must be at intervals of not more than 2m and must be at every joint so that the whole length of pipe may be inspected.

Pipes positioned in walls

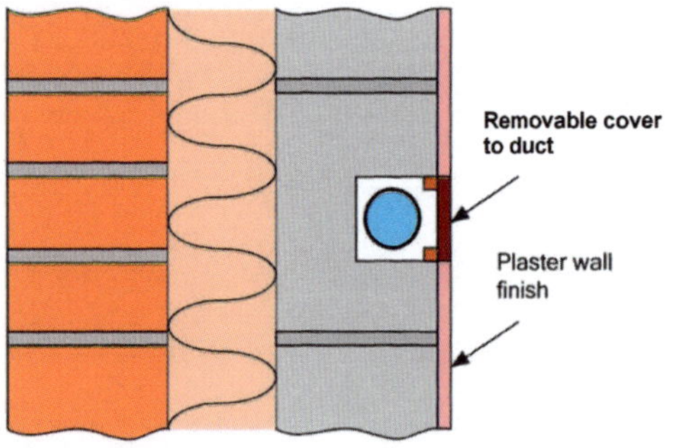

Pipe in a purpose made duct with a removable cover in an insulated cavity wall

Pipe in a purpose made duct with a removable cover in an insulated cavity wall

The enclosure within pipe ducts is acceptable providing that leaks would become apparent and the section of pipe can be exposed by the removal of superficial surface finishes such as tiles or plaster. Frost protection may be required

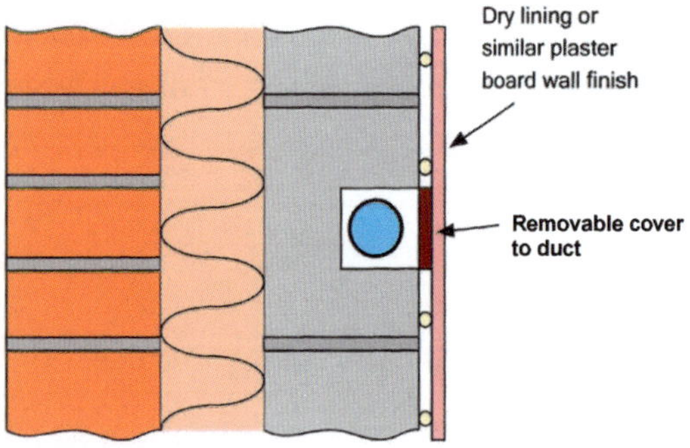

Pipe in a purpose made duct with a non-removable cover in an insulated cavity wall

Pipe in a purpose made duct with a non-removable cover in an insulated cavity wall

This method is only acceptable when there are no joints that are enclosed in the duct. Frost protection may be required.

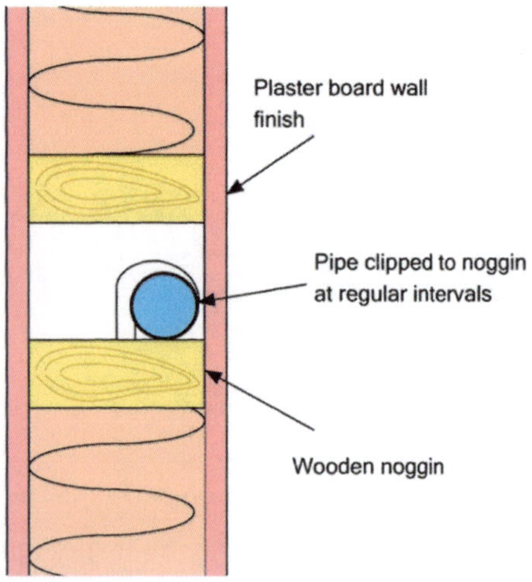

Pipe in a studded wall

Pipe in a studded wall

Generally acceptable. Compression fittings should not be placed in studded walls and there should be as few joints as possible.

Pipework in areas of the building subject to frost

Refer to AC2.6.

Pipework that is installed in vulnerable positions within the dwelling must have insulation protection:

- Unheated cellars
- Roof spaces
- Under-ventilated suspended floors
- Garages
- Outbuildings

Weight distribution of cisterns and heavy components

Refer to AC2.5.

By placing components such as cisterns on platforms that have a larger surface area distributes the load and places less stress on the building structure. Take a look at the example below:

A plumber is installing a cistern measuring 1m x 1.1m x 0.9m in a roof space. The cistern has a volume of $0.99m^3$.

$$1m\ x\ 1.1m\ x\ 0.9m\ =\ 0.99m^3$$

To convert this into litres capacity:

$$0.99\ x\ 1000\ =\ 990litres$$

Since 1 litre of water weighs 1kg @ 4°C, then 990litres = 990kg

So, the cistern has a mass 990kg. But what is the weight of the cistern on the roof trusses? Weight is measured in Newtons, so we must first convert the kg to N

$$Kg\ x\ 9.81\ =\ N$$

So:

$$990\ x\ 9.81\ =\ 9711.9\ N$$

Therefore, the cistern has a force of 9711.9 N

The formula for finding the pressure exerted is:

$$\text{Pressure} = Force\ \div\ Area = Pressure$$

Area of the ***bottom of the cistern***:

$$1m\ x\ 1.1m = 1.1m^2$$

$$9711.9\ \div\ 1.1 = 8829N\ pressure$$

But what if the area of the cistern was enlarged by placing on a platform that covers a greater surface area?

The plumber decides to build a 1.5m x 1.5m platform to stand the cistern on.

$$1.5m\ x\ 1.5m\ =\ 2.25m^2$$

$$9711.9\ \div 2.25 = 4316.4N$$

Therefore, by increasing the surface area, the weight of the cistern on the trusses is less than half of what it was. This puts less strain on the roof structure.

AC2.8 Identify sources of information required when undertaking work on cold water systems

The Water Act 2003 (The Water Industry Act 1991)

The Water Act 2003 combines two previous pieces of UK legislation, the Water Industry Act 1991 and the Water Resources Act 1991. It introduced a series of changes to the regulation of the water industry across the UK. Enforced by the Environment Agency, the Water Act governs the following:

- The appointment, licencing and regulation of water and sewage companies by the Water Services Regulation Authority OFWAT.
- The obligations of the Water Companies and licenced water suppliers to supply water that is fit for human consumption.
- The enforcement of those obligations by Department for the Environment, Food and Rural Affairs and the Drinking Water Inspectorate.
- The control of charges made by the Water Companies and licenced water suppliers to consumers by OFWAT.
- The protection of consumers by OFWAT

The Water Industry Act 1991, however, remains the primary national legislation for all water supplies in England and Wales and defines the powers and responsibilities of local authorities in relation to private water supplies and the definition of a relevant person(s).

Water supplied by a Water Undertaker

The supply of wholesome water to the majority of consumers is the responsibility of the UK water authorities. They are known as Water Undertakers. It is their responsibility to collect, clean, distribute and maintain the water supply in such a manner that the water is fit for human consumption under section 67 of the Water Industry Act 1991. Their responsibility ends at the external (boundary) stop tap to the property. From here, the responsibility rests with the owner of the property or the plumber.

Private water supplies

Private water supplies are water supplies that are not supplied by the Water Undertaker and, as such, are solely the responsibility of the owner and user of the water supply. This could be a single dwelling or multiple properties supplied from the same source. Private water may be supplied from a borehole, stream, river, spring or well.

The quality of the water supplied from private sources can be variable. Some sources offer good quality water where little treatment is necessary but others present a risk to health due to the water quality and require careful monitoring.

The responsibility for regulating private water supplies rests with the local authority but it is also vital that owners are aware of the potential risks. Owners should:

- Ensure that the source is protected against contamination from grazing animals and any contamination that may come from upstream.
- Install and maintain sufficient equipment that will maintain a satisfactory water quality.
- Ensure that the water is adequately disinfected before use.
- Ensure that any stored water is stored in such a way as to prevent contamination after storage but before use.
- Ensure that the water quality is tested regularly.

Ensuring water quality in the property

Under provisions laid down by the Water Industry Act 1991 and then later the Water Act 2003, the UK government introduced two pieces of legislation that regulate how plumbing systems are installed and maintained in domestic buildings in the UK. These are:

- The Water Supply (water fittings) Regulations 1999; and,
- The Private Water Supply Regulations 2016

The Water Supply (water fittings) Regulations 1999

On July 1st, 1999, the government issued the Water Supply (Water Fittings) Regulations 1999 that standardised working practices across all areas of the United Kingdom. These were initially linked to BS6700 and then later to BSEN806: Specification for installations inside buildings conveying water for human consumption: PARTS 1 to 5 and BS 8558:2015: Guide to the design, installation, testing and maintenance of services supplying water for domestic use within buildings and their curtilages. Complementary guidance to BS EN 806.

The Water Supply (water fittings) Regulations were written under section 74 of the Water Industry Act 1991 to ensure that the systems of plumbing installed in properties prevent:

- Contamination of water
- Wastage of water
- Misuse of water
- Undue consumption of water
- Erroneous metering of water

It should be remembered that these regulations only apply to water that has been supplied by a Water Undertaker and do not apply to water supplied from a private source.

A free copy of the regulations can be downloaded from www.opsi.gov.uk/si/si1999/19991148.htm.

The Private Water Supply Regulations 2016

These ensure that water supplied from a private source is safe for people to drink. Changes from the 2009 Private Water Supply Regulations now classify private water supplies into four categories:

- Large supplies serving commercial and public buildings such as hotels, restaurants, nursing homes, bed and breakfast establishments, village halls and large domestic users using more than $10m^3$/day or supplying more than 50 people per day on average.
- Small supplies serving two or more domestic properties but with no commercial usage, using less than $10m^3$/day or fewer than 50 people per day on average.
- Single property supplies where water is supplied for domestic purposes only.
- Private distribution networks. The scope of the regulations now includes networks where the water is distributed further beyond the responsibility of the licenced water company. These include such properties as caravan parks, industrial estates or country estates or where water is distributed by tanker.

The Role of local authorities in regulating private water supplies

The local authority acts as the regulator for the private water supplies in their area and have statutory duties under the PWSR. Local Authorities must:

- Conduct a risk assessment on every property with private water supply in their area.
- Undertake monitoring in order to determine compliance with water standards.
- Provide monitoring data to the Drinking Water Inspectorate (DWI).

The local authority has enforcement powers to ensure that a water supply is improved, should it fall below the minimum cleanliness requirement, by the relevant person(s) who control that supply.

The Role of Drinking Water Inspectorate (DWI) in regulating private water supplies

The DWI plays a statutory role in acting as technical advisors to Local Authorities in relation to the implementation of the Private Water Supplies Regulations 2016. This includes technical and scientific assistance on all aspects of drinking water quality on both public and private supplies.

The Water Supply (water quality) Regulations 2016

In the United Kingdom, all water, whether from a private source or supplied by a water undertaker, must comply with the EU Drinking Water Directive 98/83/EC. It is the duty of every member state to implement rules that reflect the directive, which must meet the minimum European Legislation. This legislation reflects the recommendations of the World Health Organisation (WHO).

The law requires that all drinking water is wholesome and clean. It sets down minimum acceptable concentrations of potential contaminants. It states that:

> *"Water is free from any micro-organisms and parasites and from any substances which, in numbers or concentrations, constitute a potential danger to human health."*

More information can be found at http://www.water.org.uk/

Building Regulations Approved Document G - Sanitation, hot water safety and water efficiency 2010 – 2015 edition with 2016 amendments

In 2010 the Building Regulations were updated and amended. Approved Document G - Sanitation, hot water safety and water efficiency was extended to bring new areas under the control of the Building Regulations, most notably, the installation of systems and water efficiency.

The amended Regulation **Document G** is broken down into 7 parts:

1. **G1 – Cold Water Supply** - Requirements on supply of wholesome water for drinking, washing or food preparation. G1 also deals with the provision of water of a suitable quality to sanitary appliances fitted with a flushing device.
2. **G2 – Water Efficiency** - G2 and Regulations 17K and 20E of the Building Regulations 2000 set out new requirements on water efficiency in NEW dwellings.
3. **G3 – Hot Water Supply and Systems** - Enhanced and amended provisions on hot water supply and safety, applying safety provisions to all types of hot water systems and a new provision on scalding prevention.
4. **G4 – Sanitary Conveniences and Washing Facilities** - Requirements for sanitary conveniences and hand washing facilities.
5. **G5 – Bathrooms** - Requirements for bathrooms, which apply to dwellings and to buildings containing one or more rooms for residential purposes.
6. **G6 – Kitchens and Food Preparation Areas** - New provision requiring sinks to be provided in areas where food is prepared.
7. **Water efficiency Calculator for new dwellings** - Sets out the methodology required to allow the calculation of water consumption in dwellings, limiting water use to 125 litres (optional 110 litres) per person per day

Water efficiency Calculator for new dwellings

The requirement under G2 of the Building Regulations states that:

> ***G2. Reasonable provision must be made by the installation of fittings and fixed appliances that use water efficiently for the prevention of undue consumption of water.***

Water efficiency of new dwellings (Regulation 17K)

36. (1) The potential consumption of wholesome water by persons occupying a new dwelling must not exceed the requirement in paragraph (2).

(2) The requirement referred to in paragraph (1) is either—

a) 125 litres per person per day; or
b) in a case to which paragraph (3) applies, the optional requirement of 110 litres per person per day, as measured in either case in accordance with a methodology approved by the Secretary of State.

Wholesome water consumption calculation (Regulation 20E)

37.— (1) Where regulation 36 applies, the person carrying out the work must give the local authority a notice which specifies—

a) which of the requirements in regulation 36(2)(a) or (b) applies to the dwelling; and
b) the potential consumption of wholesome water per person per day in relation to the completed dwelling.

Requirement G2 applies only when a dwelling is—

a) erected; or
b) formed by a material change of use of a building

The methodology of the water efficiency calculator

The water efficiency calculator uses the water consumption figures that are provided from manufacturers' data. These must be obtained before the assessment can be attempted. The figures are then entered into a series of tables to determine the water consumption per person. Water consumption/flowrate figures are required for:

- WCs
- Taps
- Baths
- Dishwashers
- Washing machines
- Showers
- Water softeners
- External taps.

Bidets are exempt from the calculation due to their minimal water consumption.

Example of the Water efficiency calculator

Installation type	Unit of measure	(1) Capacity/flow rate	(2) Use factor	(3) Fixed use (litres/person/day)	(4) Litres/person/day = [(1) × (2)] + (3)
WC (single flush)	Flush volume (litres)		4.42	0.00	
WC (dual flush)	Full flush volume (litres)		1.46	0.00	
	Part flush volume (litres)		2.96	0.00	
WCs (multiple fittings)	Average effective flushing volume (litres)		4.42	0.00	
Taps (excluding kitchen/utility room taps)	Flow rate (litres/minute)		1.58	1.58	
Bath (where shower also present)	Capacity to overflow (litres)		0.11	0.00	
Shower (where bath also present)	Flow rate (litres/minute)		4.37	0.00	
Bath only	Capacity to overflow (litres)		0.50	0.00	
Shower only	Flow rate (litres/minute)		5.60	0.00	
Kitchen/utility room sink taps	Flow rate (litres/minute)		0.44	10.36	
Washing machine	Litres/kg dry load		2.1	0.00	
Dishwasher	Litres/place setting		3.6	0.00	
Water softener	Litres/person/day		1.00	0.00	
(5) Total calculated use = (Sum column 4)					
(6) Contribution from greywater (litres/person/day) from Table 4.6 Appendix A, App Doc G					
(7) Contribution from rainwater (litres/person/day) from Table 5.5 Appendix A, App Doc G					
(8) Normalisation factor					0.91
(9) Total water consumption = [(5) – (6) – (7)] = (8)					
(10) External water use					5.0
(11) Total water consumption = (9) + (10) (litres/person/day)					

British Standards BSEN806:2012 Pts 1 to 5

BSEN806: Specifications for installations inside buildings conveying water for human consumption

- BSEN806 – 1: 2000. General. This document specifies potable water installation requirements and gives recommendations on the following aspects of hot and cold water installations:
    - Design
    - Installation
    - Alteration
    - Testing
    - Maintenance and operation.
    - Part 1 of BSEN806 also covers the pipework installation from the point of entry when the property is supplied with water from a private water source.

- BSEN806 – 2: 2005. Design. This document provides technical information so that the following points may be achieved from the design and installation:
    - Appropriate pressures and flow rates
    - Water quality at the tap is not contaminated or affected by the location or environment
    - The system avoids wastage of water and leakage.
    - The system is efficient, convenient, reliable and safe.
    - The system has a reasonable working life span.
- BSEN806 – 3: 2006. Pipe sizing. Simplified method. This document describes the simplified method for pipesizing drinking water standard installations as defined in section 4.2. However, the document does not engage in pipesizing for domestic firefighting systems.
- BS EN 806 – 4:2010. Installation. Specifies the requirements of water installations within buildings and gives recommendations for their correct installation. It also covers pipework outside buildings but within the premises as outlined in BSEN806 – 1:2000. It applies to new, altered and repaired installations.
- BS EN 806 – 5:2012. Operation and maintenance. This document takes the form of a practice specification. It specifies the requirements for the correct operation and maintenance of potable water supply installations within buildings and for pipework outside buildings but within premises in accordance with BSEN806 – 1:2000.

BSEN806 Parts 1 to 5 completely supersede BS6700 in all aspects of hot and cold water supply. **BS 8558: 2015 Guide to the design, installation, testing and maintenance of services supplying water for domestic use within buildings and their curtilages** - complementary guidance to BS EN 806, now becomes the lead document for potable water supply in premises.

BS 8558: 2015 Guide to the design, installation, testing and maintenance of services supplying water for domestic use within buildings and their curtilages - complementary guidance to BS EN 806

BS8558 was developed primarily to provide complimentary guidance to BSEN806 Pts 1 to 5. Together, this complete suite of British Standards provides recommendations into best practice for the design, installation, testing, operation, maintenance and alteration of hot and cold water systems for domestic buildings.

BS8558 bridges the gap between BSEN806 and its predecessor BS6700 and provides UK guidelines for, not just plumbers, but the water supply industry as a whole.

PD 855468:2015 Guide to the flushing and disinfection of services supplying water for domestic use within buildings and their curtilages

Flushing and disinfection of systems used to be part of BS8558:2011. However, the latest version of BS8558:2015 excludes flushing and disinfection. Instead, flushing and disinfection of systems is now a separate document PD 855468:2015.

PD 855468:2015 provides guidance on the cleaning, flushing and disinfection of cold water systems to control microbiological growth and the removal of debris. This includes guidance on:

- Deployment of the correct tools and personnel
- The use of the correct disinfectants
- How to respond if a microbiological problem is identified
- Keeping records of cleaning and disinfection

This new document applies to systems supplying water to domestic purposes within buildings and their curtilages, and includes water used in food preparation.

Manufacturer technical instructions

The manufacturer's instructions are probably the most important document to read and consult when installing, servicing and maintaining appliances, components and equipment, because they instruct us on the best methods to use whilst keeping to current legislation and regulations. In some cases, it may appear that these instructions contradict the regulations. This occurs because regulations and codes of practice are only updated periodically, whereas manufacturers are constantly reviewing and updating their literature in line with modifications and current good practice. Where a conflict exists, manufacturer's literature should always be followed. If not:

- The warranty of the equipment may be void.
- Regulations may be inadvertently broken
- The installation may be dangerous.

AC2.9 Explain backflow risk and required methods of prevention

Schedule 2 paragraph 15 of the Water Supply (water fittings) Regulations 1999, deals with backflow prevention. Backflow is simply the reversing of the normal direction of the water flow and can be as a result of backpressure (where the water pressure downstream is greater than the water pressure upstream) or back syphonage (where the water is sucked backwards through the pipework due to sudden loss of pressure). This can lead to contamination of the water undertaker's mains cold water supply. See diagrams:

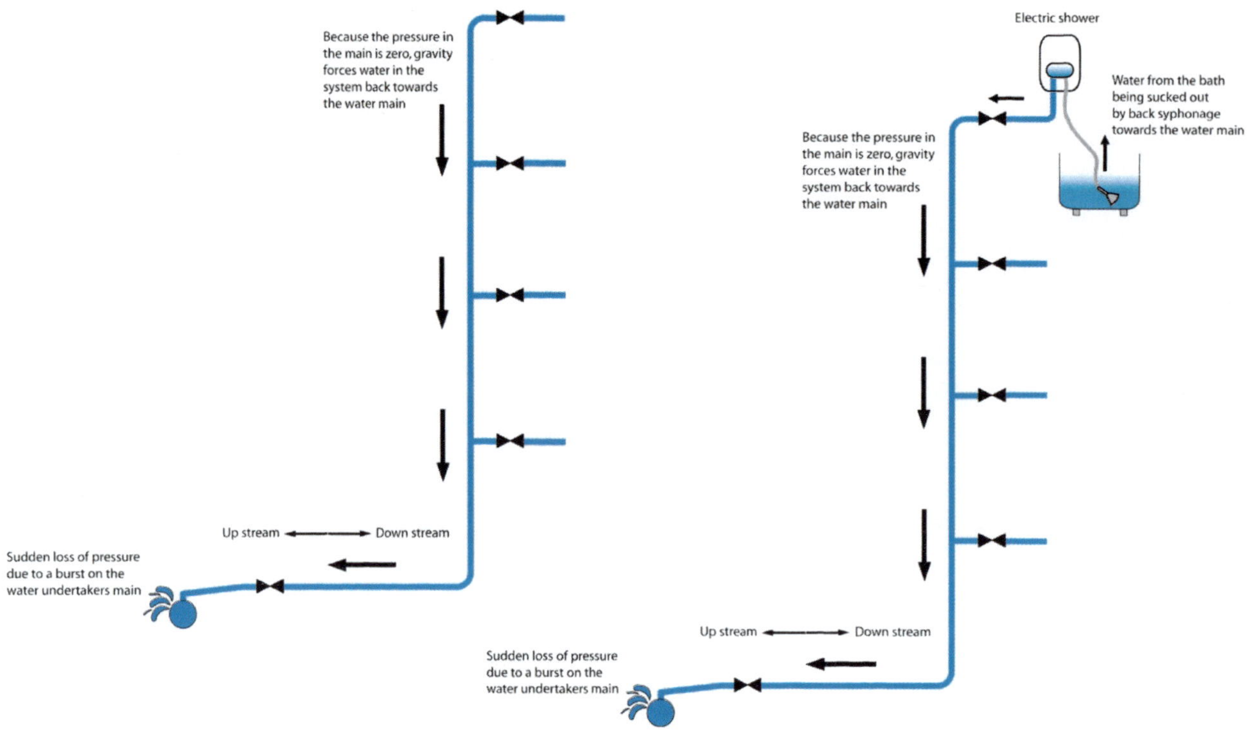

Back pressure (image left) Back syphonage (image right)

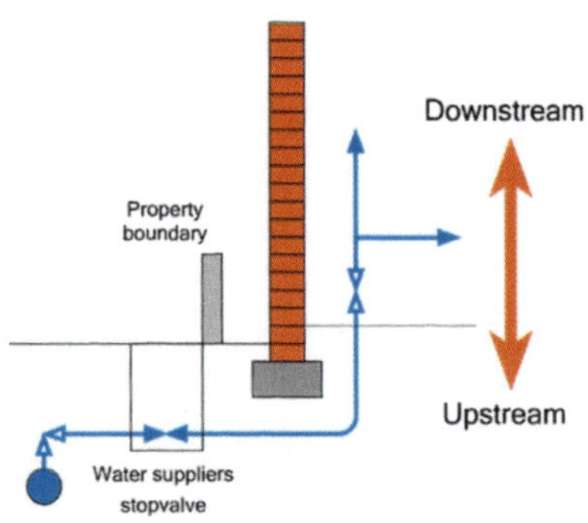

Upstream and downstream

Cross connection of wholesome water with water from other sources can be another major source of contamination by back pressure and back syphonage. Cross connections can occur in the home easily simply by connecting mixer taps on hot and cold water supplies. If the mixing tap or valve is a true mixer (not a bi-flow mixer), then a cross connection between hot and cold supplies is deemed a cross connection between a fluid cat. 1 water (cold) to a fluid cat. 2 water (hot). In this instance, the simple installation of a single check valve on the cold supply will prevent contamination. However, not all back flow situations are so easily sorted out.

Preventing contamination of the water supply by back pressure, back syphonage and cross connection

Air gaps are the most effective method of preventing contamination between water undertakers' mains cold water supply and water from possible unwholesome sources. An air gap is the only method of protecting against contamination from fluid category 5. However, for fluid categories 2 to 4, mechanical devices, such as Type EA/ED single check valves (fluid cat. 2), Type EC/ED double check valves (fluid cat. 3) and Type BA Reduced Pressure Zone valves or RPZ valves (fluid cat. 4) can often be used.

The tables below indicate the air gaps and mechanical devices.

Non-Mechanical back flow prevention arrangements

Table: Schedule of *non-mechanical backflow prevention arrangements (air gaps)* and their respective fluid category protection

<table>
<tr><th rowspan="2">Type</th><th rowspan="2" colspan="2">Description of backflow prevention arrangements and devices.</th><th colspan="2">Suitable for protection against fluid category</th></tr>
<tr><th>Back pressure</th><th>Back syphonage</th></tr>
<tr><td>AA</td><td colspan="2">Air gap with unrestricted discharge above spill over level</td><td>5</td><td>5</td></tr>
<tr><td>AB</td><td colspan="2">Air gap with weir overflow</td><td>5</td><td>5</td></tr>
<tr><td>AD</td><td colspan="2">Air gap with injector</td><td>5</td><td>5</td></tr>
<tr><td>AG</td><td colspan="2">Air gap with minimum size circular overflow determined by measure or vacuum test</td><td>3</td><td>3</td></tr>
<tr><td>AUK1</td><td colspan="2">Air gap with interposed cistern (For example, a WC suite)</td><td>3</td><td>5</td></tr>
<tr><td rowspan="5">AUK2</td><td colspan="2">Air gaps for taps and combination fittings (tap gaps) discharging over domestic sanitary appliances, such as a washbasin, bidet, bath or shower tray shall not be less than the following:</td><td rowspan="5">X</td><td rowspan="5">3</td></tr>
<tr><td>Size of tap or combination fitting</td><td>Vertical distance of bottom of tap outlet above spill-over level of receiving appliance</td></tr>
<tr><td>Not exceeding G ½</td><td>20 mm</td></tr>
<tr><td>Exceeding G ½ but not exceeding G ¾</td><td>25 mm</td></tr>
<tr><td>Exceeding G ¾</td><td>70 mm</td></tr>
<tr><td>AUK3</td><td colspan="2">Air gaps for taps or combination fittings (tap gaps) discharging over any higher risk domestic sanitary appliances where a fluid category 4 or 5 is present, such as:
a) Any domestic or non-domestic sink or other appliance; or
b) Any appliances in premises where a higher level of protection is required, such as some appliances in hospitals or other health care premises,
shall be not less than 20 mm or twice the diameter of the inlet pipe to the fitting, whichever is the greater.</td><td>X</td><td>5</td></tr>
<tr><td>DC</td><td colspan="2">Pipe interrupter with permanent atmospheric vent</td><td>X</td><td>5</td></tr>
</table>

Notes:

1. X Indicates that the backflow prevention arrangement or device is not applicable or not acceptable for protection against backpressure for any fluid category within water installations in the UK.
2. Arrangements incorporating Type DC devices shall have no control valves on the outlet of the device; they shall be fitted not less than 300 mm above the spillover level of a WC pan, or 150 mm above the sparge pipe outlet of a urinal, and discharge vertically downwards.
3. Overflows and warning pipes shall discharge through, or terminate with, an air gap, the dimension of which should satisfy a Type AA air gap.

Type AA air gap

The Type AA air gap provides protection against fluid category 5. Water is discharged through an air gap into a cistern that has an unrestricted spill-over to atmosphere. The air gap is measured vertically downwards from the lowest of discharge to the cistern spill-over level:

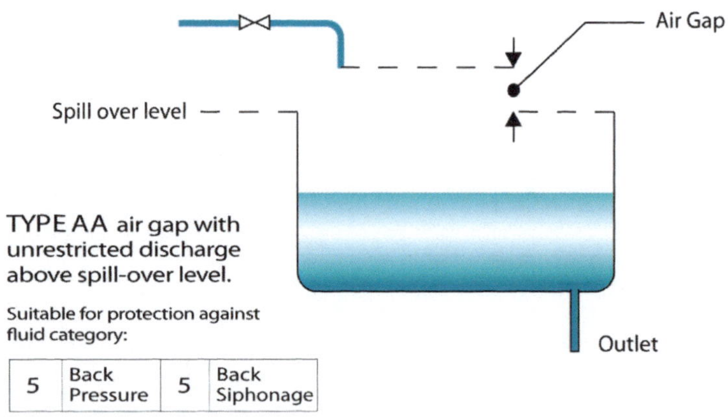

- It is suitable for all fluid categories
- The size of the air gap is dependent on the size of the inlet (see table below)
- The inlet flow must not be more than 15° from vertical

Air gaps at taps, valves, fittings and cisterns		
Situation	**Nominal size of inlet, tap, valve or fitting**	**Vertical distance between tap or valve outlet and the spill over level of the receiving appliance or cistern**
Domestic situation with fluid categories 2 and 3 (AUK2)	Up to and including G ½	20mm
	Over G ½ and up to G ¾	25mm
	Over G ¾	70mm
Non-domestic situation with fluid categories 4 and 5 (AUK3)	Any size of inlet pipe	Minimum diameter of 20mm or twice the diameter of the inlet pipe, whichever is the greater of the two.

Animal drinking troughs use the Type AA air gap where the discharge of the water into the trough is through a raised housing. The housing is covered to prevent the animals from having direct access to the water supply (see diagram on the next page).

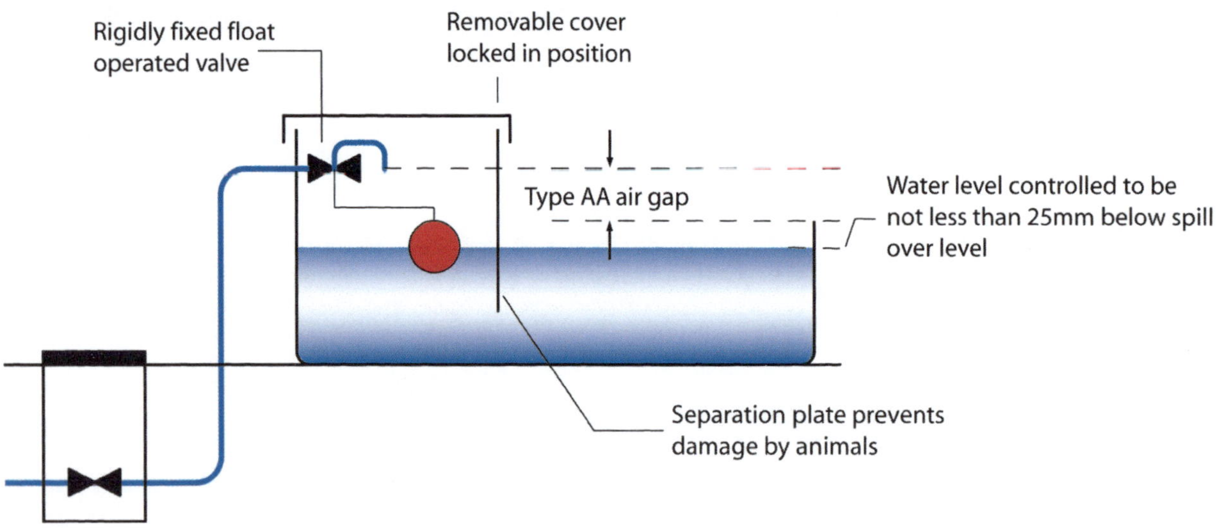

Animal trough with Type AA air gap

Type AB air gap with weir overflow

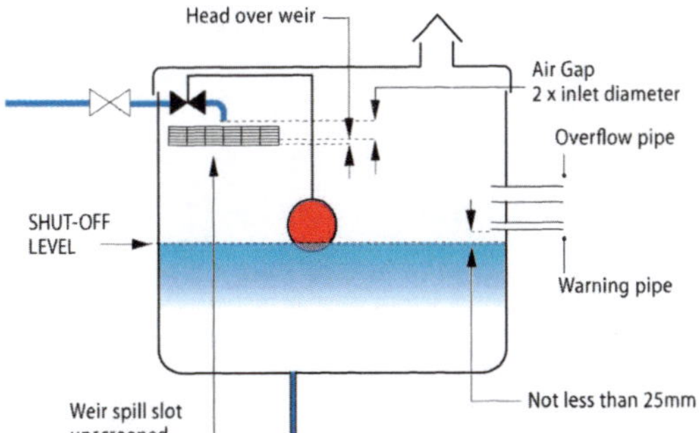

TYPE AB air gap with weir overflow.

Suitable for protection against fluid category:

5	Back Pressure	5	Back Siphonage

The Type AB air gap is a non-mechanical backflow prevention arrangement. It provides protection up to Fluid Category 5.

This arrangement is suitable for high risk situations and it particularly suited where protection is required from insects, dust and vermin. It also gives protection in various medical situations such as dental practices and GP Doctors' surgeries.

Type AD air gap with injector

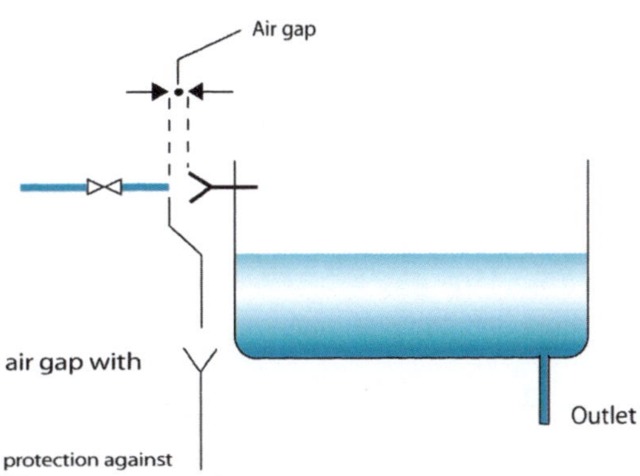

TYPE AD air gap with injector.

Suitable for protection against fluid category:

5	Back Pressure	5	Back Siphonage

Again, the Type AD air gap with injector provides protection up to Fluid Category 5. It has a horizontal injector with a 20mm air gap known as a 'jump jet'. It is primarily used in commercial clothes washing machines and dish washers.

Type AG gap with minimum size circular overflow determined by measure or vacuum test

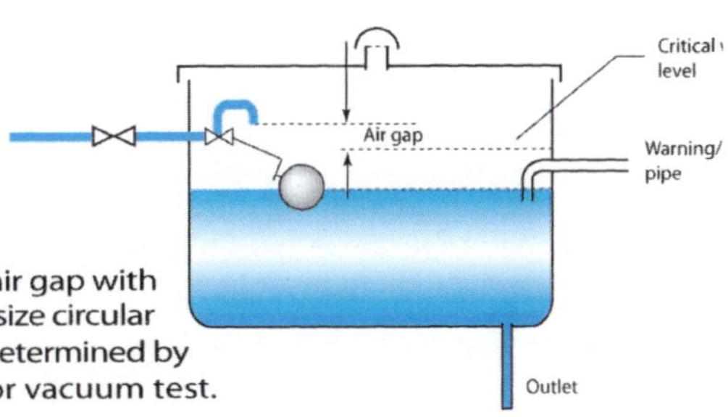

TYPE AG air gap with minimum size circular overflow determined by measure or vacuum test.

Suitable for protection against fluid category:

3	Back Pressure	3	Back Siphonage

The Type AG air gap is provided when cold water cisterns are installed correctly:

- The Type AG air gap is fitted in the process, provided that a **BS1212 part 2** Float operated Valve is fitted to the cistern,
- The overflow is installed **25mm below the Float operated Valve**, and;
- The water level is set to **25mm below the invert of the overflow**.

Points to remember:

- The air gap between the float operated valve outlet and the critical water level should be sufficient to provide back syphonage protection.
- Type AG air gaps **MUST comply with BS 6281**
- The overflow must be of sufficient size to be able to cope with the flow of water when the Float operated Valve is running at full bore, without submerging the outlet of the Float operated Valve.

It is suitable for backflow protection up to Fluid Category 3.

AUK1 air gap with interposed cistern

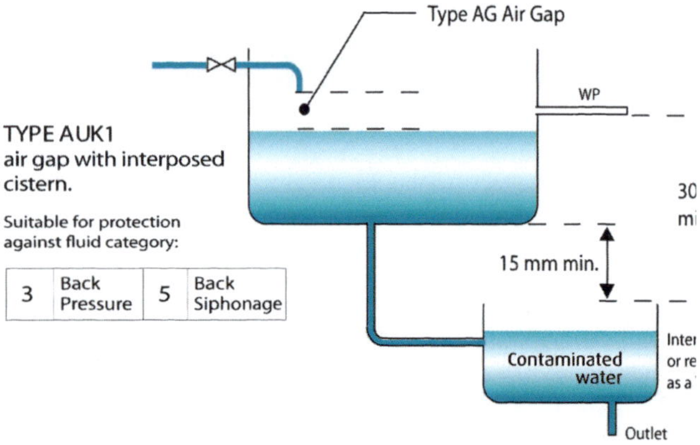

The AUK1 air gap can be found on WC Cistern and pan installations, where fluid category 1 water is discharged into a fluid category 5 vessel. The cistern incorporates a Type AG air gap and overflow arrangement.

The spill over level of the receiving vessel or cistern must not be less than 300mm below the invert of the overflow pipe and not less than 15mm below the bottom of the supplying cistern.

AUK2 air gap for taps/combination fittings (tap gaps) discharging over sanitary appliances

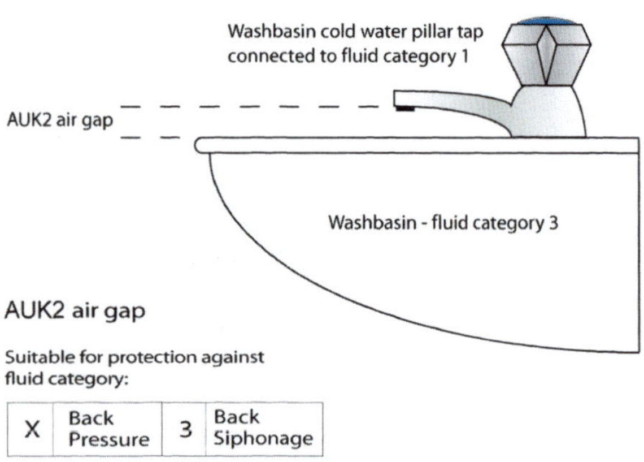

An AUK2 air gap is the gap located between the bottom of the outlet of a tap or other fitting discharging water over an appliance and the spill over level of the sanitary appliance, such as a washbasin as shown in the diagram. The vertical distance depends on the size of the outlet (see table on page 104). It can only be used where there is a back syphonage of Fluid Category 2 or 3 risk present downstream.

AUK3 air gap for taps/combination fittings where fluid category 4 or 5 risk exists

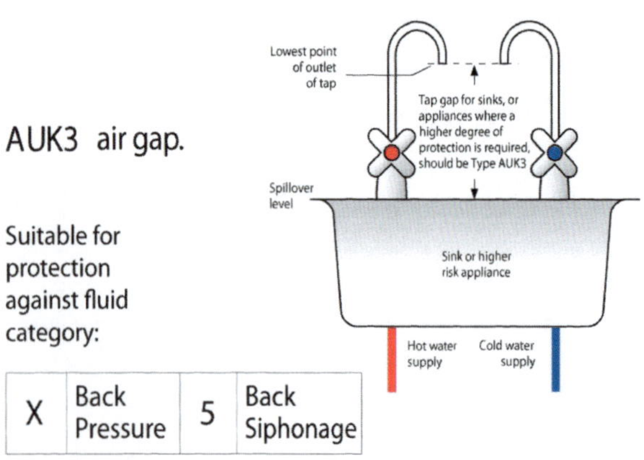

As with the AUK2 air gap, the AUK3 air gap is the gap located between the bottom of the outlet of a tap or other fitting discharging water over an appliance and the spill over level of the appliance. The difference is that the required gap is much bigger, because of the greater risk. (see table on page 104). AUK3 air gaps are present at kitchen sinks, through high neck pillar taps and monobloc mixer taps.

AUK3 air gaps can be found on appliances at all places where a fluid category 4 or 5 exists such as hospitals, doctors' surgeries, healthcare premises and veterinary practices.

Type DC Pipe Interrupter with a permanent atmospheric vent

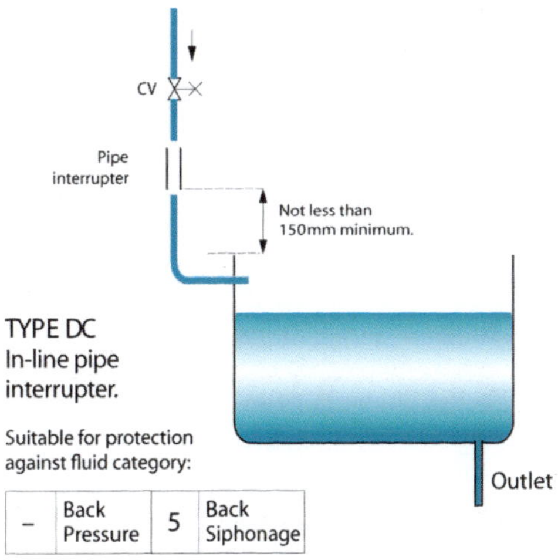

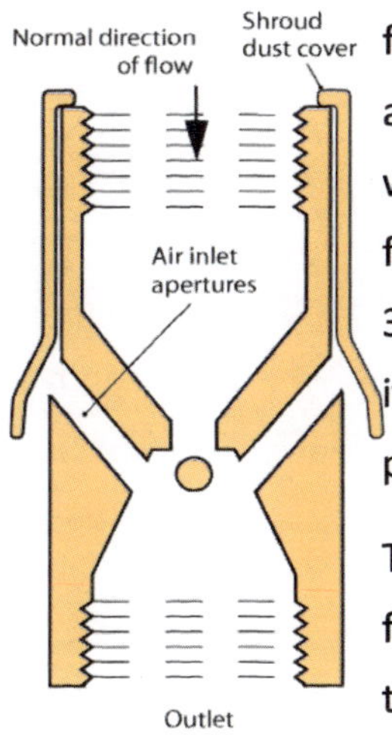

The DC pipe interrupter is a non-mechanical fitting that prevents back syphonage by allowing air into the stream of back flowing water to break the syphonic action. It must be fitted in the vertical position, no less that 300mm above the over spill level of the vessel it is serving. They must comply to BS 6281 part 3.

They are usually fitted to urinals that use a flushing valve discharging water direct from the mains cold water supply.

Mechanical back flow prevention arrangements

Table: Schedule of *mechanical backflow prevention arrangements* and fittings and their respective fluid category protection

Type	Description of backflow prevention arrangements and devices.	Suitable for protection against fluid category	
		Back pressure	Back siphonage
BA	Verifiable backflow preventer with reduced pressure zone	4	4
CA	Non-verifiable disconnector with difference between pressure zones not greater than 10%	3	3
DB	Pipe interrupter with atmospheric vent and moving element	X	3
EA/EB	Verifiable and non-verifiable single check valves	2	2
EC/ED	Verifiable and non-verifiable double check valves	3	3
HA	Hose union backflow preventer. Only permitted for use on existing hose union bib-tap in house installations.	2	3
HUK1	Hose union bib-tap incorporating a double check valve arrangement. Only permitted as a replacement for existing bib-taps in house installations.	3	3
HC	Diverter with automatic return (normally integral with some domestic appliance applications only).	X	3

Notes:
'X' indicates that the backflow prevention device is not acceptable for protection against backpressure for ANY fluid category.
Arrangements incorporating a Type DB device shall have no control valves on the outlet of the device. The device shall not be fitted less than 300mm above the spill over level of an appliance and must discharge vertically downwards.
Relief ports from BA and CA devices should terminate with an air gap, the dimension of which should satisfy a Type AA air gap.

Type BA Verifiable backflow preventer with reduced pressure zone

TYPE BA or reduced pressure zone valve (RPZ valve).

Suitable for protection against fluid category:

4	Back Pressure	4	Back Siphonage

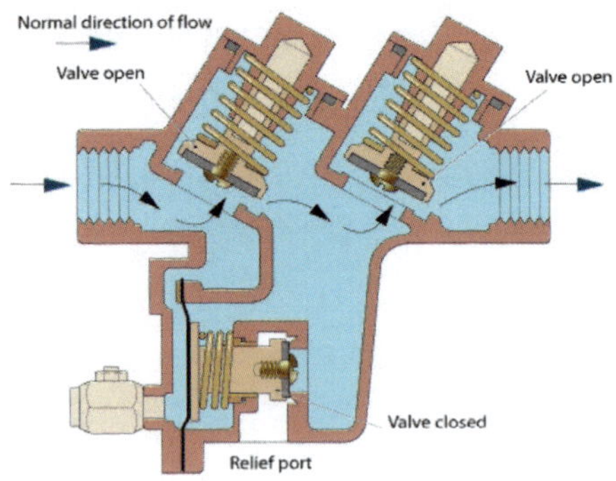

Known as the Reduced Pressure Zone or RPZ valve, the Type BA Verifiable backflow preventer is used where a fluid category 4 risk exists. It consists of three separate elements – (1) two single check valves, (2) a differential relief valve and (3) test points. They should only be installed horizontally.

These are specialist valves and should only be fitted and maintained by approved installers who have been trained in RPZ valve installation. It is advised that learners read Water Regulations Advisory Scheme Information and Guidance Note No. 9-03-02.

Type CA Non-verifiable disconnector with difference between pressure zones not greater than 10%

Type CA

Non-verifiable disconnector with different pressure zones

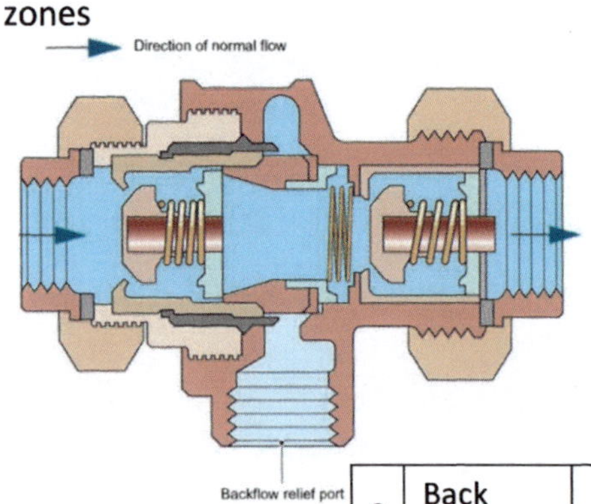

3	Back pressure	3	Back syphonage

Type CA valves provide a positive disconnection space between the upstream and downstream water. Like the Type BA valve, these contain 2 single check valves and a path for water to be released. In the event of pressure loss, the backflow relief port opens to release the water to drain via a tundish.

These are often used as permanent connections to sealed central heating systems, negating the need for a filling loop between the heating system and mains cold water supply.

Type DB Pipe interrupter with atmospheric vent and moving element

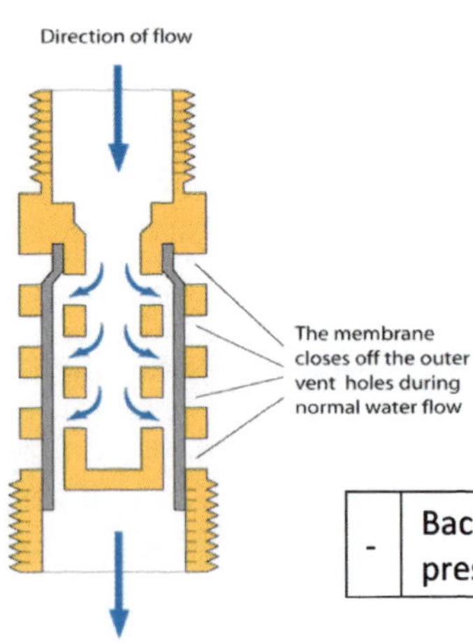

-	Back pressure	4	Back syphonage

Designed for fluid category 4 installations, the Type DB pipe interrupter works by diverting the flow of water through a series of holes covered by a membrane. The water pressure forces the membrane against the sides of the valve allowing water to flow. In a back syphonage occurrence, the membrane collapses blocking off the path back to the mains supply. The fluid category 4 water then diverts away via another series of holes on the outside of the valve, eliminating the back syphonage risk.

Type EA/EB Verifiable and non-verifiable single check valves

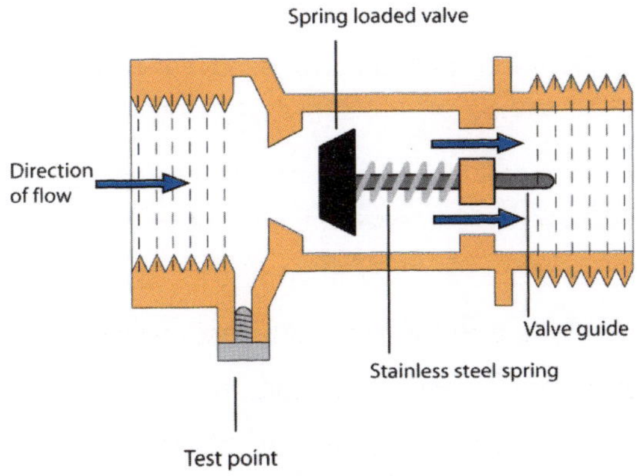

2	Back pressure	2	Back syphonage

Also known as 'one-way valves' or 'non-return valves', the Type EA/EB single check valves simply stop the flow of water upstream in the event of a back pressure or back syphonage situation. They are spring loaded to aid closure in the event of a mains cold water failure.

Type EA valves are verifiable. This means they have a test point where they can be checked for the operation of the valve. Type EB do not have a test point. Used for fluid category 1 protection against fluid category 2 risk.

Type EC/ED Verifiable and non-verifiable double check valves

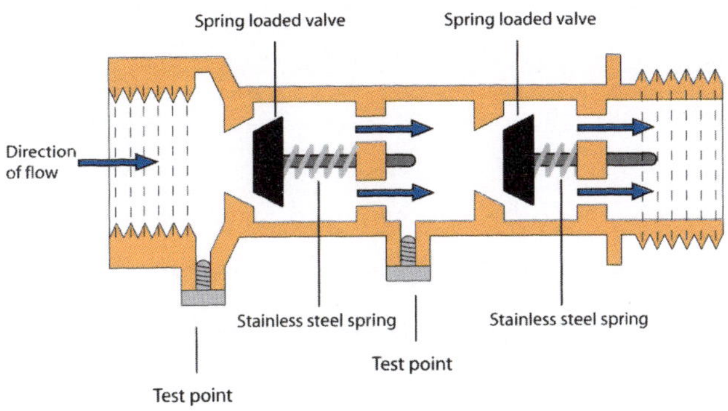

3	Back pressure	3	Back syphonage

Type EC/ED double check valves are used in exactly the same way as single check valves. They are basically two single check valves in tandem with one another.

Type EC valves are verifiable. This means they have a test point where they can be checked for the operation of the valve. Type EC do not have a test point. Used for fluid category 1 protection against fluid category 3 risk i.e. outside tap installations and central heating temporary filling loops (in conjunction with an air gap)

Type HA Hose union backflow preventer. Only permitted for use on existing hose union bib-tap installations

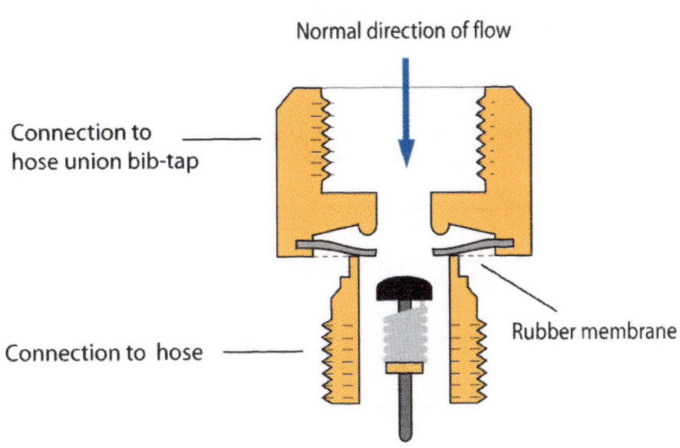

2	Back pressure	3	Back syphonage

These simply screw on to the outlet of existing outside hose union bib tap installations to bring existing taps into line with the Water Supply (water fittings) Regulations, where no backflow protection exists. They must not be used on new installations.

Type HUK1 Hose union bib-tap incorporating a double check valve arrangement. Only permitted as a replacement for existing domestic bib-tap installations.

Back pressure	3	Back siphonage	3

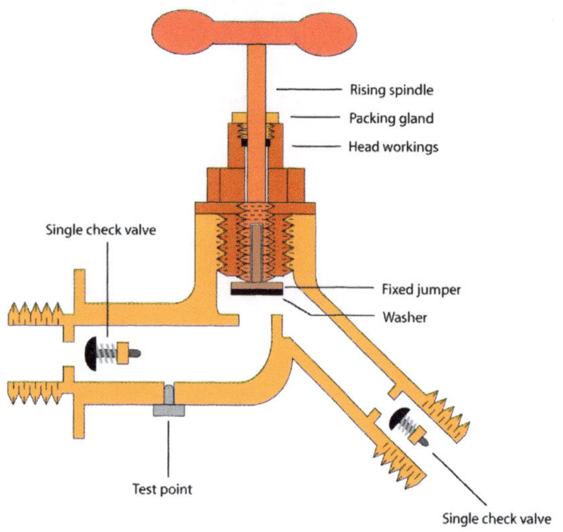

This is a hose union bib tap that incorporates two single check valves – one at the inlet to the tap and one at the outlet to provide back flow protection up to fluid category 3. They must not be used on new installations and are only recommended as replacements where a hose union bib tap already exits. This is because the check valves are liable to freezing in the open position during very cold weather, eliminating the backflow protection.

Type HC Diverter with automatic return (normally integral with some domestic appliance applications only)

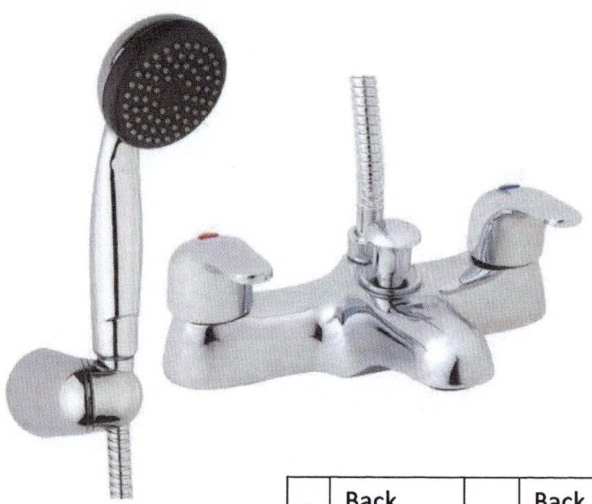

-	Back pressure	3	Back syphonage

Rather than being a separate backflow prevention device, the Type HC is integrated within bath/shower mixer combination taps.

When a bath/shower mixer is in shower mode, the pressure of the water keeps the shower diverter lever in the up position allowing water to flow to the shower head. When the water is turned off, the lever drops and the water in the shower hose is diverted out of the tap. In this way, the water in the hose does not become a risk if the pressure suddenly drops from the shower.

They are suitable up to fluid category 3 and for back syphonage protection only.

AC2.10 Install cold water systems

Cisterns – **should be installed in accordance with the Water Supply (water fittings) Regulations and British Standards BS EN 806 and BS 8558.**

Sanitary appliances – including baths, wash basins, WCs etc. The installation of sanitary ware was covered on page 58. Great care should be exercised when installing sanitary ware to ensure that damage to the appliance does not occur. Positioning needs to be completed in accordance with BS 6465–2:1996 Code of practice for space requirements for sanitary appliances to ensure that the correct spacing requirements between appliances are observed.

Booster sets – these are often used in domestic properties where the water pressure and flow rate are low. They are also used where the water supply is drawn directly from a bore hole, in the case of private water supplies. In multi-storey installations, large and often multiple boosting sets are installed to boost the water to heights that are unreachable by the water undertaker's mains supply. These are often installed alongside a break cistern to ensure that large amounts of water are not drawn from the water undertaker's mains cold water supply.

Shower – There are many different shower types available and each one will have its own installation requirements. Some can be installed when the pressure of the hot and cold water supplies are different, some require the same hot and cold water pressures and some are electric and only require a cold mains water supply. Whichever shower is chosen, the manufacturer's instructions should be thoroughly read and understood before any installation takes place.

Pipework – There are two very common materials used for installation of hot and cold water supply in domestic properties and small commercial buildings. These are copper tubes and plastic pressure pipe:

- Copper – this is the traditional material for the installation of hot and cold pipework in domestic and small commercial buildings. It is manufactured to BS EN 1057 and comes in a range of sizes to suit all applications and installation types. Copper is light, easy to bend, cut and joint using a variety of fitting types. Remember to keep the pipework as neat as possible and use the recommended clip types and distances.
- Plastic pressure pipe – manufactured from either polybutylene (PB) or cross-linked polyethylene (PE-x), plastic pressure pipe is becoming the material of choice for new build housing installations. it is very easy to install, bend and joint. As with copper tubes, remember to keep the pipework as neat as possible and use the recommended clip types and distances.

bpec LEARNING FOR LIFE

Learning Outcome 3

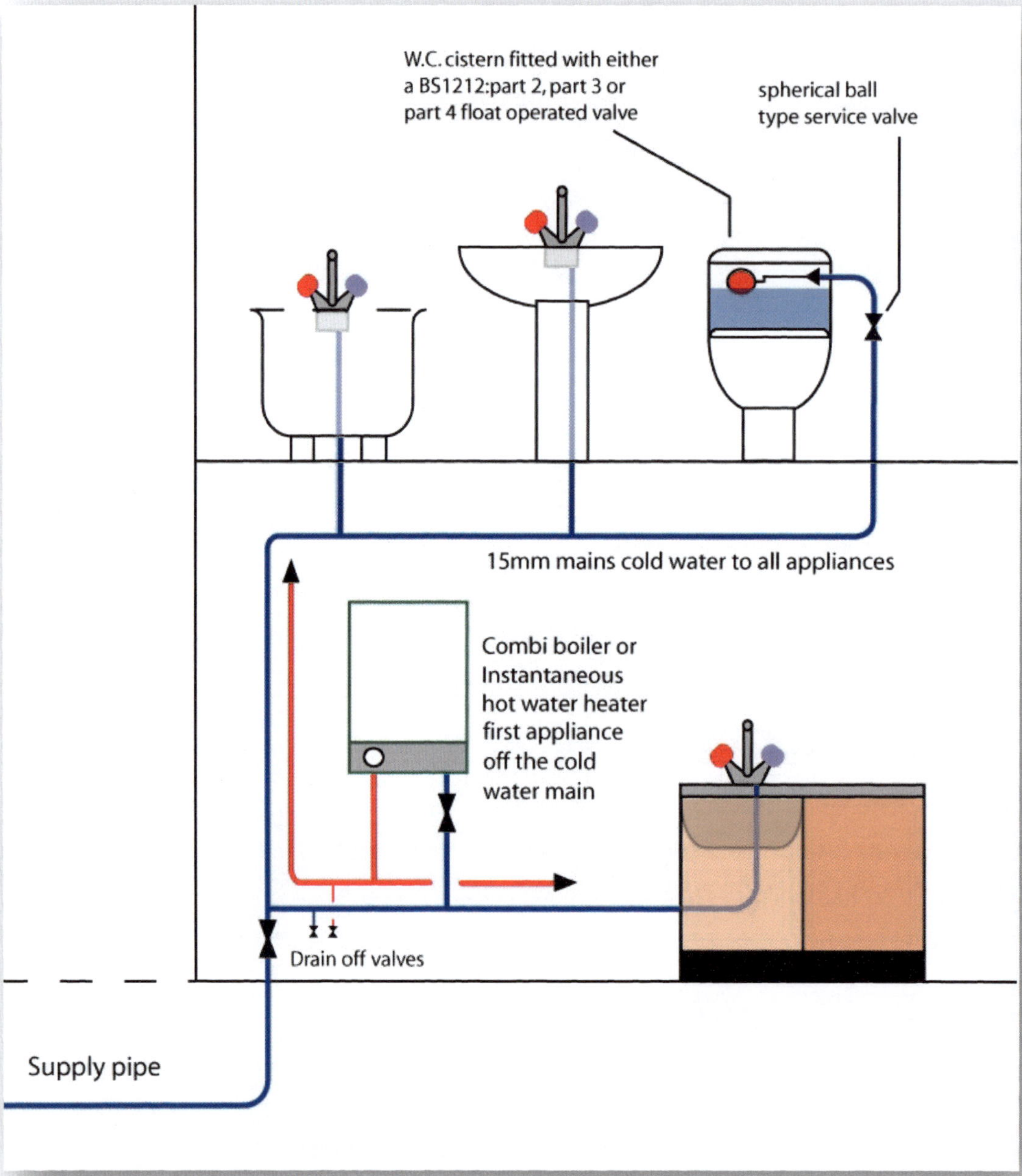

Decommission Cold Water Systems

Learning Outcome 3 - Decommission Cold Water Systems

In Learning Outcome 3, there are two Assessment Criteria:

AC3.1 Explain **procedures** for **decommissioning** systems

AC3.2 Carry out decommissioning **procedures**

There are often times when systems are required to be taken out of service, either for service and repair or extension or renewal. There are also times when systems are taken out of service permanently. Both of these procedures are called decommissioning.

In Learning Outcome 3, we will look at the decommissioning process and the tasks that must be performed to ensure that this is done safely.

AC3.1 Explain procedures for decommissioning systems

There are two types of decommissioning that we may be asked to perform:

- Permanent decommissioning – This is performed when a system is no longer required and involves completely draining down the system and the removal of all pipework, fittings, clips and appliances. The cold water supply should be turned off at the external (boundary) stop tap and the internal stop tap capped off as near to the stop tap as is possible. This is to avoid the stagnation of water.
- Temporary decommissioning – Temporary decommissioning is conducted when a system is either being serviced, repaired, renewed or extended. It involves isolating the water supply either to the whole property or to a particular section of pipework, while the work is performed. There are points to remember:
    - **Notify relevant person** – Before the decommissioning process commences, notify the home owner or any other relevant person that the water will be isolated for a period of time. Advise them to save water, i.e. fill saucepans and kettle etc., to use while the system is off. If there are other trades on site that require water to work, such as plasterers, bricklayers etc. let them know in advance so they can make alternative water arrangements.
    - **Isolate fuel/electricity supply to the system as appropriate** – isolating the cold water supply may also isolate the hot supply as well. You must ensure that any electricity using appliances or components, such as immersion heaters etc. are isolated also.
    - **Isolate water supply** – if possible, only isolate the section of the system you are working on. If this is not possible, then the whole system may need isolating at the mains internal stop tap.
    - **Apply warning notices and signs** – leave a warning notice that the system is turned off at the point where you have turned it off. This will prevent people from turning it on and causing problems. If possible, make sure the notice gives an approximate time that the supply will be reinstated and leave a phone number so that you may be contacted.
    - **Drain system to a suitable location** – using the lowest possible drain point, drain the system water to a safe location. Avoid discharging the water onto the pathway of the property or the road, especially in cold weather. If it freezes it may cause an ice hazard for passers-by. If there is no drain point, then drain the system at the lowest draw-off point. While the system is drained, it may be a good idea to install a drain point for future use.
    - **Appropriately dispose of contents and any additives** – in most cases, water drained from a cold water system can be disposed of down the foul water drain.

You should always be aware of top water drains as these may not be appropriate drain points.

- **Continuity bonding as required** – when cutting into an existing system, always remember to apply temporary continuity bonding clips first.
- **Temporary capping of pipework sections as required** – any pipework being left drained for a period of time will require capping to prevent the ingress of dirt and vermin/insects. This can be done with cap ends of the appropriate tube size.

AC3.2 Carry out decommissioning procedures

You will be asked to perform decommissioning procedures as part of your practical assessment. It is advisable to follow the procedure as stated in AC3.1.

Learning Outcome 5

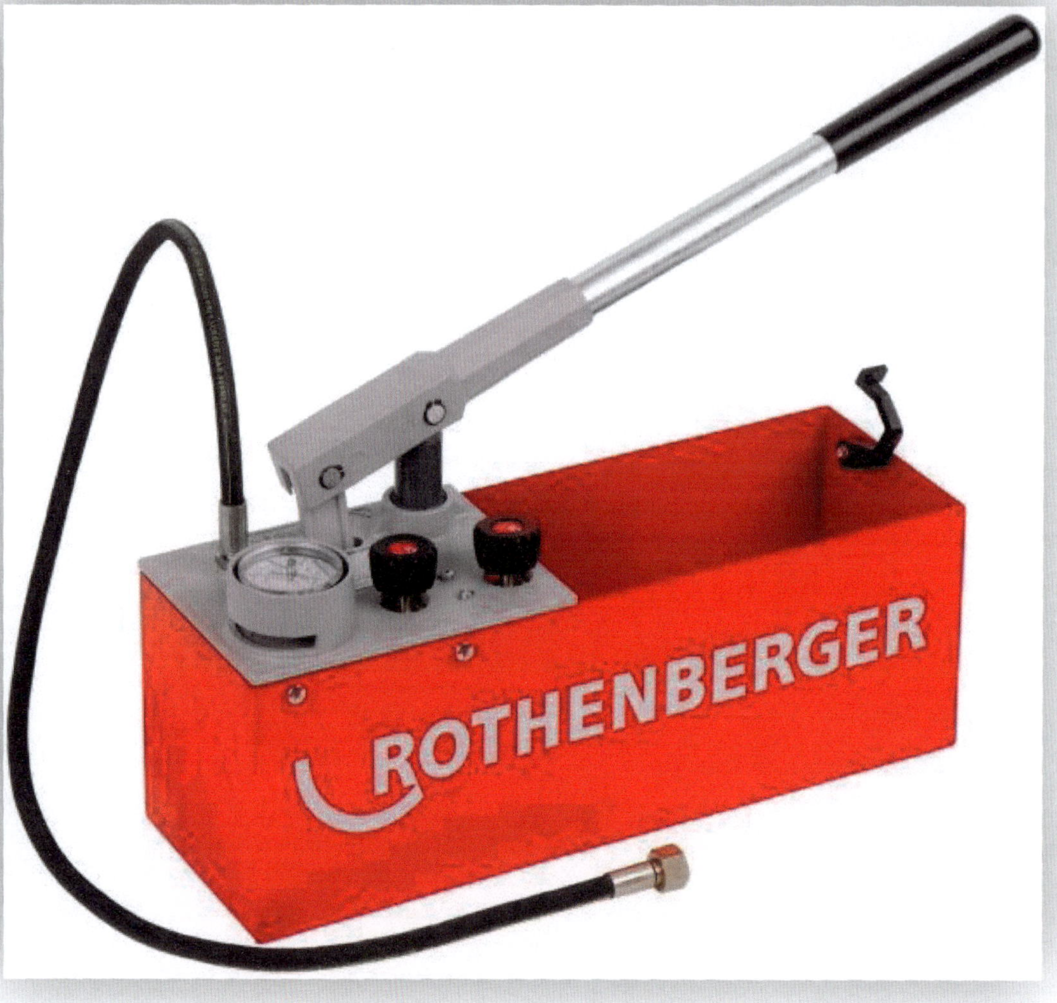

Perform A Soundness Test and Commission Cold Water Systems and Components

Learning Outcome 5 - Perform A Soundness Test and Commission Cold Water Systems and Components

In Learning Outcome 5, there are six Assessment Criteria at Phase 2:

AC5.1. Identify information sources required to complete testing and commissioning

AC5.2. Describe how to fill and vent cold water systems

AC5.3. Describe a visual inspection of a cold water system to confirm that it is ready to be soundness tested

AC5.4. Describe a **soundness test** to industry requirements on cold water systems **pipework** and components

AC5.5. State the **flushing requirements** including the use of **system additives** for new and existing cold water systems

AC5.6. Carry out a soundness test

AC5.7 to AC5.11 will be completed at Phase 3

Pressure testing systems is vital, especially if pipework is to be hidden in walls or buried in floors and ceilings. Pressure testing ensures that the pipework is sound and leak free. It is usually conducted at the end of the first fix stage in any installation.

In this Learning Outcome, we will look at how we perform pressure testing, the procedures we must follow and the rules and regulations in place to ensure that it is done correctly.

AC5.1 Identify information sources required to complete testing and commissioning

The testing of hot and cold water installations is detailed in the following documents:

- The Water Supply (water fittings) Regulations 1999: Schedule 2: Paragraph 5 and Guidance G5.1
- British Standard BS EN 806 – 4: 2010 **Specifications for installations inside buildings conveying water for human consumption. Installation**

AC5.2 Describe how to fill and vent cold water systems, and

AC5.3 Describe a visual inspection of a cold water system to confirm that it is ready to be soundness tested

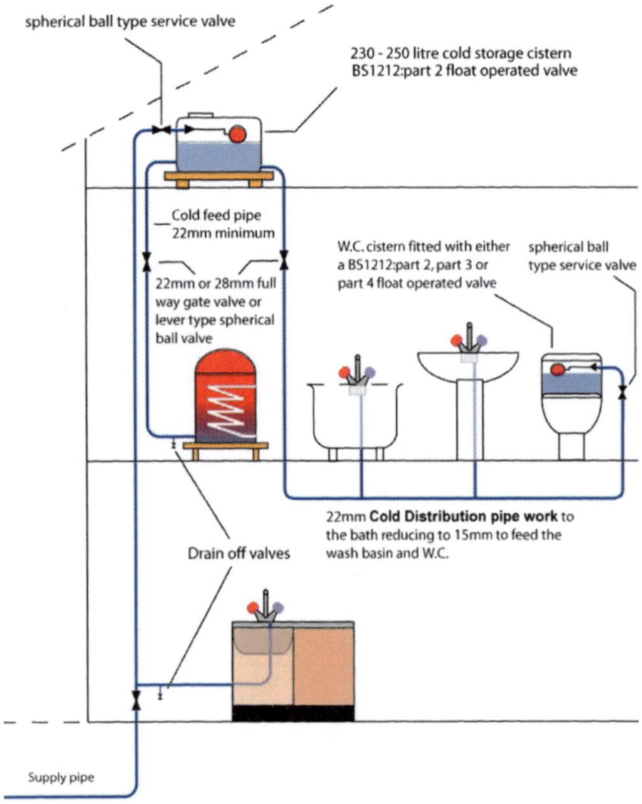

Before pressure testing of the completed installation can take place, we must first ensure that the system is water tight under normal operating conditions. To complete this, the system must be filled and vented and the system checked visually to ensure that there are no leaks.

Filling and venting cold water systems varies with each installation. No two installations are ever the same. Add to this the fact that there are at least 8 different systems types and the subject of filling and venting can become confusing.

Filling and venting the system

The system talked about here, is the indirect cold water system containing a storage cistern in the roof space to feed the bathroom with low pressure water supply, as shown in the drawing. The indirect system has been chosen because it is more difficult to fill than the direct system and requires more than just turning on the water and checking the system. There are four considerations:

1. Are there any leaks on the mains cold water supply to the kitchen sink and the storage cistern?
2. Are there any leaks on or around the storage cistern?
3. Is the water level in the storage cistern set correctly?
4. Are there any leaks on the low pressure cold water supply from the cistern to the other sanitary appliances?

Filling the system

Point 1: walk the system. This is to ensure that all taps and isolation valves are in the 'off' position so we can conduct a controlled fill, testing the installation section by section. We do not want a situation to develop where ALL parts of the system are filling at the same time. If leaks begin to develop, it makes it extremely difficult to cope with.

Point 2: **walk the system part 2.** Whilst you are checking the isolation valves as described in point 1, take the time to check that you are happy with the installation in general:

- Is the pipework neatly installed?
- Are all the compression joints tight?
- Are the tap connectors tight?
- Has any residual flux been wiped off capillary joints?
- Have enough pipe clips been used to prevent excessive movement?
- Are the float operated valves set to a provisional water level?

If the answer to these questions is 'yes', then you can proceed to point 3.

Point 3: Turning on the water. Open the kitchen tap first BEFORE you turn on the stop tap. This ensures that any debris finds its way to the kitchen tap. We do not want a situation to develop where dirt and debris are being sent all over the system. Even the smallest piece of grit can block the orifices of float operated valves and ruin tap washers. Turn the water on at the main internal stop tap **VERY SLOWLY** and wait for the water to come out of the kitchen sink tap. There will be quite a bit of flux residue and other swarf and dirt initially, so it's advisable to wait until this has cleared a little before turning on the stop tap fully.

When the water runs clear and the stop tap is fully open, turn off the kitchen tap. If there are no leaks, turn on the isolation valve to the storage cistern and fill the cistern until it is full and the FoV has shut off. Now check and adjust if necessary, the water level in the cistern. Check around the cistern for any leaks, especially the tank connectors. If there are no leaks, then go to point 4.

Point 4: Filling the low-pressure pipework. With the high-pressure side of the installation full of water, its now time to turn our attention to the low-pressure pipework. Turn on the gate valve slowly and listen for the water filling the pipework. Open the gate valve fully and let the system stand for a moment. Are there any signs of leakage? The indirect system runs very quietly and so leaks may not be apparent at first.

Go to the bath and turn on the tap. This will vent any air out of the system, and it may splutter a little until all of the air has cleared. Run the bath for a moment to remove any debris that may have accumulated in the pipework. The bath is connected in 22mm pipe and more likely to cope with any debris that has entered the system. The wash basin may be connected in 10mm pipe, which is much easier to block than 22mm.

Now turn on the isolation valve to the WC and fill the WC cistern. This may take a little longer than a WC connected to the mains supply, even if the low-pressure orifice is fitted to the FoV. Once filled, check the water level is correct and flush the WC. This will ensure that the WC works as it should. Let the WC cistern fill and shut off.

Turn on the tap to the washbasin and let it run for a moment. Once you are satisfied and the water runs clear, turn it off.

Point 5: Final system walk. With the system full of water and under normal operating pressures, walk the system again and check for any signs of leakage.

AC5.4 Describe a soundness test to industry requirements on cold water systems pipework and components

The type of testing procedure will depend on the pipe work material used. Hot and cold water systems testing is detailed in BSEN806 pts 2 and 4, central heating systems testing is detailed in **BSEN14336:2004 Heating systems in buildings — Installation and commissioning of water based heating systems**. Above ground sanitation systems should be tested in accordance with Document H of the Building Regulations.

The recommendations of BSEN806 Pts 2 and 4

For many years the recommendations of BS6700 and the Water Supply (water fittings) Regulations 199 stated that a hot/cold water system should be tested at 1.5 times the maximum operating pressure of the system. However, since the introduction of BSEN806, the test pressures for hot and cold water systems has been updated and is no longer as straightforward as it once was.

Calculating the test pressure

BSEN806 pt 1 gives a definition of the Maximum Design Pressure or MDP of a system. It states that the MDP is:

> *'the maximum hydrostatic pressure at which the potable water installation is designed to work.'*

BSEN806 pt 2, clause 3.4.2 states that:

> *'To ensure adequate strength, all components of the system shall be designed to meet the test pressure requirements of the local and national laws and regulations. The test pressure shall be **at least** 1.5 times the allowable Maximum Operating Pressure.'*

So, the maximum design pressure as stated in BSEN806 pt 1 is:

$$Max. Operating\ Pressure\ x\ 1.5 = Max. Design\ Pressure\ (MDP)$$

BSEN806 pt 4, which covers the installation and commissioning of hot and cold water systems, identifies in clauses 6.1.2 and 6.1.3 that the test pressure shall be:

$$MDP\ x\ 1.1\ or\ MOP\ x\ 1.65$$

How does this work in practice?

If a system is designed to operate at a Maximum Operating Pressure of 2 bar pressure, then

$$2\ bar\ (MOP)\ x\ 1.5 = 3bar\ (MDP)$$
$$3\ bar\ (MDP)\ x\ 1.1 = 3.3\ bar\ (test\ pressure)$$

Or

$$2\ bar\ (MOP)\ x\ 1.65 = 3.3\ bar\ (test\ pressure)$$

This results in a test pressure that is slightly higher than that required by the Water Supply (water fittings) Regulations 1999 in Schedule 2, Paragraph 5 (1.5 x MOP).

The testing procedures

BSEN806 Part 4 states that there are 3 systems for testing pipework and these testing procedures are determined by the material from which the pipework is made. The testing procedures are classed as:

a) **Test A**
b) **Test B**
c) **Test C**

The table below illustrates which pipe material is associated with each of the tests:

Table taken from BSEN806 Part 4 page 20	
Type of material	**Hydrostatic test procedure**
Linear elastic materials (i e. metals: copper, LCS and stainless-steel)	**Test procedure A**
Elastic materials (PVC-U, PVC-C, etc.) and multi-layer materials	
Visco-elastic materials (i. e. PP, PE, PEX, PA, PB, etc.) with DN/OD < 63mm	
Visco-elastic materials with DN/OD > 63mm (i.e. PP, PE, PEX, PA, PB, etc.)	**Test procedure B or C**
Combined system with DN/OD < 63mm (metals and plastics)	**Test procedure A**
Combined systems with DN/OD > 63mm (metals and plastics)	**Test procedure B or C**

Test A requirements page 21, BSEN806 pt 4

The system should be filled with water and all air vented to atmosphere. All outlet valves and vents should be sealed once the air is removed. Apply the test pressure calculated (equal to 1.1 times MDP) for a period of 10 minutes. The test pressure must stay constant for 10 minutes. If there is any pressure loss, the system must be maintained at the test pressure until all leaks have been found. The test is shown in graph form right.

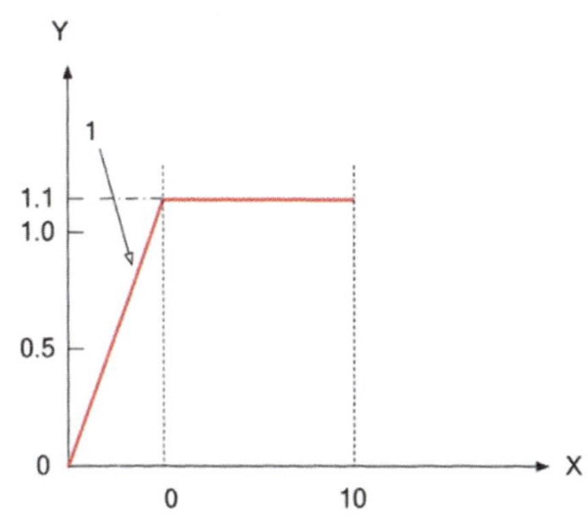

X = Time in minutes
Y = Test pressure
1 = Pumping

Test B requirements page 21, BSEN806 pt 4

The system should be filled with water and all air vented to atmosphere. All outlet valves and vents should be sealed once the air is removed. Apply the test pressure calculated (equal to 1.1 times MDP) for a period of 30 minutes. Inspect the system to identify any leaks.

Reduce the pressure by bleeding water from the system until the pressure reads 0.5 times the test pressure.

The pressure must remain constant for a period of 30 minutes after pressure reduction. Check the system visually for leaks. If, after the pressure has been reduced there is a further

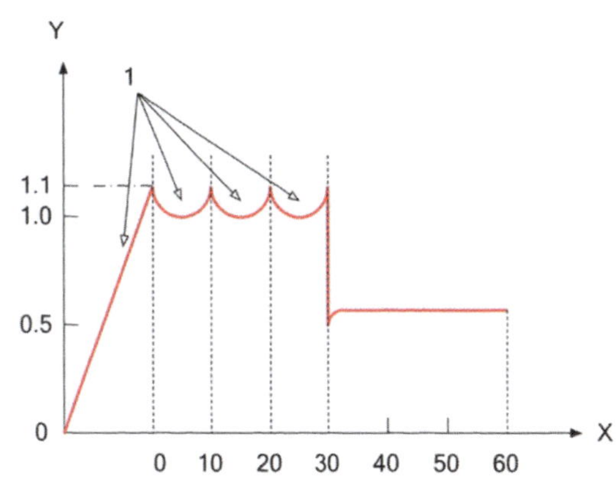

X = Time in minutes
Y = Test pressure
1 = Pumping

pressure drop, there is a leak on the system. The test pressure must be maintained while the leak is identified. Again, the test is shown in graph form.

Test C requirements page 22, BSEN806 pt 4

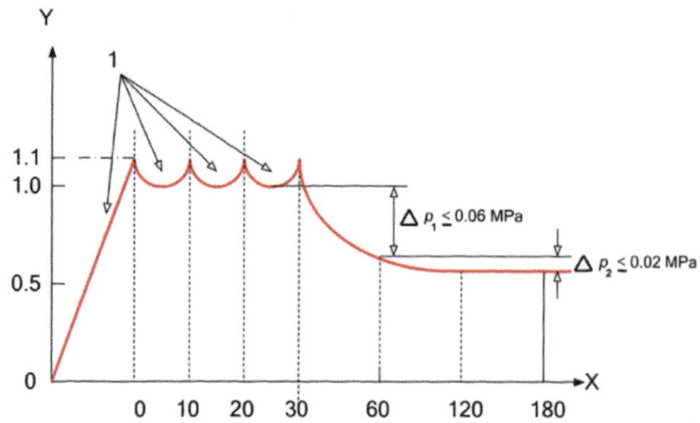

X = Time in minutes
Y = Test pressure
1 = Pumping
Δ P1 = Maximum pressure drop between 30 mins and 60 mins of test pressure
Δ P2 = Maximum pressure drop between 30 mins and 60 mins of test pressure

The system should be filled with water and all air vented to atmosphere. All outlet valves and vents should be sealed once the air is removed. Apply the test pressure calculated (equal to 1.1 times MDP) for a period of 30 minutes. Inspect the system to identify any leaks.

Note the pressure after a period of 30 mins and recheck for leaks.

Note the pressure after a further 30 mins. If the pressure drop is less than 0.06 MPa or 0.6 bar, the system can be considered to be sound. Continue the test without further pumping.

Continue the test for a further 120 minutes (2 hours). If the pressure drops more than 0.02 MPa or 0.2 bar in this time, then this would indicate that the system has a leak. Maintain the pressure and identify the leak.

AC5.5 State the flushing requirements including the use of system additives for new and existing cold water systems

PD 855468:2015 Guide to the flushing and disinfection of services supplying water for domestic use within buildings and their curtilages

Flushing and disinfection of systems used to be part of BS8558:2011. However, the latest version of BS8558:2015 excludes flushing and disinfection. Instead, flushing and disinfection of systems is now a separate document PD 855468:2015.

PD 855468:2015 provides guidance on the cleaning, flushing and disinfection of cold water systems to control microbiological growth and the removal of debris. This includes guidance on:

- Deployment of the correct tools and personnel
- The use of the correct disinfectants
- How to respond if a microbiological problem is identified
- Keeping records of cleaning and disinfection

This new document applies to systems supplying water to domestic purposes within buildings and their curtilages, and includes water used in food preparation.

System Flushing

All systems must be flushed with cold water taken directly from the water undertaker's mains cold water supply before handing over to the end user. This is a requirement of the British Standards. It should be completed as soon as possible on completion of the installation and commissioning process. System flushing removes any dirt and debris, such as flux residues, swarf, jointing compounds etc. from the system that may be classed as contaminates.

The flushing process

The system should be filled with water, all cisterns should be full, and all float operated valves should be off. Starting with the cold tap that is nearest to the incoming cold water supply, open the tap until the water runs clear and there is no visible dirt, debris or discolouration present. Work away from the tap, visiting each tap and appliance in turn, performing the same process each time.

Cisterns must be checked for debris and cleaned as necessary and float operated valves checked and adjusted. Any components, such as showers, should also be flushed thoroughly. If a break cistern is connected to the system, this should be checked thoroughly, as it is the first major point on the system.

It is accepted practice, that once a system has been flushed, it should not be left charged with water if the property is not going to be occupied straight away. Unused cold water systems can become stagnant if left for more than 4-6 weeks. However, draining a system can often be detrimental, since it's almost impossible to completely drain a system of water, especially if there are long, horizontal runs on the system. The water left in the system becomes stagnant, which immediately raises contamination issues. Also corrosion can set in, which often goes undetected. It is recommended that, in these instances, the system should be left fully charged with water and the system flushed through completely twice weekly. Again, this will involve opening all taps and outlets and should be completed until the build is permanently occupied.

Disinfection of cold water systems

At the present time, there is no requirement that domestic cold water systems be disinfected. However, large systems should be disinfected using a special water system disinfecting agent that contains either sodium hypochlorite or calcium hypochlorite. The British Standard tells us that systems should be disinfected in the following circumstances:

- In all new installations, except those used as private dwellings
- Where extensions and alterations have been made to large cold water systems
- Where it is suspected that the system has been compromised by contamination
- Where the system has not been in regular use
- Where underground pipework has been installed (this does not include localised repairs or where a junction has been installed after the fittings have been immersed in a solution of sodium hypochlorite diluted to 200ppm)(1ppm = 1mg/l)

Systems should be cleaned and disinfected as part of a good maintenance regime. It should also be remembered that, where large amounts of water are stored, these systems must also be disinfected to prevent the onset of legionella pneumophila bacteria leading to Legionnaire's disease.

Disinfection types

There are two disinfectants approved by the Drinking Water Inspectorate:

- ⇒ EN900 – Calcium Hypochlorite
- ⇒ EN901 – Sodium Hypochlorite

Most disinfection processes for cold water installations use EN901 – Sodium Hypochlorite.

The disinfection process

Before disinfection can take place, the system must be first thoroughly flushed with fresh water from the water undertaker's cold water main and then completely drained. It should then be refilled adding the disinfecting agent as the system fills. As already stated, most systems are disinfected using a solution of EN901 – Sodium Hypochlorite and this should be added at a concentration of 50ppm. Once the system is full and all Float Operated Valves have shut off, the system must be allowed to stand for 1 hour. At the end of this period, the furthest outlet point should be tested to ensure that at least 30ppm of chlorine are present.

All person either working in or using the building MUST be informed that disinfection is going to be conducted BEFORE it takes place. This includes those persons who are not normally present during working hours. During disinfection, WARNING NOTICES must be posted to warn people that the system is being disinfected and that the water MUST NOT be used. These should be placed at ALL outlets.

Household bleaches, toilet cleaners and chemicals MUST NOT be allowed to come into contact with the disinfected water as this can produce highly reactive toxic fumes. All operatives must be made aware of the COSHH requirements of the disinfecting solution and PPE must be issued to all operatives beforehand. This should include goggles, face mask and protective gloves or gauntlets.

Calculating the disinfection solution required

To calculate the correct amount of solution to administer, we must first calculate the exact capacity of the system including cistern and pipework capacities.

Sodium Hypochlorite contains around 5% chlorine, which is equivalent to 50,000ppm. So, to calculate the amount of solution to add to the system that is equal to 50ppm:

$$50 \div 1{,}000{,}000\, x\, 100 \div 5 =$$
$$0.001\ part\ of\ Sodium\ Hypochlorite\ solution\ to\ 1\ litre\ of\ water$$

In this form, the calculation doesn't mean very much because we need to know how much Sodium Hypochlorite to add. So, it is generally accepted that a ration of 1 litre of Sodium Hypochlorite (containing 5% chlorine) for every 1000 litre of water in the system be used.

Calculating the capacity of the cold water system

To add the correct amount of solution, we must first calculate the total system capacity:

⇒ Calculate the capacity of the cistern up to the water line (in litres)

⇒ Calculate the capacity of the pipework

Calculate the capacity of the cistern up to the water line (in litres)

Cistern capacity is calculated by:

$$Length\ of\ the\ cistern\ x\ width\ of\ cistern\ x\ height\ of\ the\ cistern\ to\ the\ water\ line\ x\ 1000 = cistern\ capacity$$

Let's assume:

Length of cistern = 2m

Width of cistern = 1.25m

Height to water line = 1.25m

$$2\ x\ 1.25\ x\ 1.25 = 3.125m^3\ x\ 1000\ (to\ convert\ to\ litres) = 3125\ litres$$

Calculate the capacity of the system pipework (in litres)

Again, let us assume the following amounts of copper tube are used in the installation. In reality these should be measured accurately. The table gives the capacity of the various sizes of copper tube:

⇒ 50m of 42mm copper tube

⇒ 25m of 35mm copper tube

⇒ 25m of 28mm copper tube

⇒ 10m of 22m copper tube

⇒ 30m of 15mm copper tube

Calculating the capacity of the installation pipework

The capacity of copper tube per metre		
Standard/Spec	Tube Dimension (OD)	Carrying Capacity (L/m)
BS EN 1057 R250 half-hard straight lengths	6	0.018
	8	0.036
	10	0.061
	12	0.092
	15	0.145
	22	0.320
	28	0.539
	35	0.835
	42	1.232
	54	2.091

Pipe	Size (mm)	Length (m)	Capacity per metre (l)	Capacity of the pipe (l)
1	42	50	1.232	61.6
2	35	25	0.835	20.875
3	28	25	0.539	13.475
4	22	10	0.320	3.2
5	15	30	0.145	4.35
Total pipework capacity (l)				103.5 litres

$$103.5 + 3125 = 3228.5\ litres\ total\ capacity$$

So, to calculate the amount of disinfectant:

$$3228.5 \ x \ 0.001 = 3.228 \ litres \ of \ sodium \ hypochlorite$$

How to disinfect a large cold water system

This method presumes that capacity calculations have already been conducted and that the amount of disinfectant is known:

Stage 1 Thorough flush the system with clean cold water from the water undertaker's cold mains supply and drain down completely. This will remove any system installation debris.

Stage 2 Refill the cold water cistern adding the disinfectant to the water as it enters the cistern. This will ensure that the solution of disinfectant is completely and evenly diluted into the system water. Add the disinfectant slowly allowing the water to reach all parts of the system.

Stage 3 Working away from the cistern, open each cold water outlet in turn until the smell of chlorine is detected.

Stage 4 Once the system is full and the cistern Float Operated Valves are shut down, allow the system to stand for 1 hour. A free residual chlorine content of 30ppm must be detected at the furthest outlet point. If it is below this, then the system must be drained down, flushed and the process started again from the beginning.

Stage 5 If the disinfection process has been successful, then the system should be completely drained and thoroughly flushed with clean water until the free residual chlorine content of the water is no higher than that of the water undertakers' mains cold water supply.

Water colour	Chlorine level (ppm or mg/l)
Clear	**None**
Faint pink/pink	**0.2 – 1**
Pink/red	**1 – 5**
Red/purple	**5 – 10**
Purple/blue	**10 – 20**
Blue/grey-green	**20 – 30**
Grey-green/yellow	**30 – 50**
Muddy brown	**Over 50**

Testing for free residual chlorine content

The test involves taking a sample of system water from the furthest outlet and testing it using a specialised kit. The kit contains a test tube and special chemicals in tablet form that are dropped into the test tube of water. As the tablet dissolves, the water changes colour, the colour indicating what the free residual content is. The colours and their representative free residual contents are shown in the table left:

Flushing the system after disinfection has been completed

It is extremely important to remember that Sodium Hypochlorite and Calcium Hypochlorite are extremely dangerous to wild life because of the chlorine content. Because of this, we cannot simply drain the contents to a foul water or surface water drain. Permission must be sought from the water undertaker so that they can expect water with a high chlorine content arriving at the water treatment plant when the system is flushed. Alternatively, a neutralising agent can be added to the water.

The neutralising agent is known as **Sodium Thiosulphate** and must be added at the following strength:

$$System\ volume\ x\ ppm\ of\ chlorine\ in\ the\ water\ x\ 2\ =\ grams\ of\ sodium\ thiosulphate\ required$$

At the end of the process, the system should be completely drained and thoroughly flushed with clean water until the free residual chlorine content of the water is no higher than that of the water undertaker's mains cold water supply at all outlets.

General points to remember

- ⇒ Warn all persons before disinfection is conducted
- ⇒ Handle all chemicals with care and always where the appropriate PPE
- ⇒ Check the free residual chlorine content at the end of the disinfection test period
- ⇒ NEVER leave the disinfected water in the system overnight
- ⇒ NEVER discharge chlorinated water directly down the drainage system without seeking advice from the water undertaker or the Environment Agency first
- ⇒ Always complete the disinfection record paperwork as appropriate

AC5.6 Carry out a soundness test

During your Phase 2 practical assessment, you will be asked to carry out a soundness test to industry standards.

Unit 6

Hot Water – Phase 2

Unit 6 – Hot Water – Phase 2

In Phase 2, there 3 Learning Outcomes:

LO1 - Install Hot Water Systems:

- **AC1.1** Identify types and layout features of **hot water systems (completed at Phase 1 – please refer to Phase 1 textbook)**
- **AC1.2** State advantages and disadvantages of hot water systems
- **AC1.3** Identify types and typical pipe sizes used in hot water systems within dwellings
- **AC1.4** Describe working principles of hot water systems, positioning, fixing, connection and operation of components
- **AC1.5** Describe insulation requirements and system frost protection
- **AC1.6** Identify the positioning and fixing of pipework within the **building fabric**
- **AC1.7** Explain expansion and contraction in hot water systems and negative effects
- **AC1.8** Identify location and function of **unvented system components**
- **AC1.9** Describe secondary circulation and how trace heating can be used
- **AC1.10** Identify **sources of information** required when undertaking work on hot water systems
- **AC1.11** Explain **backflow** risk and required **methods** of prevention
- **AC1.12 Install** hot water systems

LO2 - Decommission hot water systems:

- **AC2.1** Explain **procedures** for **decommissioning systems**
- **AC2.2** Carry out decommissioning **procedures**

LO4 - Perform a soundness test and commission Hot Water systems and components:

- **AC4.1** Identify information sources required to complete testing and commissioning
- **AC4.2** Describe how to fill and vent hot water systems
- **AC4.3** Describe a visual inspection of a hot water system to confirm that it is ready to be soundness tested
- **AC4.4** Describe a **soundness test** to industry requirements on hot water systems **pipework** and components
- **AC4.5** State the **flushing requirements** including the use of **system additives** for new and existing hot water systems
- **AC4.6** Carry out a **soundness test**

Learning Outcome 1

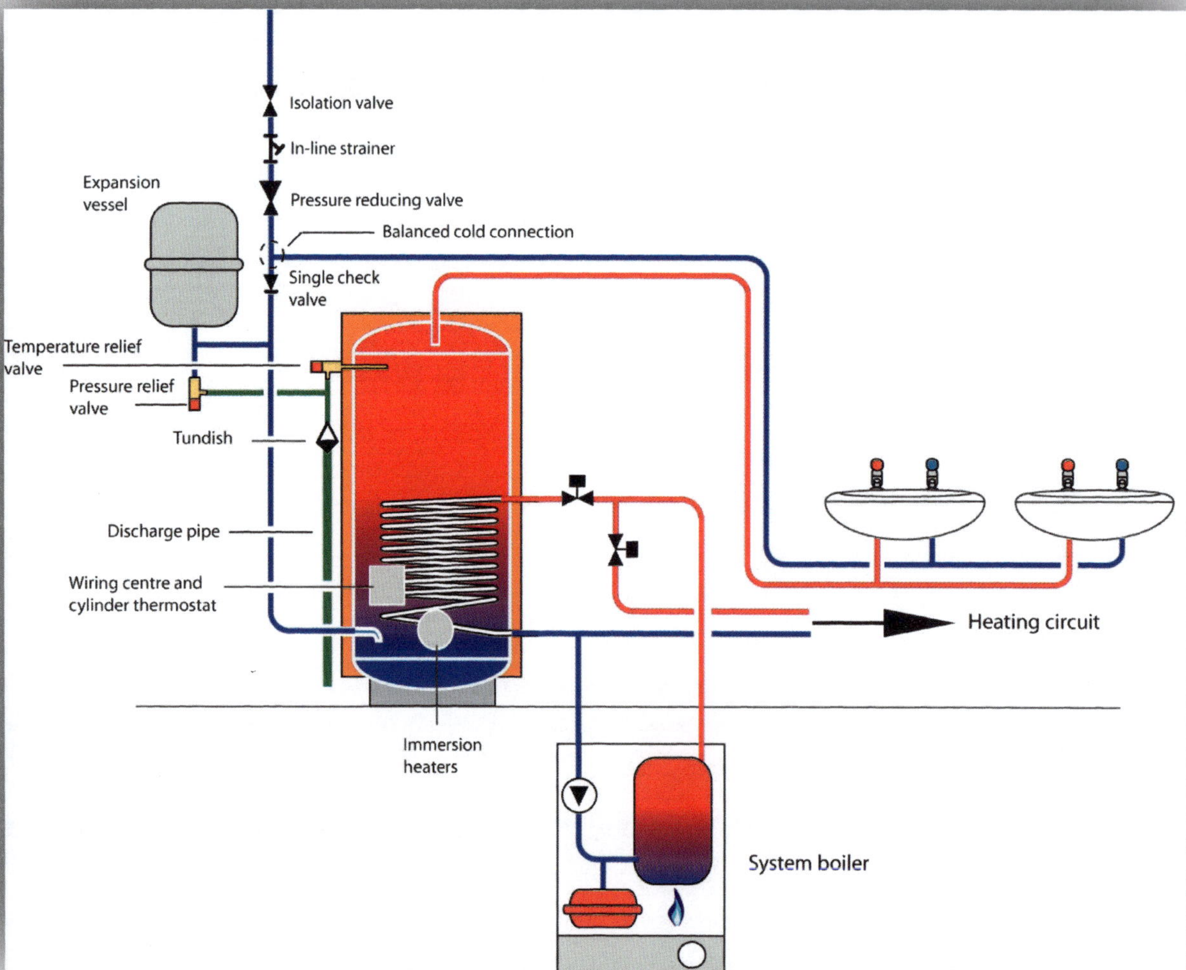

Install Hot Water Systems

Learning Outcome 1- Install Hot Water Systems

In Phase 2, you will be investigating the following Assessment Criteria from Learning Outcome 1:

AC1.2 State advantages and disadvantages of hot water systems

AC1.3 Identify types and typical pipe sizes used in hot water systems within dwellings

AC1.4 Describe working principles of hot water systems, positioning fixing, connection and operation of components

AC1.5 Describe insulation requirements and system frost protection

AC1.6 Identify the positioning and fixing of pipework within the **building fabric**

AC1.7 Explain expansion and contraction in hot water systems and negative effects

AC1.8 Identify location and function of **unvented system components**

AC1.9 Describe secondary circulation and how trace heating can be used

AC1.10 Identify **sources of information** required when undertaking work on hot water systems

AC1.11 Explain **backflow** risk and required **methods** of prevention

AC1.12 Install hot water systems

AC1.2 State advantages and disadvantages of hot water systems

In Phase 1, we looked at the system layouts for hot water systems. Now, in phase 2 we will investigate the advantages and disadvantages of those systems. Not all system types will be discussed as many have similar advantages and disadvantages.

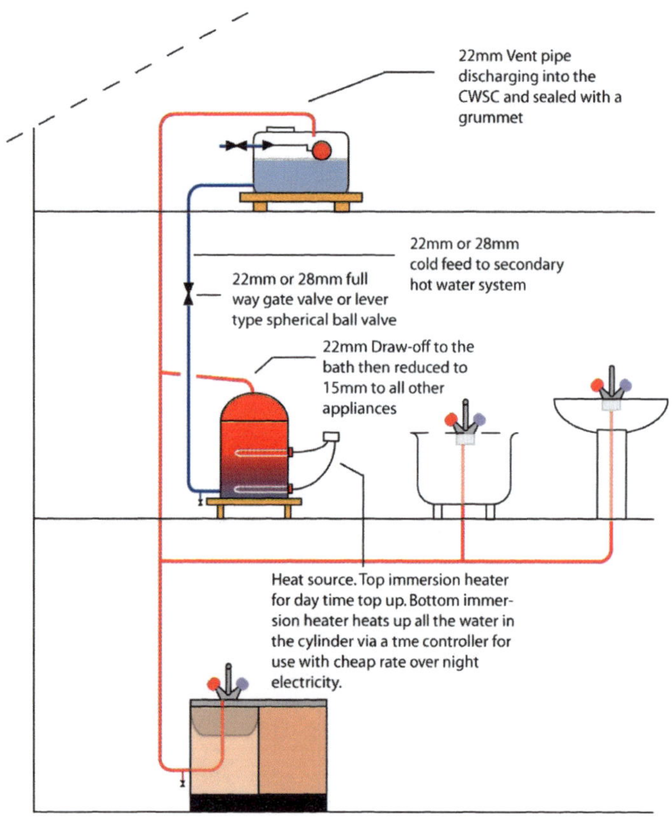

Direct open vented system using immersion heaters

Open vented Hot Water Storage Systems

Open Vented Direct System

Direct open vented hot water storage systems are named after the type of hot water storage cylinders that the systems use. They use a direct-type hot water storage cylinder, which is heated either by **a)** a small hot water only boiler, or **b)** an electric immersion heater.

The direct hot water storage cylinder does not contain a heat exchanger. The water is heated directly by either the hot water only boiler or an electric immersion heater. This type of cylinder is NOT suitable for use on central heating systems.

Table 1: Direct Open Vented Hot Water Storage Systems

Advantages	Disadvantages
Direct water heating means quicker hot water heating times	Greater risk of frost damage of pipework and cisterns in the roof space
When used with an immersion heater it is the cheapest of all storage hot water systems	A large amount of water storage is needed
Low pressure system is quiet in operation	Cisterns and hot water cylinder use a lot of storage space in the roof and airing cupboard
Pipework can be sized to give very good flow rates despite the lack of pressure	More structural support required to carry the weight of the cisterns/cylinder when full of water
Shower valves on low pressure supplies may be supplied with equal pressures and flow rates	Cannot be used with central heating systems
Reserve water in the cistern should the cold water mains fail	Immersion heaters can be expensive to run
Less risk of water hammer and reverberation noise	When used with a back boiler, the boiler must be made of non-ferrous metal, these can be expensive

Open Vented Double Feed Indirect System

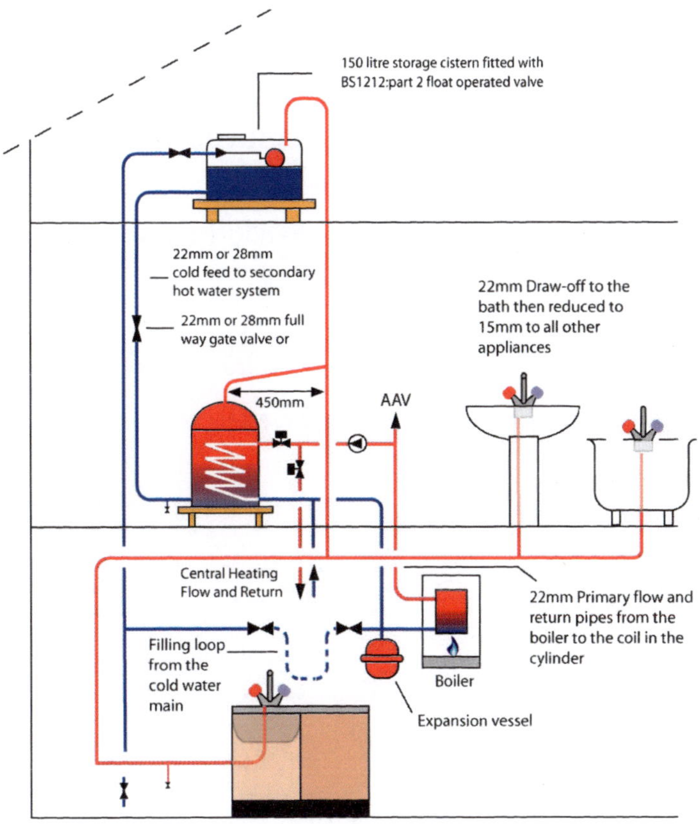

The double feed indirect hot water storage cylinder

This system utilises a double feed indirect hot water storage cylinder that contains a heat exchanger in the form of a coil.

The coil, which is made from copper tube (stainless steel storage cylinders with stainless steel coils are also available), is the primary heat exchanger for the hot water supply. This type of hot water storage cylinder is suitable for use with central heating systems. It is called indirect because the water is heated indirectly by the primary water in the boiler via the heat exchanger.

The double feed indirect system has two cisterns. A large cold feed cistern that supplies the cold water to the hot water system via the cold feed pipe and a smaller feed and expansion cistern to supply water to the boiler, hot water heat exchanger and central heating system. Here the hot water circulates via the primary flow and return pipes, with the aid of a circulating pump or without a circulating pump through gravity (see Phase 1 textbook Unit 3 Scientific Principles).

Table 2: Indirect open vented hot water storage systems	
Advantages	**Disadvantages**
Low pressure system is quiet in operation	Greater risk of frost damage of pipework and cisterns in the roof space
Pipework can be sized to give very good flow rates despite the lack of pressure	A large amount of water storage is needed
Shower valves on low pressure supplies may be supplied with equal pressures and flow rates	Cisterns and hot water cylinder use a lot of storage space in the roof and airing cupboard
Reserve water in the cistern should the cold water mains fail	More structural support required to carry the weight of the cisterns/cylinder when full of water
Less risk of water hammer and reverberation noise	More pipework required, so the system is expensive to install
Can be used with central heating systems	***Important note:*** *The single feed self-venting indirect system has not been mentioned, since this is an obsolete system and is not recognised by Approved Document L: Conservation of fuel and power as an energy efficient system.*
Can be used with sealed heating systems eliminating the F&E cistern	
Can be used with underfloor heating systems	
Good hot water temperature control	

Unvented Hot Water Storage Systems

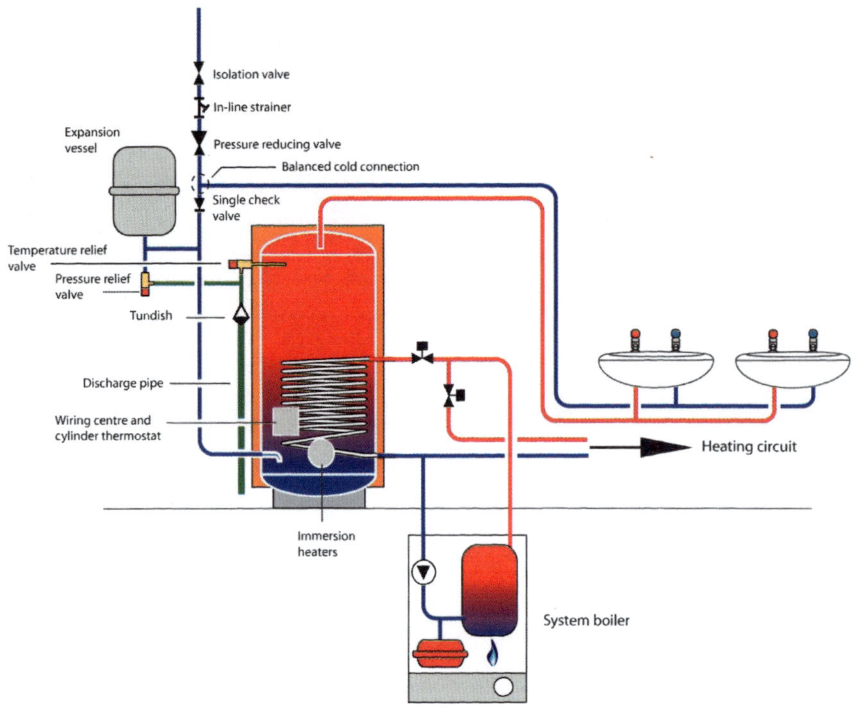

An indirect Unvented Hot Water Storage System with system boiler

Unvented Hot Water Storage Systems (UVHWSS) differ from open vented systems because they do not require a vent pipe or storage cistern. They are filled directly from the cold water mains supply, giving much better flow rates and pressures at the hot water outlets and taps. However, filling a hot water storage system from a pressurised water supply can cause problems with the boiling point of the water and subsequently excessive water expansion.

Unvented Hot Water Storage Systems are available in both direct and indirect types. The advantages/disadvantages below relate to the method of hot water supply, since direct/indirect differences are highlighted with the open vented systems above.

Table 3: Unvented hot water storage systems	
Advantages	**Disadvantages**
High pressure and flow rate of hot water at all outlets	Hot Water storage cylinders are expensive
Smaller hot water pipework sizes can be used in some instances.	Storage cylinders are heavy and require structural support when fitted in a roof space
Shower valves may be supplied with equal pressures and flow rates	Requires more safety controls than other hot water systems
Less risk of frost damage	Can cause explosion if fitted incorrectly
No cold water storage required	Potential for water temperature to exceed 100°C
Can be used with all wet central heating systems (indirect type only)	Requires extra pipework for discharge pipework system
Good hot water temperature control	Requires regular maintenance
Less pipework required compared with Open Vented systems	Cannot be used with solid fuel heating
Ideal for flats and apartments with no roof space	
Can be used with solar thermal hot water systems and heat pumps	

Instantaneous Hot Water Systems

Thermal store

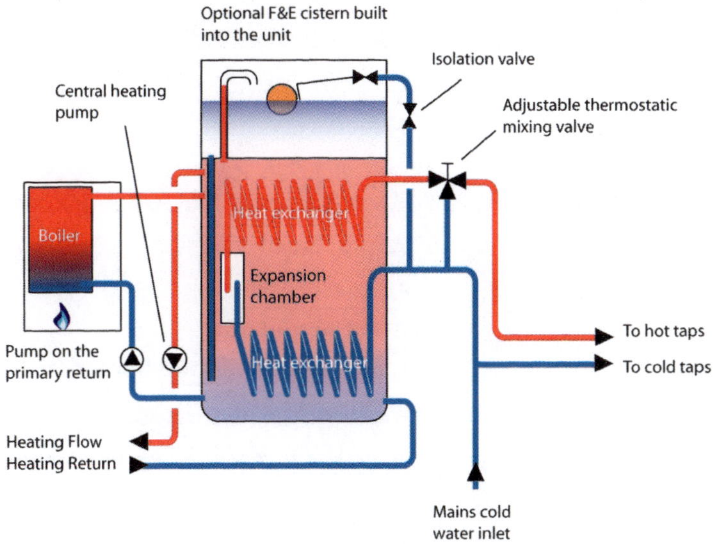

A thermal store also known as a water jacketed tube heater

Thermal stores (also known as water jacketed tube heaters) are centralised instantaneous multipoint water heaters that work by passing water direct from the cold water main through a series of heat exchangers. These heat exchangers are completely surrounded by water at around 80°C inside a large hot water vessel that is directly heated by a boiler. They look very similar to indirect hot water storage cylinders but actually work in reverse. Water from the mains cold supply enters the thermal store, passing through the first heat exchanger where the water is warmed. After passing through the expansion chamber, the water flows through the second heat exchanger where it is heated to full hot water temperature. Because hot water should not exceed a temperature of 60 – 65°C, the water is blended with cold water through an adjustable thermostatic blending valve on the outlet of the thermal store.

The water in the hot water vessel is completely separate from the domestic hot water and is heated by a boiler. The system can be used for properties with central heating systems.

Table 4: Thermal store hot water systems

Advantages	Disadvantages
High pressure and flow rate of hot water at all outlets	Hot Water storage cylinders are expensive
Smaller hot water pipework sizes can be used in some instances.	Storage cylinders are heavy and require structural support when fitted in a roof space
Shower valves may be supplied with equal pressures and flow rates	Requires regular maintenance
Less risk of frost damage	Fuels that can used with thermal stores: • Conventional boiler systems (gas or oil) • Solar panels • Wood burning stoves & back boilers (or other uncontrollable heat sources)
Good hot water temperature control	
Less hot water pipework required compared with storage hot water systems	
Ideal for flats and apartments with no roof space	
Can be used with solar thermal hot water systems and heat pumps	

Water can be heated up by a secondary heating source such as a conventional boiler or electric immersion heater	• Gravity or pumped heat sources • Heat recovery systems (Ground source & Air source) • Electricity (from grid or renewable sources
Enables a renewable heating system to work more efficiently	
Reduces the need to buy expensive fossil fuels to meet on-demand hot water or space heating	
Allows the use of a wood burning boiler stove or a stove with a back boiler at maximum efficiency without overheating the room	

Multipoint heaters

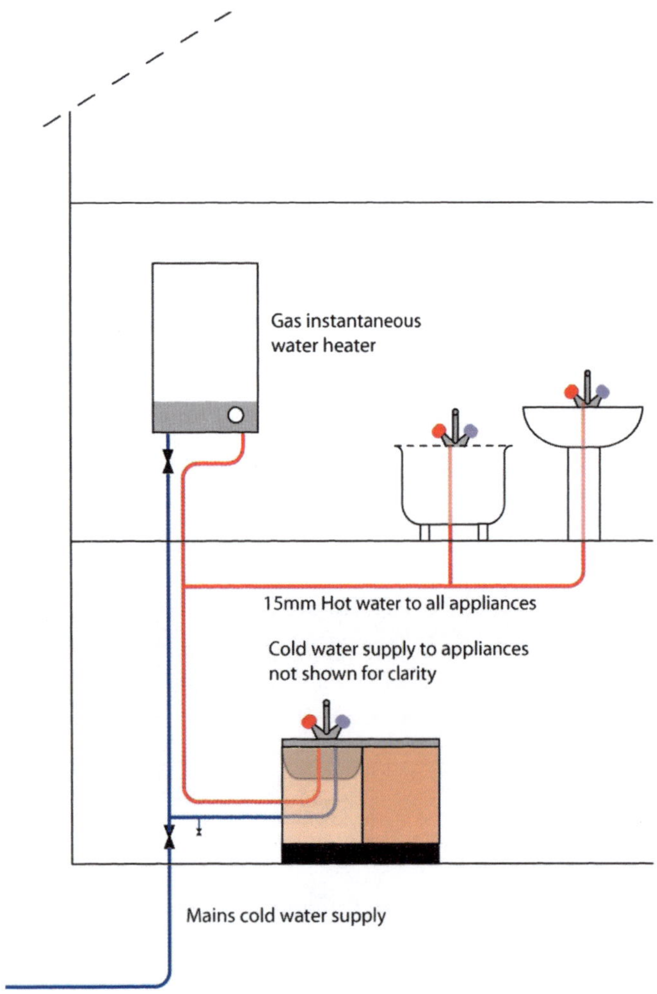

A gas fired instantaneous water heater

Fired by gas, instantaneous multipoint water heaters use the principle of pressure difference. When a tap is opened, the movement of water through a venturi tube creates a difference in pressure over a rubber diaphragm inside the pressure differential valve. The difference in pressure flexes the diaphragm inside the valve and this opens a gas valve. The gas passes through to the gas burner where it is ignited by either a pilot light or piezo ignitor. The water is then heated. When the tap is closed, the pressure across the pressure differential valve equalises and a spring in the valve closes the valve, shutting off the flow of gas.

Instantaneous gas water heaters can provide enough hot water for a small domestic dwelling but the hot taps should only be used one at a time as these hot water heaters cannot generate enough flow rate to supply two open hot taps simultaneously.

Electric-type multipoint water heaters are also available.

Table 5: Multipoint water heaters	
Advantages	Disadvantages
Smaller hot water pipework sizes can be used in some instances.	Requires regular maintenance
Shower valves may be supplied with equal pressures and flowrates	No central heating facility

Less risk of frost damage	Although classed as multipoint heaters, most types will only allow one tap to be open at once
No cold water storage required	Flow rate can be poor
Less hot water pipework required compared with storage hot water systems	
Ideal for flats and apartments with no roof space	

Combination boilers

Combination boilers are fast becoming one of the main sources of hot water supply in small to medium sized dwellings in the UK. They combine hot water supply and central heating in a single appliance that can be sited virtually anywhere in the property.

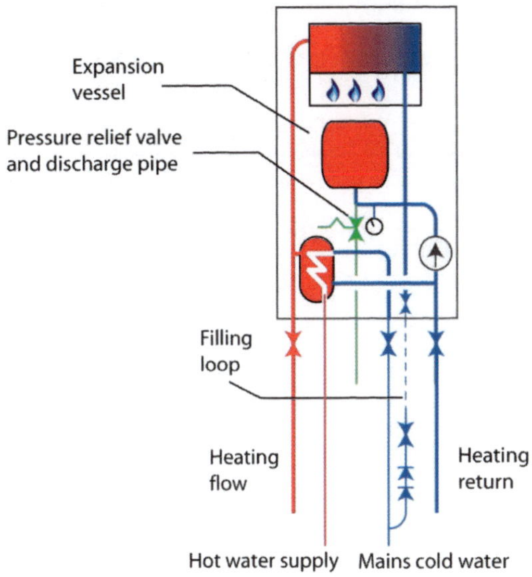

A combination boiler

'Combi' boilers use the same principle as thermal stores by heating the water through a water-to-water heat exchanger. Known as a 'plate heat exchanger', the heat exchanger can alternate the flow of water through a diverter valve to heat either the hot water instantaneously or the central heating system. They are hot water priority appliances, which means that if the central heating is on and a hot tap is opened, all energy created by the appliance will be transferred into generating hot water supply to the open tap.

There are many different types of combination boilers available ranging from 24kW to 50kW in output. Some combis even have a small amount of hot water storage (less than 15 litres), which means hot water is delivered quickly to the hot taps. They are fitted in the same way as multipoint instantaneous hot water heaters.

Table 6: Combination boilers

Advantages	Disadvantages
Smaller hot water pipework sizes can be used in some instances.	Requires regular maintenance
Shower valves may be supplied with equal pressures and flow rates	Although classed as multipoint heaters, most types will only allow one tap to be open at once
Less risk of frost damage	Flow rate can be poor with older combi types
No cold water storage required	Requires extra condensate pipework, which is vulnerable to freezing in cold weather
Less hot water pipework required compared with storage hot water systems	Not suitable for domestic properties where more than one bathroom is installed
Ideal for flats and apartments with no roof space	Secondary hot water circulation not possible with combination boilers
Central heating available	
Small amount of hot water storage available on larger combination boilers	

Single point (point of use) heaters

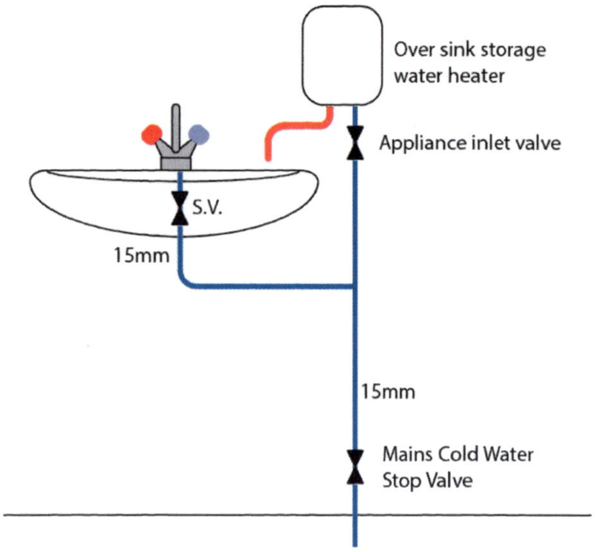

A point of use water heater

Instantaneous single point water heaters

These provide instant hot water to a single appliance and are usually fuelled by electricity. There are two basic types:

Hand wash type – small water heaters, usually around 3kW output, that are fitted over a single washbasin. They are inlet controlled. The temperature of the water depends on the flow rate through the heater. The faster the flow rate, the cooler the water. Flow rates are generally poor but adequate for hand washing.

Electric showers – outputs up to 12Kw, many electric showers feature sophisticated microchip technology allowing temperature stabilisation at low and high flow rates. Electric showers feature a low pressure cut-out to guard against scalding if the pressure or flow rate suddenly drop. Flow of water is controlled by an electrically operated solenoid valve that operates when the electricity supply is on to the heater.

Table 7: Instantaneous single point water heaters

Advantages	Disadvantages
Very little pipework required.	Only one appliance can be served
Cheap to install	No central heating facility
Cost effective, especially if the main hot water system is some distance away from where the hot water is required.	Flow rate is usually poor (except electric showers)
Cheap to run	
Ideal for isolated washing facilities	

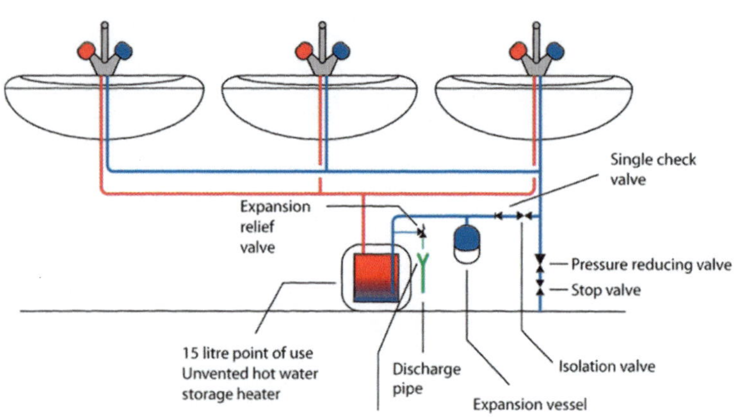

The unvented under sink storage water heater

Storage type single point water heaters

Storage point of use heaters are small water heaters that store a small amount of hot water, usually of less than 15 litres capacity, for use at a single appliance or small range of appliances. They can be either outlet or inlet controlled depending on the type. There are 3 basic types:

1. Over sink type
2. Undersink type
3. Unvented hot water type

Table 8: Storage type single point water heaters	
Advantages	**Disadvantages**
Very little pipework required.	Only one appliance can be served (in most cases)
Cheap to install	No central heating facility
Cost effective, especially if the main hot water system is some distance away from where the hot water is required.	Flow rate is usually poor (except unvented types)
Cheap to run	
Ideal for isolated washing facilities	
Small amount of storage available	
Quick water heat up times	

AC1.3 Identify types and typical pipe sizes used in hot water systems within dwellings

The pipe sizes used for hot water systems depends on:

- The type of system installed
- The size of the property
- How many hot water outlets there are on the system
- The distance from the heat source/hot water storage point to the outlets

There are three materials that can be used for hot water systems in domestic properties:

- Copper tubes and fittings
- Plastic pressure pipe and push-fit fittings
- Stainless steel tubes and fittings

Pipe sizes vary according to the system being installed. Generally, open vented systems, because of the low pressure of the hot water delivery, tend to have larger pipe sizes:

Open vented hot water storage systems – the cold feed from the cistern to the cylinder should be of 28mm size pipe. In smaller properties, this may be reduced to 22mm if there is no ensuite bathroom in the property. The hot water draw-off should be a minimum of 22mm, although larger systems will require a 28mm draw-off because of the number of hot water outlets. At least 22mm should be taken to the bath, with 15mm serving washbasin, bidet and kitchen/utility sink. The Vent pipe must have a vertical rise and discharge over the cistern in the roof space. The height of the vent above the water line in the cistern must be at least 40mm per meter of static head plus 150mm. The draw-off from the top of the cylinder must have a gentle rise to the vent pipe and must be at least 450mm in length to prevent parasitic circulation. Secondary circulation may be required with large systems.

Unvented hot water storage systems – unvented hot water storage systems are fed directly from the water undertaker's cold water main. Because of this, generally smaller pipe sizes can be used. Most manufacturers insist that the unvented hot water storage cylinder is fed by 22mm cold water supply pipe, with 22mm hot water draw-off from the top of the storage vessel, although with these systems, baths can usually be supplied by 15mm pipe. Again, if the system is large with multiple hot water outlets, it may be advisable to use 28mm pipe to ensure a good flow rate. Secondary circulation may be required with large systems.

Instantaneous hot water systems – generally these are fed via 15mm cold water supply direct from the water undertaker's cold water main. It is usual to feed all appliances in 15mm pipe. However, to aid the flow rate with larger systems, it is advisable to install 22mm from the hot water appliance (combination boiler, instantaneous hot water heater or thermal store). It advisable to seek manufacturer's advice with instantaneous supplies, because larger systems are not suited to instantaneous hot water systems. Secondary circulation is not possible with these systems.

AC1.4 Describe working principles of hot water systems, positioning, fixing, connection and operation of components

In this Assessment Criteria, you may be referred to **Unit 5 – Cold Water Systems**, as some of the components are generic to both hot AND cold water systems.

Hot Water Storage Vessels

Open vented hot water cylinders

Open vented hot water cylinders are manufactured to BS 1566-1:2002+A1:2011: **Copper indirect cylinders for domestic purposes. Open vented copper cylinders. Requirements and test methods.**

Table showing sizes of open vented hot water cylinders

Diameter	300mm	350mm	375mm	400mm	450mm	500mm
Height	**litres**	**litres**	**litres**	**litres**	**litres**	**litres**
900mm	55	72	86	95	122	150
975mm	59	81	95	105	130	163
1050mm	63	90	100	113	145	177
1125mm	68	95	110	122	154	190
1200mm	72	100	118	130	168	204
1350mm	81	113	130	150	190	231
1425mm	90	122	140	160	204	245
1500mm	95	130	150	168	213	263
1575mm	100	136	154	177	227	277
1650mm	105	145	163	188	236	290
1725mm	110	150	72	195	250	304
1800mm	113	160	181	204	263	322

There are 3 grades. These are shown in the drawing below.

Grade 1: 2.5 bar operating pressure
Grade 2: 1.5 bar operating pressure
Grade 3: 1.0 bar operating pressure

They are available in the sizes shown in the table above. Only the most common sizes are shown.

Below are the types of open vented hot water cylinder:

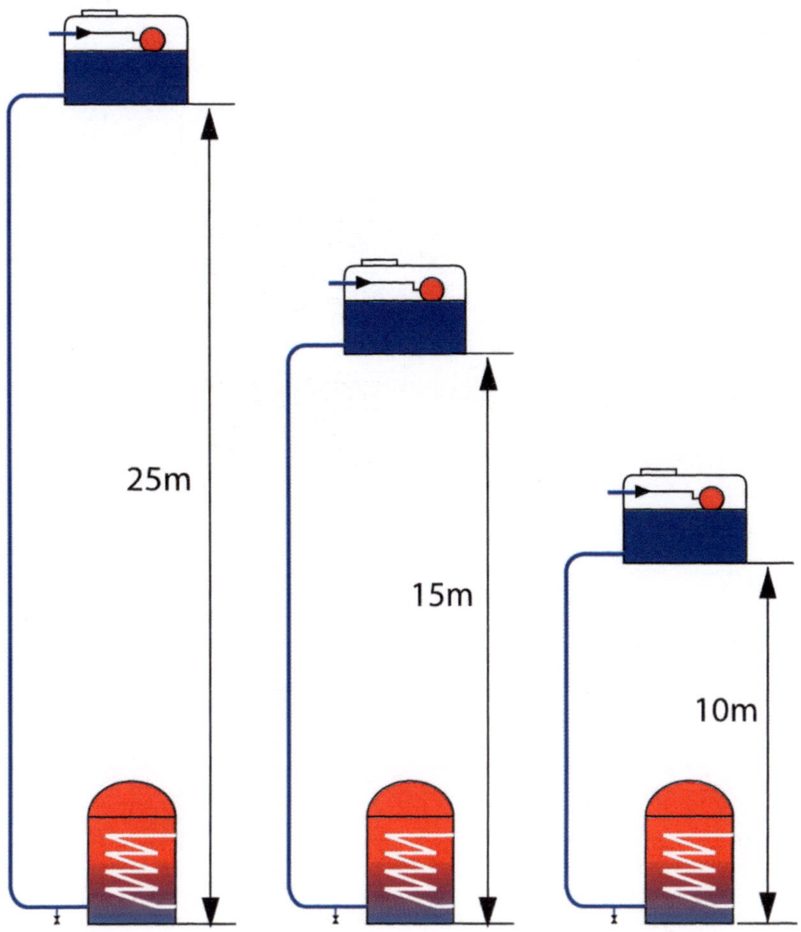

Grade 1	Grade 2	Grade 3
2.5 bar operating pressure	1.5 bar operating pressure	1.0 bar operating pressure
3.65 bar test pressure	2.20 bar test pressure	1.45 bar test pressure

Direct

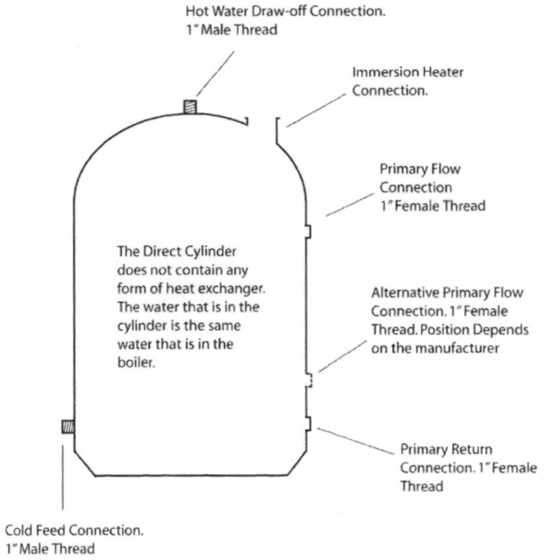

The direct hot water storage cylinder does not contain a heat exchanger. The water is heated directly by either the hot water only boiler or an electric immersion heater. This type of cylinder is NOT suitable for use on central heating systems. The connections to the cylinder are usually male-type threads for the cold feed pipe and the hot water draw-off connections and female-type threads for the primary flow and primary return connections. A drawing of the hot water storage cylinder is shown left.

Double feed indirect

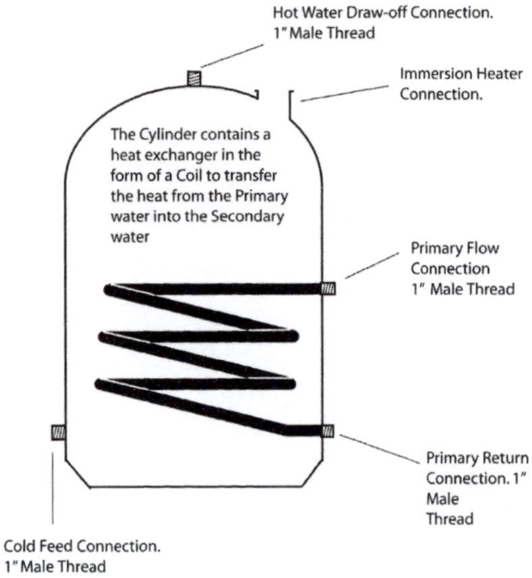

The double feed indirect hot water storage cylinder incorporates a coil heat exchanger.

The coil, which is made from copper tube (stainless steel storage cylinders with stainless steel coils are also available), is the primary heat exchanger for the hot water supply. This type of hot water storage cylinder is suitable for use with central heating systems. It is called indirect because the water is heated indirectly by the primary water in the boiler via the heat exchanger

Single feed self-venting indirect

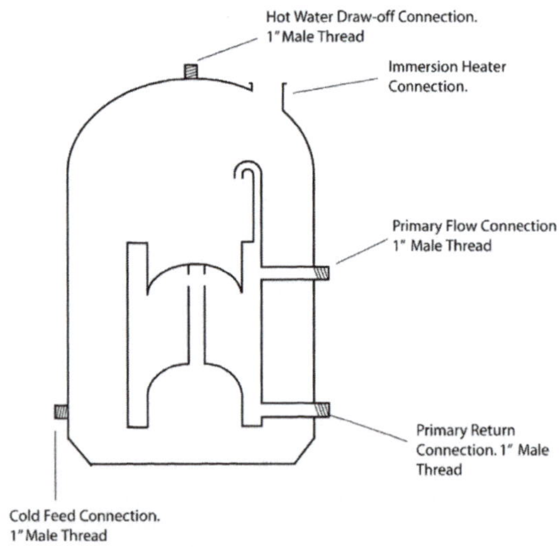

The Single Feed cylinder incorporates a special heat exchanger that uses air entrapment to separate the secondary water from the primary (heating) water.

Quick recovery Cylinders

– also known as 'superduty' cylinders, these have a multi-coil that is made up of many small coils to transfer as much heat as possible as quickly as possible into the stored water in the cylinder.

Installing a superduty cylinder increases the operating efficiency of the boiler by reducing cycling times resulting in savings of up to 40% compared with standard cylinders.

The return pipe on the superduty cylinder remains fairly cool which maintains full hot water heating load for the period of water heating. This eliminates wasteful boiler cycling.

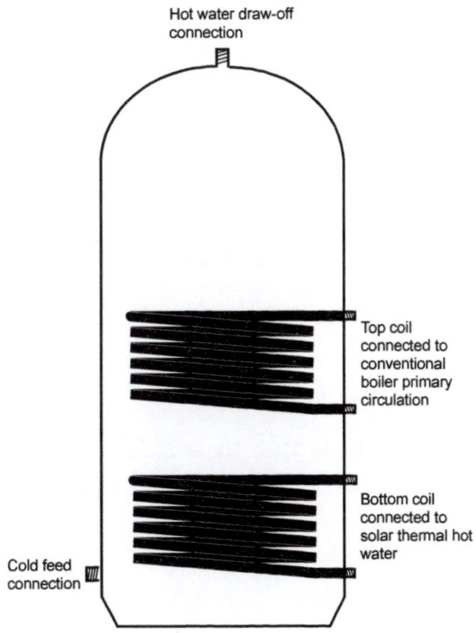

Dual Coil Cylinders – These are used when there is more than one heat source being used to heat the water. They are typically used where solar thermal hot water systems are installed.

- The bottom coil is fed from the solar heating system as this provides the greater volume of hot water.
- The top coil is used when solar hot water heating is unavailable

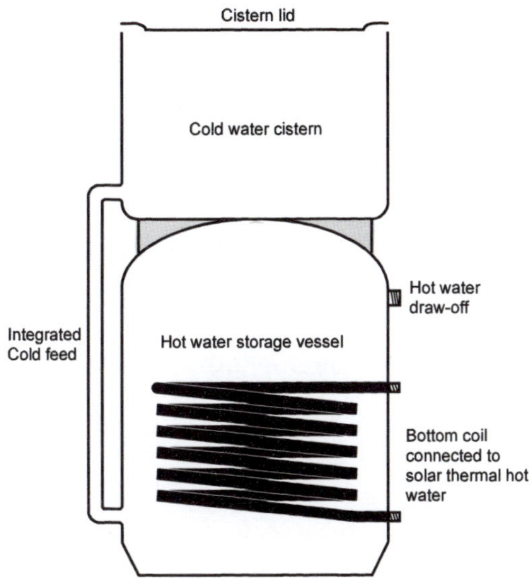

Combination Cylinders – These are hot water storage vessels and feed cistern combined into one unit. The unit has an integrated cold feed pipe already installed. Direct and indirect models are available.

Combination units are often fitted where there is no access to a roof space such as flats and apartments. They should be fitted as high as possible but still allow for maintenance of the float operated valve in the cistern. The main disadvantage with this unit is the lack of suitable pressure at the hot taps.

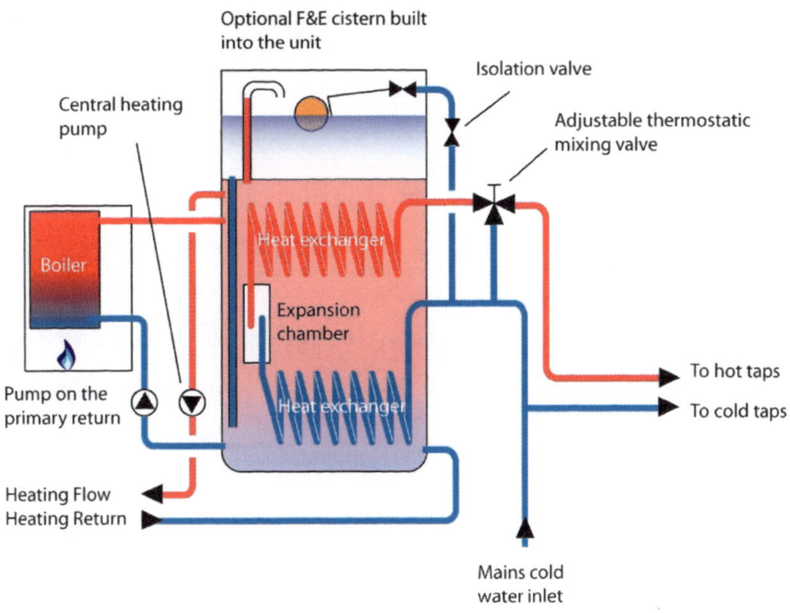

Thermal stores – also known as water jacketed tube heaters, these are centralised instantaneous multipoint water heaters that work by passing water direct from the cold water main through a series of heat exchangers. These heat exchangers are completely surrounded by water at around 80°C inside a large hot water vessel that is directly heated by a boiler. They look very similar to indirect hot water storage cylinders but actually work in reverse. Water from the mains cold supply enters the thermal store, passing through the first heat exchanger where the water is warmed. After passing through the expansion chamber, the water flows through the second heat exchanger where it is heated to full hot water temperature. Because hot water should not exceed a temperature of 60 – 65°C, the water is blended with cold water through an adjustable thermostatic blending valve on the outlet of the thermal store.

The water in the hot water vessel is completely separate from the domestic hot water and is heated by a boiler. The system can be used for properties with central heating systems.

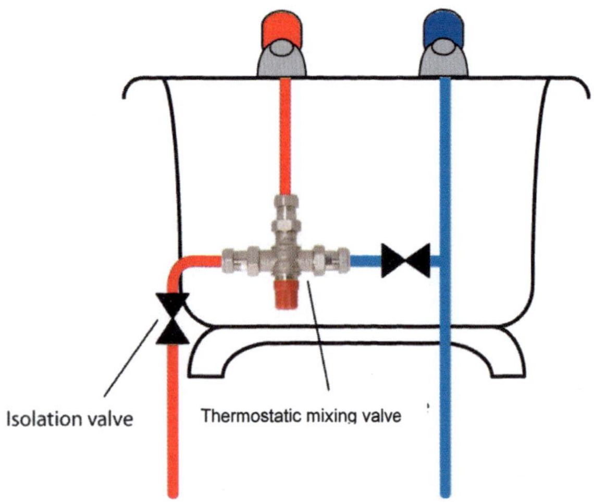

Thermostatic mixing valve

Thermostatic mixing valve – These are valves that mix hot and cold water to a safe temperature – usually 38 – 42°C – before the water is discharged out of the outlet or tap. This is very important in public buildings such as schools, health care premises, or any building where the public have access to prevent scalding occurring. There are two different types:

TMV2 – These are primarily for domestic use.

TMV3 – These offer a better control of water temperature especially when the water is subject to temperature variations and pressure fluctuations.

It is a mandatory requirement that a TMV is fitted to all new bath installations, including domestic baths.

Flow limiting taps and valves – these are designed to limit the flow of water to a specific flow rate. Most have interchangeable cartridges allowing different flow rates from the same valve body. They are often used when an appliance or outlet has a very specific flow rate requirement

Secondary circulation pump – these circulating pumps are very similar to central heating circulating pumps. The difference is that they are made from non-rusting bronze to prevent the hot water being contaminated due to rust formation. They are primary designed to circulate hot water in a secondary circulation system. Most pumps are fitted on the return as close to the hot water storage vessel as possible.

For information on the following components please refer to Unit 5 – Cold Water Systems:

Taps, outlets and valves

Mixer taps Outside taps Pillar taps Bib taps Mixer taps Bi-flow mixer taps Ceramic disc taps Infra-red operated taps Concussive taps Stop valves Spray taps Servicing valves Full way gate valves Spherical plug valves Drain valves Float-operated valves (part 1–4)	Unit 5 – Cold Water Systems, pages 60 to 65

Appliances

Baths WCs Over the rim bidets Wash hand basins Sinks Washing machines Dishwashers	Unit 5 – Cold Water Systems, pages 58 to 60

Showers

Gravity Digital shower valves Bath shower mixer Pumped (single and twin impeller) Mixer valve	Unit 5 – Cold Water Systems, pages 66 to 70

Cisterns

Unit 5 – Cold Water Systems pages 73 to 85

AC1.5 Describe insulation requirements and system frost protection

Frost protection and insulation is covered in detail in **Unit 5 – Cold water systems** on page 85. However, there are further requirements for the insulation of hot water systems that we need to be aware of.

Insulating hot water systems

There are two points to consider:

- The insulation of hot water storage vessels – hot water storage cylinders and vessels are manufactured with polyurethane insulation that is sprayed on at the factory to a thickness of 50mm for those vessels installed in new properties and 35mm for replacement cylinders. The thickness of the insulation is covered in **Building Regulations Approved Document L1: Conservation of fuel and power**.
- The insulation of hot water pipework – new hot water installation pipework requires insulating in line with the recommendations in the **Domestic Building Services Compliance Guide**. There are four considerations:
    - Whenever hot water pipework or primary circulation pipes pass outside of the living space, such as those pipes installed under unheated floor spaces or in roof spaces, those pipes should be lagged adequately. This is for frost and freezing protection.
    - Primary circulation pipes should be insulated along their entire length, except where the pipes pass through obstructions such as floor joists and walls.
    - Pipework connected to hot water storage vessels must be insulated for at least 1m from the connection point on the cylinder or at least up to the point where the pipe becomes concealed. This requirement includes the vent pipe.
    - Secondary circulation pipework must be lagged along its entire length to prevent excessive heat loss from the pipework.

AC1.6 Identify the positioning and fixing of pipework within the building fabric

This subject is covered in detail in **Unit 5 – Cold water systems** on page 87.

AC1.7 Explain expansion and contraction in hot water systems and negative effects

When pipework is subjected to heat from hot water, it expands. As the pipework cools down, it contracts. This constant expansion and contraction process is accommodated within the system pipework so that, in domestic installations, it does not cause major problems such as leaks. However, if the pipework passes through wooden joists in floors and timber studs in internal walls, there is the possibility of noise in the form of 'ticking' or 'creaking', especially if not enough room has been made for the pipe to pass through the structure.

The rate of expansion depends on the pipe material. Generally, plastic pipes such as polybutylene and cross-linked polyethylene expand more than copper or steel. It is known as the coefficient of thermal linear expansion and affects the linear length of the pipework:

- For plastic pipe, the coefficient of thermal linear expansion is 0.00018m/m/°C. This means that for every 1°C rise in temperature, the pipe will expand 0.18mm for every metre of pipe.
- For copper pipe, the coefficient of thermal linear expansion is 0.000016m/m/°C. This means that for every 1°C rise in temperature, the pipe will expand 0.016mm for every metre of pipe.

AC1.8 Identify location and function of unvented system components

Controls for unvented hot water storage systems

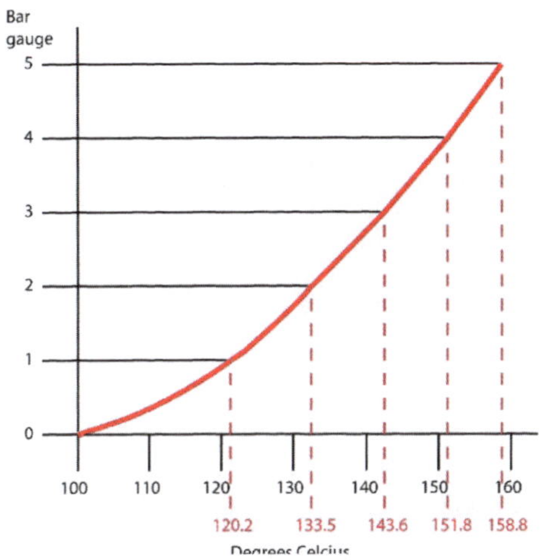

Boiling Points of water under pressure

There are two groups of controls for unvented hot water storage systems:

a) Safety controls

b) Functional controls

Safety Controls

Heating water that is under pressure can create unwanted side effects. We know that water at atmospheric pressure boils at 100°C. In other words, the water goes from a liquid state to a gaseous state known as steam.

When water is pressurised, the boiling point of the water increases. This is shown in the chart left. As you can see from the chart, at 1bar pressure, the boiling point of water is 120.2°C, at 2 bar it's 133.5°C and so on.

Obviously, to heat the water, heat is introduced in the form of an immersion heater or a boiler to make it hot. If that heat source was not controlled, then the temperature of the water would increase beyond the 100°C point. As the water temperature increased, the pressure would also increase further increasing the boiling point. Eventually, the storage vessel would rupture because of the increased stress, with disastrous consequences! Here's why:

$$1cm^3 \ of \ water \ expands \ by \ 1604\% \ when \ changing \ state \ from \ water \ to \ steam.$$

$$1 \ litre \ of \ water \ (l) \ = \ \mathbf{1,000} \ cubic \ centimeters \ of \ water \ (cm^3 \ - \ cc)$$

$$The \ average \ capacity \ of \ an \ unvented \ hot \ water \ storage \ vessel \ is \ 200 litres$$

$$200 \ x \ 1000 \ = \ 200{,}000 cm^3 - cc$$

$$200{,}000 \ x \ 1604 \ = \ 320{,}800{,}000 \ cm^3 \ of \ steam!$$

The resulting flash from water to steam would be explosive enough to cause structural damage to the building! Hence the need for strict safety controls to prevent such a scenario from occurring. Unvented hot water storage systems use what is known as a three-tier control method:

Tier 1 This is the ordinary thermostat designed to switch off the heat when it reaches the point set by the user, usually 60 °C – 65°C. This can be the immersion heater thermostat or the cylinder or boiler thermostat.

Tier 2 In the unlikely event that the control thermostat fails, the immersion heater/boiler MUST have a second thermostat designed to shut off the supply of heat in the event of control thermostat failure. It is known as a high-energy cut-out and shuts down the heat at a predetermined level of 85°C. it is also known as the high limit thermostat.

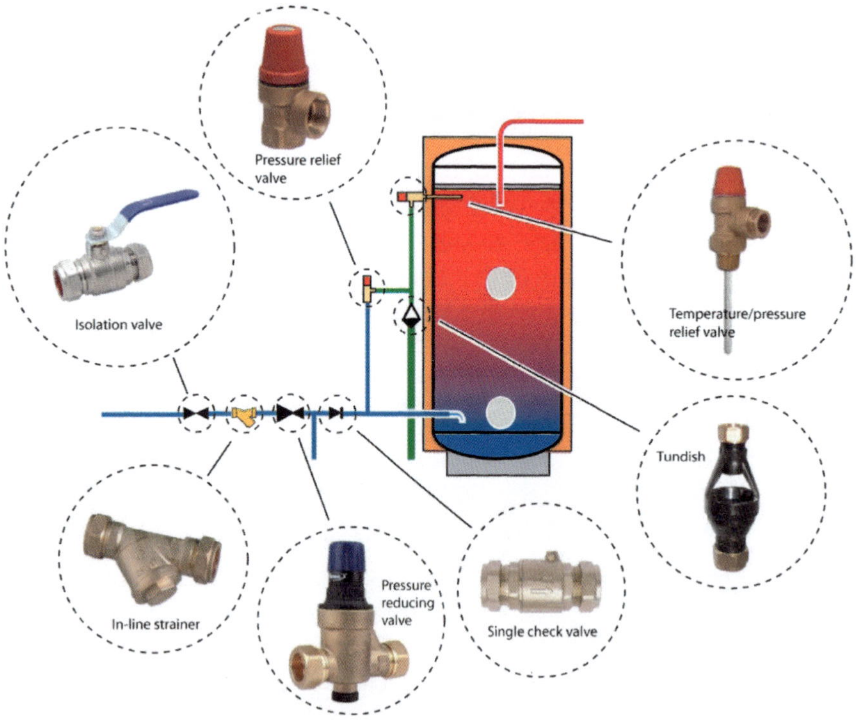

The controls of an unvented hot water storage system

Tier 3 In the event that the high limit thermostat should fail, then the water would be discharged from the temperature relief valve that is factory fitted by the manufacturer. This valve is set to open at 90°C to 95°C to discharge the very hot water through the tundish and down the discharge pipework and safely to outside. In some systems the temperature relief vale also acts as a secondary pressure relief valve to open in the even of pressure relief valve failure.

Safety controls

Control thermostat	This controls the stored water temperature to between 60 – 65°C. The immersion heater will have one built in and a separate one will be supplied for indirect cylinders. Both will have a built in Energy Cut Out (ECO).
Energy Cut-Out (ECO) (High limit thermostat)	This device is designed to isolate the electrical supply to the immersion heater via the built in non- self-resetting cut out. For indirect systems it turns off the heat source to an indirect cylinder by closing a 2-port motorized valve using a non-self-resetting cut out. It has a maximum setting of 900C but usually operates at around 850C.

Temperature & Pressure relief valve

This valve is often referred to simply as the temperature relief valve but it will also operate as a pressure relief valve in the event of the failure of the pressure relief valve. It is designed to open to discharge water when the temperature of the stored water is 90 ± 5°C .

Functional Controls

These are required to protect the mains water supply from contamination and the unvented unit from over pressure, these controls consist of the components shown in the diagram below:

In most modern systems, the pressure reducing valve, the in-line strainer and the single check valve are all incorporated into one valve. This is known as the composite valve and is shown in the annotated diagram below.

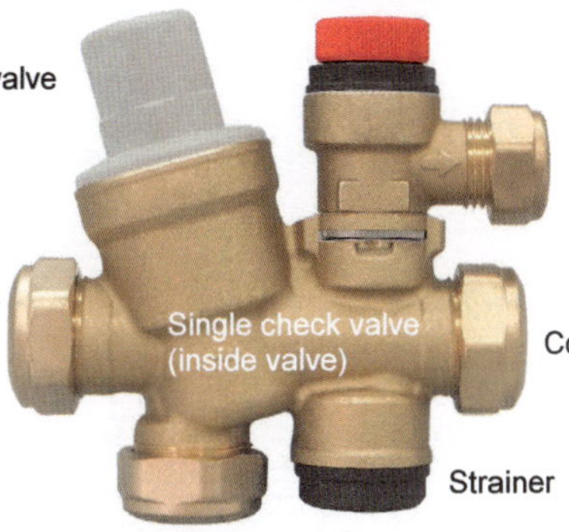

The composite valve

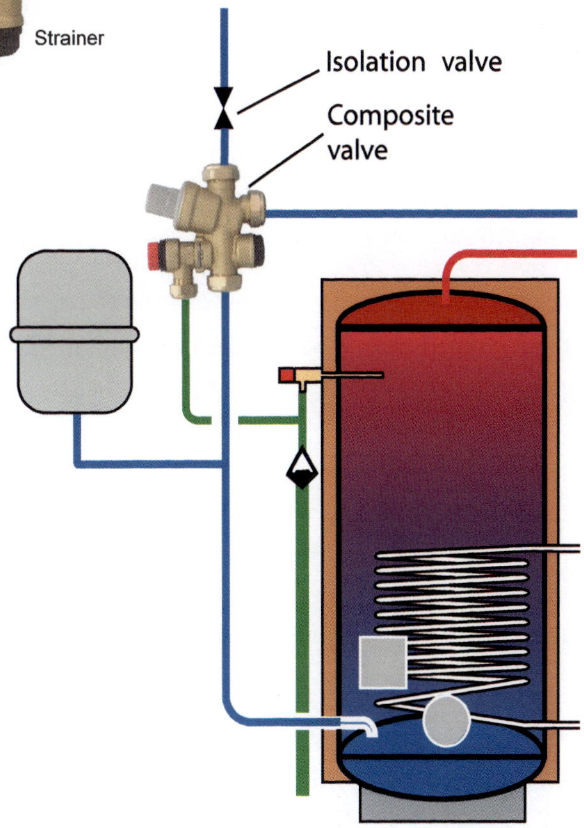

Drawing showing how the composite valve is fitted to the pipework

Functional controls

Isolating valve

This is required to turn off the water supply in order for maintenance to be carried out without undue wastage of water. At least one manufacturer includes this in the combination valve.

Composite valve

Modern unvented hot water storage systems usually have a combination valve which contains the following:

- Line strainer
- Pressure reducing valve
- Pressure relief valve
- Non-return or check valve

Older systems often have separate controls installed in the order listed.

Line strainer

This is required to prevent particles entering the pipework and controls which may prevent them operating correctly.

Pressure reducing valve

The pressure of the incoming service can vary, and this valve is used to set and maintain a constant operating pressure within the system.

Pressure relief valve

Designed to discharge water from the system at a pre-determined pressure to prevent high pressure damage to the operational controls. It will discharge in the event of failure of the pressure reducing valve or the expansion vessel or bubble.

Single check valve or non-return valve

This valve prevents stored hot water or potentially contaminated balanced cold water being siphoned or forced back into the supply and being drawn off as drinking water.

Expansion vessel

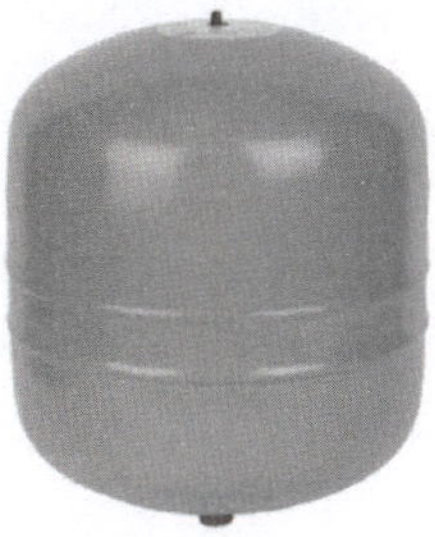

This is required to accommodate the expansion of the stored water when heated. Some systems will have a separate external expansion vessel, some systems utilise an air gap trapped at the top of the cylinder.

Tundish

Required as a visible means for the customer to identify when either the temperature or expansion relief valve is discharging and to provide a relief opening in the event that the discharge pipe becomes blocked.

Components Required by the Building Regulations

Direct heating

A directly heated unit or package should have a minimum of two temperature activated safety devices operating in sequence:

- A non-self-resetting thermal cut-out to BS 3955: 1986 - Specification for electrical controls for household and general purposes, or to BS 4201: 1984 - Specification for thermostats for gas burning appliances, and
- One or more temperature relief valves to BS 6283: Safety and control devices for use in hot water systems Part 2: 1991 - Specification for temperature relief valves from 1 bar to 10 bar, or Part 3: 1991 Specification for combined temperature and pressure relief valves for pressures from 1 bar to 10 bar.
- These devices are in addition to any thermostatic control which is fitted to maintain the temperature of the stored water (control thermostat).
- In both units and packages, the temperature relief valve(s) should be located directly on the storage vessel such that the stored water does not exceed 100°C. In most cases the valve opens at 95°C. The valves should be sized to give a discharge rating in line with BS 6283. Each valve should discharge via a short length of metal pipe (D1) of a size not less than the nominal outlet size of the temperature

relief valve either directly or by means of a manifold sized to accept the total discharge from the discharge pipes connected to it, through an air break over a tundish located vertically as near as possible to the valve(s).

- From 1 April 2004 all immersion heaters whether on unvented or vented cylinders have required a built-in non-self-resetting energy cut out.

Indirect heating

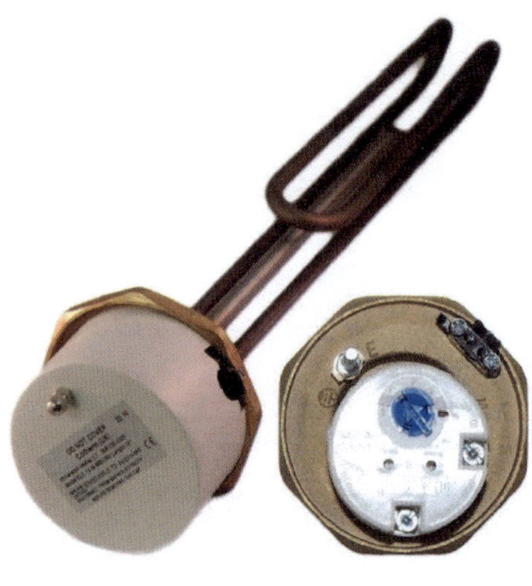

EN60335 Immersion heater and thermostat

A cylinder with an indirect heating source, i.e. a gas or oil-fired central heating boiler, will need all the controls listed for a direct system plus:

- A control thermostat to limit the temperature of the stored water usually to 6S°C with a built-in energy cut out. The energy cut out is normally wired to control a 2 port motorised valve which will close in the event of the energy cut out operating.
- If the heat source is a boiler with a high limit thermostat this may, if suitable, be used utilised as the energy cut out.

The example left shows an immersion heater suitable for an indirect unvented system, it includes a high limit thermostat (non-self-resetting thermal cut-out device). The thermostat is usually purposely designed for the cylinder and replacements should be ordered via the manufacturer.

The discharge pipework requirements

The discharge pipework is required to remove any water discharged either from the pressure relief valve (expansion relief valve) or the temperature relief valve. You must remember that water discharged from the temperature relief valve will most likely be extremely hot and is indicative of a serious problem with the temperature control of the unit/package. The discharge pipework is compiled of 3 sections:

- The D1 pipework
- The Tundish
- The D2 pipework

The two sections (D1) and (D2) are separated by a tundish.

The D1 pipework

The Regulations regarding D1 state that the safety relief valves should discharge via a short length of metal pipe directly or through a manifold to a tundish. If a manifold is to be used (this will be in almost every case) it must be factory fitted.

The D1 pipe should be at least the same diameter as the outlet of the safety device and be no longer than 600mm between the safety devices and the tundish.

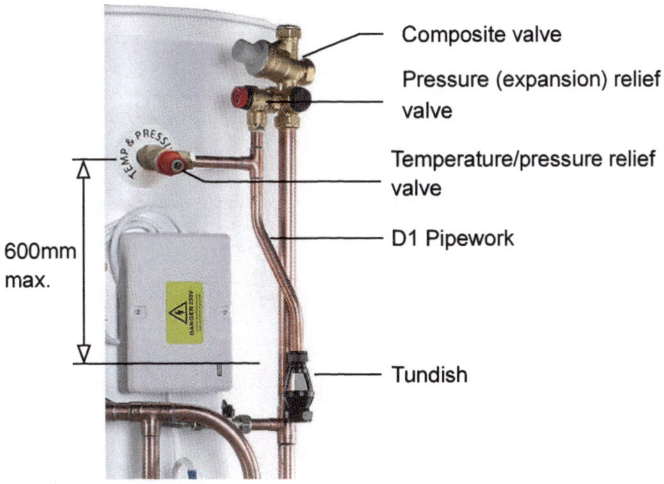

The D1 pipework

The tundish shall be located in the same space as the cylinder, as close as possible to, and lower than the outlet of the safety relief valve or valves.

The D2 pipework sizing and positioning

The D2 pipework should run via the shortest possible route to outside of the property where it should discharge safely. Once you have worked out a potential pipe run to a suitable termination point you need to calculate the pipe sizing for the D2 pipework.

One of the key factors to take into account when designing an unvented hot water storage system is the siting of the cylinder and its control components so that it will allow for safe termination of the discharge pipe. Hence, the distance from the cylinder to the discharge pipe termination point is important and will determine where the cylinder is located. Remember, it is vital to comply with Building Regulations and manufacturer's instructions.

Before we look at discharge pipe sizing we need to look at safe termination points. The Approved Document to the Regulations divides the discharge pipe into two sections Dl between the safety valve and the tundish and D2 between the tundish and the termination point.

Approved Document G3

The Approved Document G3 states that:

- Each of the temperature and expansion relief valves or combined temperature and pressure relief valves should discharge either directly or by way of a manifold via a short length of metal pipe (Dl) to a tundish. Some manufacturers provide a flexible connection from the valve to a manifold.
- The diameter of discharge pipe (Dl) should be not less than the nominal outlet size of the relief valve.
- The tundish should be vertical, located in the same space as the unvented hot water storage system and be fitted as close as possible to the valve with no more than 600mm of pipe between the valve outlet and the tundish.
- Any discharge should be visible at the tundish. In addition, where discharges from safety devices may not be apparent i.e. In dwellings occupied by people with impaired vision or mobility, consideration should be given to the installation of a suitable safety device to warn when discharge takes place e.g. Electronically operated.
- The discharge pipe (D2) from the tundish should:

- Have a vertical section of pipe at least 300mm long below the tundish before any elbows or bends in the pipework; and
- Be installed with a continuous fall of at least 1 in 200.
- The discharge pipe (D2) should be made of metal or a material capable of safely withstanding temperatures of the water discharged.

- Where the discharge pipe is connected to a soil stack discharge pipe, it should:
  - Contain a mechanical seal, not incorporating a water trap, which allows water into the branch pipe without allowing foul air from the drain to be ventilated through the tundish;
  - Be a separate branch pipe with no sanitary appliances connected to it;
  - If plastic pipes are used as branch pipes carrying discharge from a safety device, they should be either polybutylene (PB) or cross-linked polyethylene (PE-X) complying with national standards; and
  - Be continuously marked with a warning that no sanitary appliances should be connected to the pipe.
- Where a single common discharge pipe serves more than one system, it should be at least one pipe size larger than the largest individual discharge pipe (D2) to be connected.

Termination of discharge pipes

The discharge pipe (D2) from the tundish should terminate in a safe place where there is no risk to persons in the vicinity of the discharge.

Low level discharges

There are two accepted safe low-level terminations of the discharge pipe. These are:

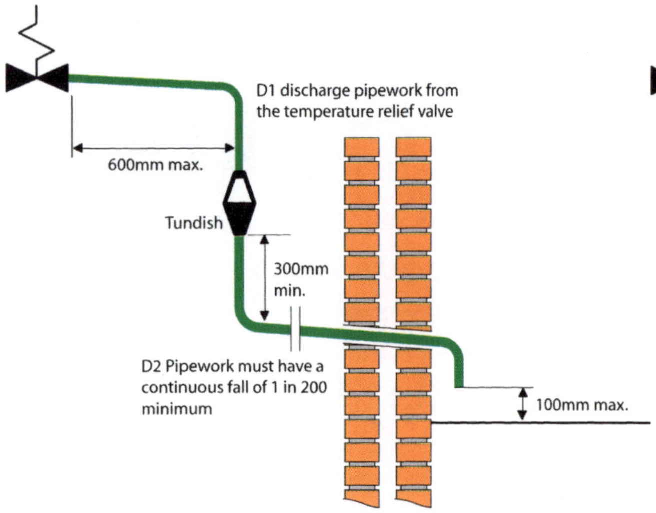

Low level discharge termination

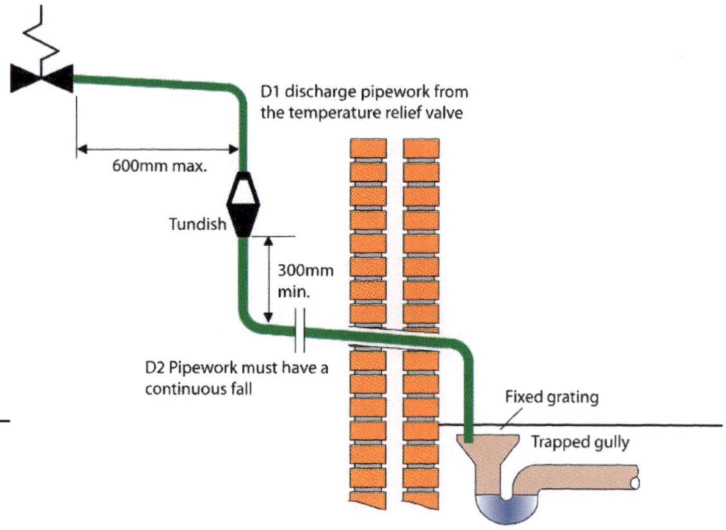

Low level discharge termination over a gully

If the low-level termination requirements cannot be met and there are concerns regarding the level of risk, the termination should be made into a trapped gully located below a fixe grating and above the water line.

When discharging into a trapped gully with a fixed grating, as illustrated above the key points to note are as follows:

- A plastic gulley can be used, as long as it is not close to the cylinder
- Termination below fixed grate but above water level
- The discharge pipe must be made of metal
- Exposed discharge pipes should be protected and covered with thermal insulation
- The termination pipe should be fixed tightly, sealed and bracketed
- The loose lid of the trap must be screwed down
- The end of the discharge pipe has to terminate below the lid
- A metal or glazed trap may have to be used

When terminating at low level, as illustrated above, the key points to note are:

- The discharge should not be of any risk to persons
- The end of the discharge pipe will require covering with guard or cover that maintains visibility.
- The discharge pipe must be made of metal
- Exposed metal discharge pipes must be covered and protected
- The discharge pipe must pass through a sleeve through the wall
- The discharge pipe must have a continuous fall of 1:200
- The discharge pipe must have a maximum termination point of 100mm from the ground
- The pipe must terminate close to the wall
- The end of the pipe does not have to be visible, provided the tundish is in full view

High level termination

The next terminations are the only safe and accepted methods for high level termination of the discharge pipe, as identified in the Building Regulations.

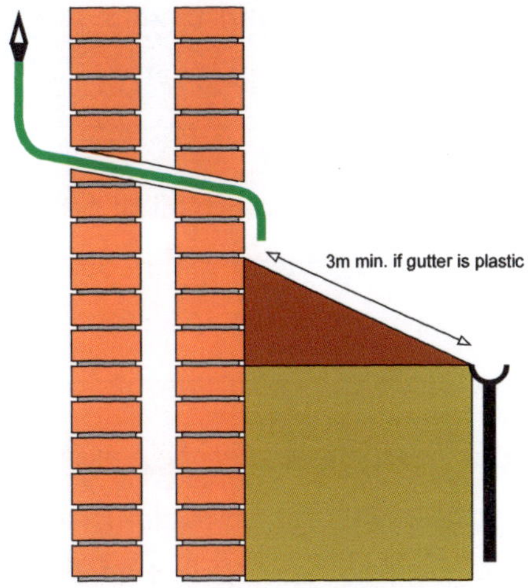

High level discharge on to a roof

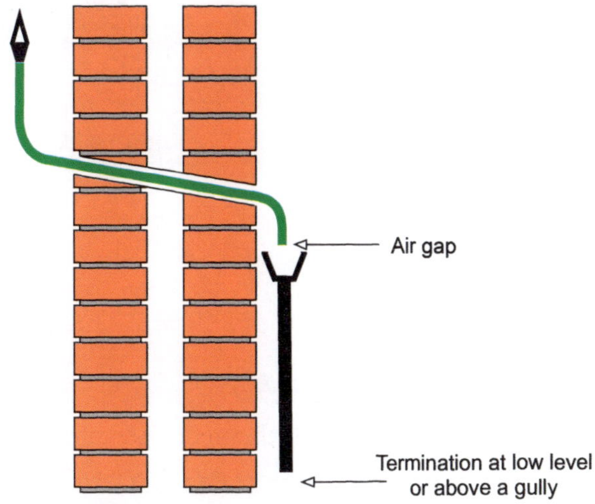

High level discharge to a hopper

When choosing to discharge at high level into a hopper, the key points to note are as follows:

- The vertical hopper pipe may be fitted to accept up to six terminations
- An air gap is required at high level for identification purposes (hopper box or additional tundishes)

- The vertical pipe has to be made of metal (02)
- The minimum size of the down-pipe is one size larger than the biggest 02 pipe going into the hopper
- Safe low level or trapped gully termination is required

The discharge may terminate to a metal hopper, where the end of the discharge pipe is clearly visible (tundish visible or not) or onto a roof capable of withstanding high temperature discharges and which is located minimum 3m away from plastic guttering that would collect such discharges (tundish must be visible).

Where the discharge pipe goes onto a roof, as shown below above left, the key points to note are:

- If termination is into a plastic gutter, the termination of the discharge pipe (02) must be at least 3m away from the gutter
- If the gutter is metal, there are no minimum measurement requirements
- The end of the pipe must face downwards
- The roof material must be capable of withstanding the high temperatures of the discharge
- It is allowed to discharge onto a flat roof without a gutter
- If the roof is a fire escape route, the discharge pipework must be treated as a low-level discharge installation for safety and protection purposes.

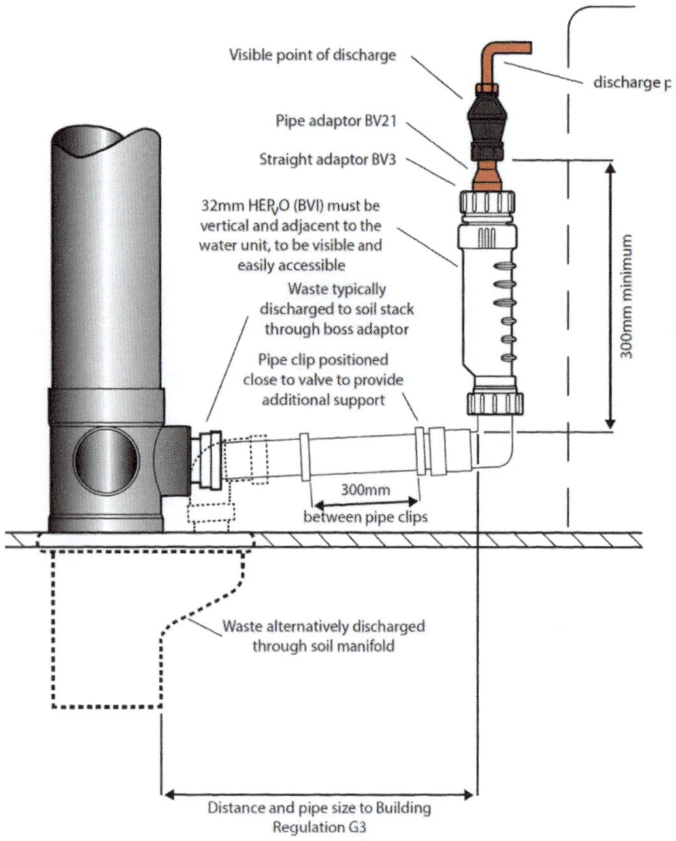

D2 pipework to a soil stack using a waterless trap

Where it is difficult to run the pipework to an external safe discharge point, an alternative method of terminating the D2 discharge pipework internally is acceptable provided the tundish is visible and the other requirements of the Building Regulations are complied with.

The first requirement is that the discharge pipework should not connect to soil stack unless it can be proved that the stack is capable of safely withstanding the temperatures of the discharged water.

If the stack is acceptable then the D2 pipework must contain a waterless mechanical seal to prevent smells, etc., entering the building from the soil stack. A normal water seal trap would dry out quickly.

The D2 discharge pipework must be a separate branch pipe to the soil stack with no sanitary appliances connected to it. If it is made of plastic, it should be of either polybutylene (PB) or cross-linked polyethylene (PE-X) and be permanently marked with a warning that NO SANITARY APPLIANCES should be connected to it.

Installation of the waterless trap and the plastic D2 pipework must be in accordance with the manufacturer's instructions.

Where the only possible location for a cylinder is in a basement or a position where the discharge pipework D2 would have to run upwards above the level of the tundish a high temperature discharge pump will be required.

The pump arrangement consists of a covered metal tank into which the initial section of the D2 pipework terminates. Inside the tank is a float switch which detects the level of water in the tank and starts the pump which lifts the discharged water up to a maximum of 8 metres. When the water level in the tank drops, the pump stops.

The pump can deal with water at up to 100°C so can be fitted to deal with both temperature and expansion relief discharges.

The termination of the D2 discharge pipe from the pump has the same requirements for safe termination as one which came directly from the safety relief valve.

Sizing the D2 pipework

We have already seen that the D1 pipework must be a minimum of the same diameter as the outlet of the safety valve it is serving. The basic formula for the D2 pipework is that it must be one size larger than the D1 pipework for maximum runs of 9 metres.

Valve outlet size	Minimum size of discharge pipe D1	Minimum size of discharge pipe D2 from the tundish	Maximum resistance allowed expressed as a straight length of pipe with no bends or elbows	Resistance created by each elbows or bend
G½	15mm	22mm	Up to 9m	0.8m
		28mm	Up to 18m	1.0m
		35mm	Up to 27m	1.4m
G¾	22mm	28mm	Up to 9m	l.0m
		35mm	Up to 18m	1.4m
		42mm	Up to 27m	1.7m
G1	28mm	35mm	Up to 9m	1.4m
		42mm	Up to 18m	1.7m
		54mm	Up to 27m	2.3m

This basic formula is OK for straight pipe runs up to 9 metres long, but as soon as a bend or elbow is fitted this adds resistance to the pipe run so the maximum pipe run is reduced.

For example, if the D2 discharge pipe is to be run in 22mm pipe and needs to have 2 elbows or bends fitted then each bend or elbow will reduce the maximum pipe run by 0.8 metres so we need to subtract 1.6 metres, meaning maximum pipe run would be 7.4 metres.

If the actual pipe length was 8 metres of 22mm pipe and had bends fitted the equivalent length would be 9.6 metres therefore you would need to increase the minimum pipe diameter by one size, i.e. to 28mm.

The table on page 155 shows how to size the D2 pipework.

A **G½** temperature relief valve has a proposed discharge pipe length **(D2) of 6m** from the tundish to its termination, the pipe will include **6 elbows**.

The calculation is as follows:

Minimum pipe size that could be used is 22mm

From table:

- Maximum resistance allowed for a straight length of 22mm copper discharge pipe from a G½ temperature relief valve is 9.0m
- Subtract the resistance for 6 number 22mm elbows at 0.8m each = 4.8m Therefore the maximum permitted length equates to 4.2m
- 4.2m is less than the actual length of 6m therefore calculate the next largest size
- Maximum resistance allowed for a straight length of 28mm copper discharge pipe from a G½ temperature relief valve is 18.0m
- Subtract the resistance for 6 number 28mm elbows at 1.0m each = 6m Therefore the maximum permitted length equates to 12m
- As the actual length is 6m, **a 28mm (D2} copper pipe will be satisfactory** from the tundish to the safe discharge point.

AC1.9 Describe secondary circulation and how trace heating can be used

Throughout this chapter, reference has been made to secondary circulation, but what is secondary circulation?

Secondary circulation systems are used to prevent system dead legs. A dead leg on a hot water system, in this context, is an outlet (tap) at the end of a long run of pipework. When the outlet is closed the water in the pipe cools down.

Outside diameter of pipe in mm	Maximum length in metres
12mm	**20m**
Over 12 and up to and including 22mm	**12m**
Over 22mm and up to and including 28mm	**8m**
Over 28mm	**3m**

When the outlet is opened, because of the length of the pipe run it takes a long time for the cool water to run out before the hot water appears. This wastes water and energy and on larger systems can lead to Legionella risks.

The Water Regulations provide recommendations for the maximum lengths of un-insulated hot water pipes as shown in the table above.

Secondary circulation systems, in simple terms, consist of a ring main from the hot water draw off which is returned about 25% down the cylinder. On unvented hot water storage systems, the secondary return pipe enters the hot water vessel at the cold feed pipe (see drawing). The circulation is pumped, and branches are taken off to the various hot water appliances.

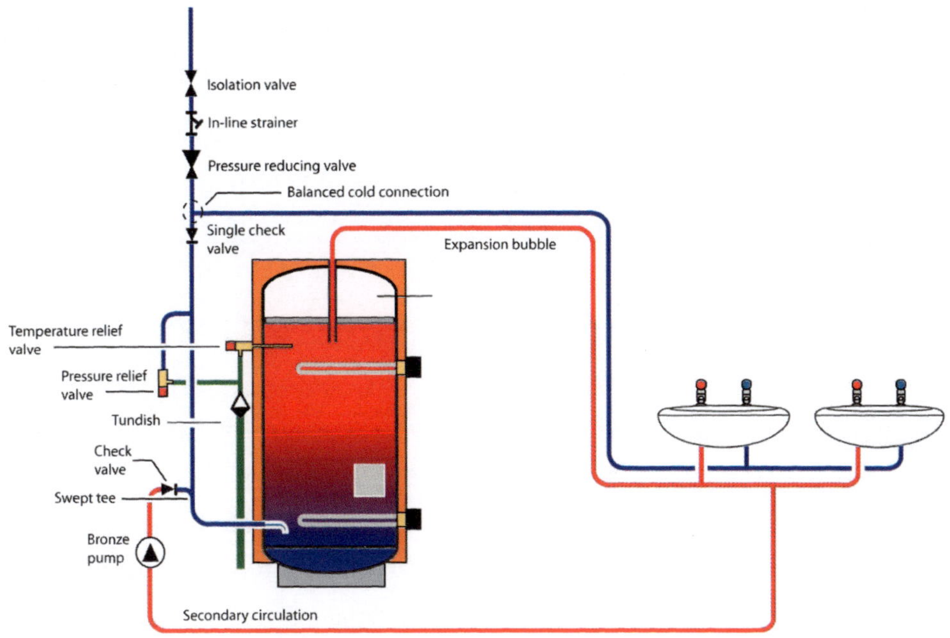

secondary circulation on an unvented system

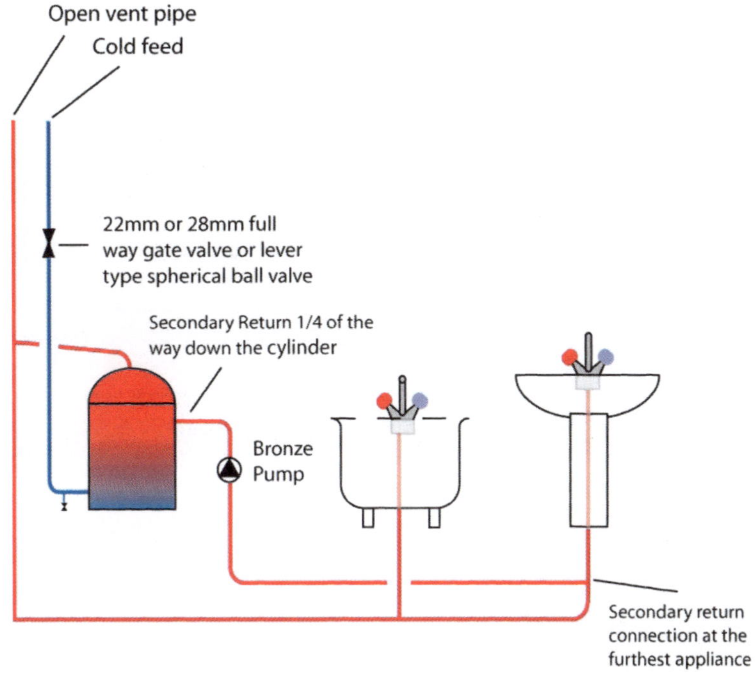

Secondary circulation on an open vented system

The secondary return pipe normally includes the installation of a non-return (check valve) near the pump, not shown in the diagram. The check valve prevents reverse circulation in the return pipe and the discharge of cooler water when a tap is opened.

The pump is normally operated by a time clock which only permits water to circulate around the circuit when the building is occupied, so conserving energy. Newer systems are installing infra-red proximity sensors so the pump only starts when a person enters a room thus saving more energy.

For secondary circulation systems, the secondary return pipe will usually be in 15mm and incorporate a check valve to prevent backflow. A suitable Water Regulations Advisory Service (WRAS) approved bronze circulation pump will be required.

On large systems incorporating either an expansion vessel or cylinder air gap, it may be necessary to fit an additional expansion vessel to the secondary circuit due to the increase in system water content. This should be done if the capacity of the secondary circuit exceeds 10 litres.

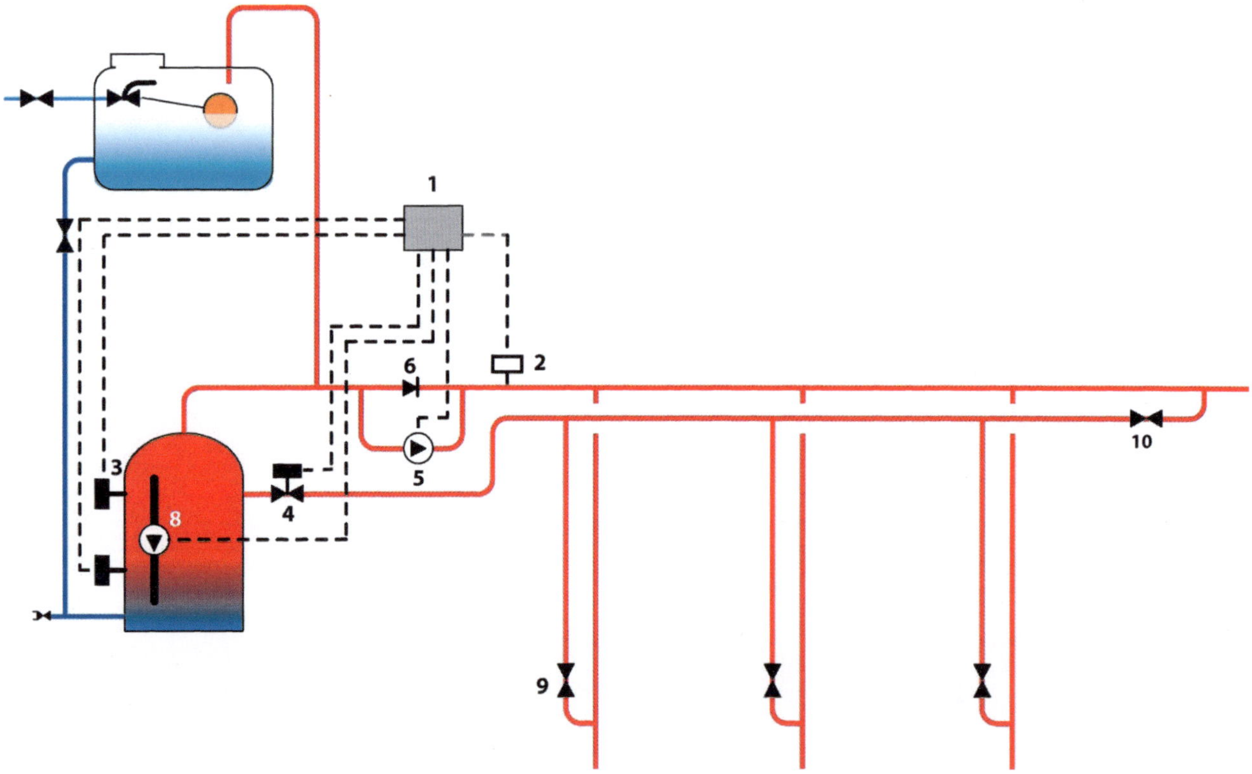

Secondary circulation on a large hot water system

The spare connection on the unvented vessel is for the secondary circulation return.

Trace heating

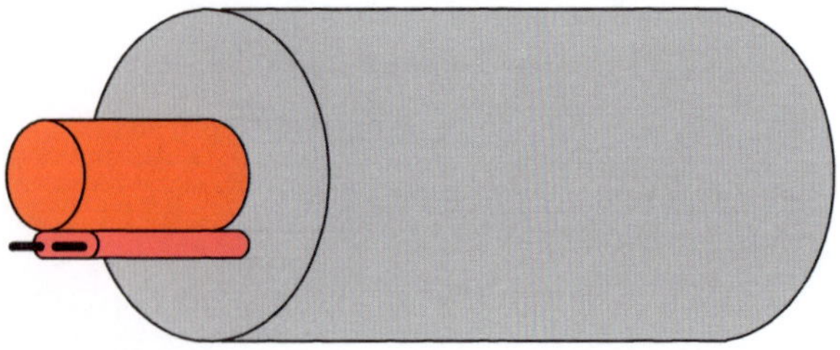

Trace heating

Another method of tackling the problem of dead legs is with the use of trace heating.

Trace heating takes the form of an electrical heating element that is in physical contact with the length of the piperun, which is then covered by thermal insulation. The insulation prevents heat losses from the pipe. Heat generated by the electrical heat element maintains the temperature of the water in the pipe.

AC1.10 Identify sources of information required when undertaking work on hot water systems

The sources of information required when undertaking work on hot water systems differs very little from those documents required for cold water systems. These were covered extensively in Unit 5 – Cold Water Systems. However, they are of such importance that they warrant a further look.

The Water Act 2003 (The Water Industry Act 1991)

The Water Act 2003 combines two previous pieces of UK legislation, the Water Industry Act 1991 and the Water Resources Act 1991. It introduced a series of changes to the regulation of the water industry across the UK. Enforced by the Environment Agency, the Water Act governs the following:

- The appointment, licencing and regulation of water and sewage companies by the Water Services Regulation Authority OFWAT.
- The obligations of the Water Companies and licenced water suppliers to supply water that is fit for human consumption.
- The enforcement of those obligations by Department for the Environment, Food and Rural Affairs and the Drinking Water Inspectorate.
- The control of charges made by the Water Companies and licenced water suppliers to consumers by OFWAT.
- The protection of consumers by OFWAT

The Water Industry Act 1991, however, remains the primary national legislation for all water supplies in England and Wales and defines the powers and responsibilities of local authorities in relation to private water supplies and the definition of a relevant person(s).

Water supplied by a Water Undertaker

The supply of wholesome water to the majority of consumers is the responsibility of the UK water authorities. They are known as Water Undertakers. It is their responsibility to collect, clean, distribute and maintain the water supply in such a manner that the water is fit for human consumption under section 67 of the Water Industry Act 1991. Their responsibility ends at the external (boundary) stop tap to the property. From here, the responsibility rests with the owner of the property or the plumber.

Private water supplies

Private water supplies are water supplies that are not supplied by the Water Undertaker and, as such, are solely the responsibility of the owner and user of the water supply. This could be a single dwelling or multiple properties supplied from the same source. Private water may be supplied from a borehole, stream, river, spring or well.

The quality of the water supplied from private sources can be variable. Some sources offer good quality water where little treatment is necessary but others present a risk to health due to the water quality and require careful monitoring.

The responsibility for regulating private water supplies rests with the local authority but it is also vital that owners are aware of the potential risks. Owners should:

- Ensure that the source is protected against contamination from grazing animals and any contamination that may come from upstream.
- Install and maintain sufficient equipment that will maintain a satisfactory water quality.
- Ensure that the water is adequately disinfected before use.
- Ensure that any stored water is stored in such a way as to prevent contamination after storage but before use.
- Ensure that the water quality is tested regularly.

Ensuring water quality in the property

Under provisions laid down by the Water Industry Act 1991 and then later the Water Act 2003, the UK government introduced two pieces of legislation that regulate how plumbing systems are installed and maintained in domestic buildings in the UK. These are:

- The Water Supply (water fittings) Regulations 1999; and,
- The Private Water Supply Regulations 2016

The Water Supply (water fittings) Regulations 1999

On July 1st, 1999, the government issued the Water Supply (Water Fittings) Regulations 1999 that standardised working practices across all areas of the United Kingdom. These were initially linked to BS6700 and then later to BSEN806: Specification for installations inside buildings conveying water for human consumption: PARTS 1 to 5 and BS 8558:2015: Guide to the design, installation, testing and maintenance of services supplying water for domestic use within buildings and their curtilages. Complementary guidance to BS EN 806.

The Water Supply (water fittings) Regulations were written under section 74 of the Water Industry Act 1991 to ensure that the systems of plumbing installed in properties prevent:

- Contamination of water
- Wastage of water
- Misuse of water
- Undue consumption of water
- Erroneous metering of water

It should be remembered that these regulations only apply to water that has been supplied by a Water Undertaker and do not apply to water supplied from a private source.

A free copy of the regulations can be downloaded from www.opsi.gov.uk/si/si1999/19991148.htm.

The Private Water Supply Regulations 2016

These ensure that water supplied from a private source is safe for people to drink. Changes from the 2009 Private Water Supply Regulations now classify private water supplies into four categories:

- Large supplies serving commercial and public buildings such as hotels, restaurants, nursing homes, bed and breakfast establishments, village halls and large domestic users using more than $10m^3$/day or supplying more than 50 people per day on average.
- Small supplies serving two or more domestic properties but with no commercial usage, using less than $10m^3$/day or fewer than 50 people per day on average.
- Single property supplies where water is supplied for domestic purposes only.
- Private distribution networks. The scope of the regulations now includes networks where the water is distributed further beyond the responsibility of the licenced water company. These include such properties as caravan parks, industrial estates or country estates or where water is distributed by tanker.

The legislation covering the installation and maintenance of hot water storage systems

There are five specific areas of Building legislation covering the installation and maintenance of unvented hot water storage systems in England & Wales:

- Building Regulations – Approved Document G section 2 Water efficiency
- Building Regulations – Approved Document G section 3 Hot water supply and systems
- Building Regulations – Approved Document L1A Conservation of fuel and power in new dwellings
- Building Regulations – Approved Document L1B Conservation of fuel and power in existing dwellings
- Building Regulations – Approved Document P1 Design and installation of electrical installations
- Domestic Building Services Compliance Guide

It should be noted that there is some overlap between both sets of legislative requirements.

Additionally, for unvented systems, the cylinder must be approved by a certification body having accreditation from the United Kingdom Accreditation Service (UKAS) or an equivalent European body. In practice this usually means BBA, KIWA or WRc approval.

- Building Regulation G3 deals primarily with the installation and safety related issues of hot water storage systems whereas the Water Regulations primarily deal with system functional issues including waste, misuse, contamination and undue consumption of the water supply.

- Building Regulation G2 deals with water efficiency.

- Building Regulations Part L1A and L1B deal with energy efficiency and prevention of wastage. The Domestic Building Services Compliance Guide is basically complementary guidance on how to comply with Part L.

- Building Regulations Part P deals with the electrical safety issues surrounding supplies and connections to hot water storage systems.

What the regulations say

Regulation G3: Unvented hot water storage systems – all systems

3.17 To minimise the danger from excessive pressure, unvented hot water storage systems should incorporate a minimum of two independent safety devices. These shall be in addition to any thermostat provided to control the temperature of the stored water. The selection of safety devices should take account of the physical location of the devices, and the design, configuration, location of components and performance characteristics of the system to which they are attached.

3.18 An acceptable approach might consist of:

a. A non-self resetting energy cut-out to disconnect the supply of heat to the storage vessel in the event of the storage system overheating; and

b. A temperature relief valve or a combined temperature and pressure relief valve to safely discharge the water in the event of serious overheating.

Alternative approaches to this are acceptable provided that they provide an equivalent degree of safety.

3.19 Water heaters with a capacity of 15 litres or less that have the appropriate safety devices for temperature and pressure will generally satisfy the requirement set out in G3(3)

Unvented hot water storage systems – systems up to 500 litres capacity and 45kW power input

3.20 Paragraphs 3.21 to 3.24 are in addition to the provisions of 3.17 above.

3.21 If an indirect supply of heat to an unvented hot water storage system incorporates a boiler, the energy cut-out may be on the boiler.

3.22 Any unvented hot water storage system up to 500 litres and less than 45kW should be in the form of a proprietary hot water storage system unit or package.

The package and components should be appropriate to the circumstances in which they are used and should satisfy any appropriate standard that will ensure the requirements of regulation G3(2) and G3(3) will be met (e.g. BS 12897: 2006 Water Supply. Specification for indirectly heated unvented (closed) hot water storage systems or BS 6700: 2006 + A1 2009 Design, installation, testing and maintenance of services supplying water for domestic use within buildings and their curtilages).

3.23 Any unvented hot water storage system unit or package should be indelibly marked with the following information:

a. The manufacturer's name and contact details;

b. A model reference;

c. The rated storage capacity of the storage water heater;

d. The operating pressure of the system and the operating pressure of the expansion valve;

e. Relevant operating data on each of the safety devices fitted; and

f. The maximum primary circuit pressure and flow temperature of indirect hot water storage system units or packages

In addition, the following warning should be indelibly marked on the hot water storage system unit or package so that it is visible after installation:

WARNING TO USER

a. Do not remove or adjust any component part of this unvented water heated; contact the installer.

b. If this unvented water heater develops a fault, such as flow of hot water from the discharge pipework, contact the installer

WARNING TO INSTALLER

a. This installation is subject to Building Regulations.

b. Use only appropriate components for installation or maintenance.

Installed by:

Name ______________________ Tel No. ______________________

Address Installation date

Unvented hot water storage systems – systems over 500 litres capacity or over 45kW power input

3.25 Paragraph 3.26 and 3.27 are in addition to the provisions of 3.17.

3.26 Systems over 500 litres capacity will generally be bespoke designs for specific projects and as such are inappropriate for approval by a third party accredited product conformity certification scheme. Where this is the case, the unvented hot water system should be designed to the safety requirements in 3.17 by an appropriately qualified engineer.

3.27 Any unvented hot water storage system having a power input of more than 45kW, but with a capacity of 500 litres or less should be in the form of a proprietary hot water storage unit or package.

The package and components should be appropriate to the circumstances in which they are used and should satisfy an appropriate standard that will ensure the requirement of regulation G3(2) and G3(3) will be met (e.g. BS 12897: 2006 Water Supply. Specification for indirectly heated unvented (closed) hot water storage systems or BS 6700: 2006 + A1 2009 Design, installation, testing and maintenance of services supplying water for domestic use within buildings and their curtilages).

Water efficiency Calculator for new dwellings – Approved Document G2

Sets out the methodology required to allow the calculation of water consumption in dwellings, limiting water use to 125 litres (optional 110 litres) per person per day.

The requirement under G2 of the Building Regulations states that:

> ***G2. Reasonable provision must be made by the installation of fittings and fixed appliances that use water efficiently for the prevention of undue consumption of water.***

Water efficiency of new dwellings (Regulation 17K)

36. (1) The potential consumption of wholesome water by persons occupying a new dwelling must not exceed the requirement in paragraph (2).

(2) The requirement referred to in paragraph (1) is either—

c) 125 litres per person per day; or

d) in a case to which paragraph (3) applies, the optional requirement of 110 litres per person per day,

as measured in either case in accordance with a methodology approved by the Secretary of State.

Wholesome water consumption calculation (Regulation 20E)

37.— (1) Where regulation 36 applies, the person carrying out the work must give the local authority a notice which specifies—

c) which of the requirements in regulation 36(2)(a) or (b) applies to the dwelling; and

d) the potential consumption of wholesome water per person per day in relation to the completed dwelling.

Requirement G2 applies only when a dwelling is—

c) erected; or

d) formed by a material change of use of a building

The water efficiency calculator was dealt with in Unit 5 – Cold Water Systems on page 95

British Standards BSEN806:2012 Pts 1 to 5

BSEN806: Specifications for installations inside buildings conveying water for human consumption

- BSEN806 – 1: 2000. General. This document specifies potable water installation requirements and gives recommendations on the following aspects of hot and cold water installations:
  - Design
  - Installation
  - Alteration
  - Testing
  - Maintenance and operation.
  - Part 1 of BSEN806 also covers the pipework installation from the point of entry when the property is supplied with water from a private water source.
- BSEN806 – 2: 2005. Design. This document provides technical information so that the following points may be achieved from the design and installation:
  - Appropriate pressures and flowrates

  - Water quality at the tap is not contaminated or affected by the location or environment
  - The system avoids wastage of water and leakage.
  - The system is efficient, convenient, reliable and safe.
  - The system has a reasonable working life span.

- BSEN806 – 3: 2006. Pipe sizing. Simplified method. This document describes the simplified method for pipesizing drinking water standard installations as defined in section 4.2. However, the document does not engage in pipesizing for domestic firefighting systems.
- BS EN 806 – 4:2010. Installation. Specifies the requirements of water installations within buildings and gives recommendations for their correct installation. It also covers pipework outside buildings but within the premises as outlined in BSEN806 – 1:2000. It applies to new, altered and repaired installations.
- BS EN 806 – 5:2012. Operation and maintenance. This document takes the form of a practice specification. It specifies the requirements for the correct operation and maintenance of potable water supply installations within buildings and for pipework outside building but within premises in accordance with BSEN806 – 1:2000.

BSEN806 Parts 1 to 5 completely supersede BS6700 in all aspects of hot and cold water supply. **BS 8558: 2015 Guide to the design, installation, testing and maintenance of services supplying water for domestic use within buildings and their curtilages** - complementary guidance to BS EN 806, now becomes the lead document for potable water supply in premises.

BS 8558: 2015 Guide to the design, installation, testing and maintenance of services supplying water for domestic use within buildings and their curtilages - complementary guidance to BS EN 806

BS8558 was developed primarily to provide complimentary guidance to BSEN806 Pts 1 to 5. Together, this complete suite of British Standards provides recommendations into best practice for the design, installation, testing, operation, maintenance and alteration of hot and cold water systems for domestic buildings.

BS8558 bridges the gap between BSEN806 and its predecessor BS6700 and provides UK guidelines for, not just plumbers, but the water supply industry as a whole.

PD 855468:2015 Guide to the flushing and disinfection of services supplying water for domestic use within buildings and their curtilages

Flushing and disinfection of systems used to be part of BS8558:2011. However, the latest version of BS8558:2015 excludes flushing and disinfection. Instead, flushing and disinfection of systems is now a separate document PD 855468:2015.

PD 855468:2015 provides guidance on the cleaning, flushing and disinfection of cold water systems to control microbiological growth and the removal of debris. This includes guidance on:

- Deployment of the correct tools and personnel
- The use of the correct disinfectants
- How to respond if a microbiological problem is identified
- Keeping records of cleaning and disinfection

This new document applies to systems supplying water to domestic purposes within buildings and their curtilages, and includes water used in food preparation.

Manufacturer technical instructions

The manufacturer's instructions are probably the most important document to read and consult when installing, servicing and maintaining appliances, components and equipment, because they instruct us on the best methods to use whilst keeping to current legislation and regulations. In some cases, it may appear that these instructions contradict the regulations. This occurs because regulations and codes of practice are only updated periodically, whereas manufacturers are constantly reviewing and updating their literature in line with modifications and current good practice. Where a conflict exists, manufacturer's literature should always be followed. If not:

- The warranty of the equipment may be void.
- Regulations may be inadvertently broken
- The installation may be dangerous.

AC1.11 Explain backflow risk and required methods of prevention

The methods of backflow protection were covered extensively in **Unit 5 – Cold Water Systems** from page 99 to page 107.

AC1.12 Install Hot Water Systems

Sanitary appliances – including baths, wash basins, WCs etc. The installation of sanitary ware was covered on page 58. Great care should be exercised when installing sanitary ware to ensure that damage to the appliance does not occur. Positioning needs to be completed in accordance with BS 6465–2:1996 Code of practice for space requirements for sanitary appliances to ensure that the correct spacing requirements between appliances are observed.

Showers – There are many different shower types available and each one will have its own installation requirements. Some can be installed when the pressure of the hot and cold water supplies are different, some require the same hot and cold water pressures and some are electric and only require a cold mains water supply. Whichever shower is chosen, the manufacturer's instructions should be thoroughly read and understood before any installation takes place.

Pipework – There are two very common materials used for installation of hot and cold water supply in domestic properties and small commercial buildings. These are copper tubes and plastic pressure pipe:

- **Copper** – this is the traditional material for the installation of hot and cold pipework in domestic and small commercial buildings. It is manufactured to BS EN 1057 and comes in a range of sizes to suit all applications and installation types. Copper is light, easy to bend, cut and joint using a variety of fitting types. Remember to keep the pipework as neat as possible and use the recommended clip types and distances.
- **Plastic pressure pipe** – manufactured from either polybutylene (PB) or cross-linked polyethylene (PE-x), plastic pressure pipe is becoming the material of choice for new build housing installations. It is very easy to install, bend and joint. As with copper tubes, remember to keep the pipework as neat as possible and use the recommended clip types and distances.

Hot water storage cylinders and vessels – There are two very common types of cylinder used for installation of hot water supply in domestic properties and small commercial buildings. These are the open vented type hot water storage cylinder and the unvented hot water storage cylinder. The configurations of these, i.e. direct/indirect type, were dealt with earlier in the unit. Here, only the hot water pipework will be dealt with:

The open vented type hot water storage cylinder – There are three pipes that we are primarily concerned with when installing open vented hot water systems:

a) The cold feed pipe to the cylinder
b) The open vent pipe
c) The hot water draw-off pipe

The cold feed from the cistern to the cylinder should be of 28mm size pipe. In smaller properties, this may be reduced to 22mm if there is no ensuite bathroom in the property. The hot water draw-off should be a minimum of 22mm, although larger systems will require a 28mm draw-off because of the number of hot water outlets. At least 22mm should be taken to the bath, with 15mm serving washbasin, bidet and kitchen/utility sink. The Vent pipe must have a vertical rise and discharge over the cistern in the roof space. The height of the vent above the water line in the cistern must be at least 40mm per meter of static head plus 150mm. The draw-off from the top of the cylinder must have a gentle rise to the vent pipe and must be at least 450mm in length to prevent parasitic circulation. Secondary circulation may be required with large systems.

The unvented hot water storage cylinder – unvented hot water storage systems are fed directly from the water undertaker's cold water main. Because of this, generally smaller pipe sizes can be used. Most manufacturers insist that the unvented hot water storage cylinder is fed by 22mm cold water supply pipe, with 22mm hot water draw-off from the top of the storage vessel, although with these systems, baths can usually be supplied by 15mm pipe. Again, if the system is large with multiple hot water outlets, it may be advisable to use 28mm pipe to ensure a good flow rate. Secondary circulation may be required with large systems. It is vitally important that the control components, i.e. the pressure reducing valve, the pressure relief valve, the single check valve, the expansion vessel etc. are installed exactly as instructed by the manufacturer's instructions.

The D1/D2 discharge pipework from unvented hot water storage systems was dealt with on pages 150 to 156.

Learning Outcome 2

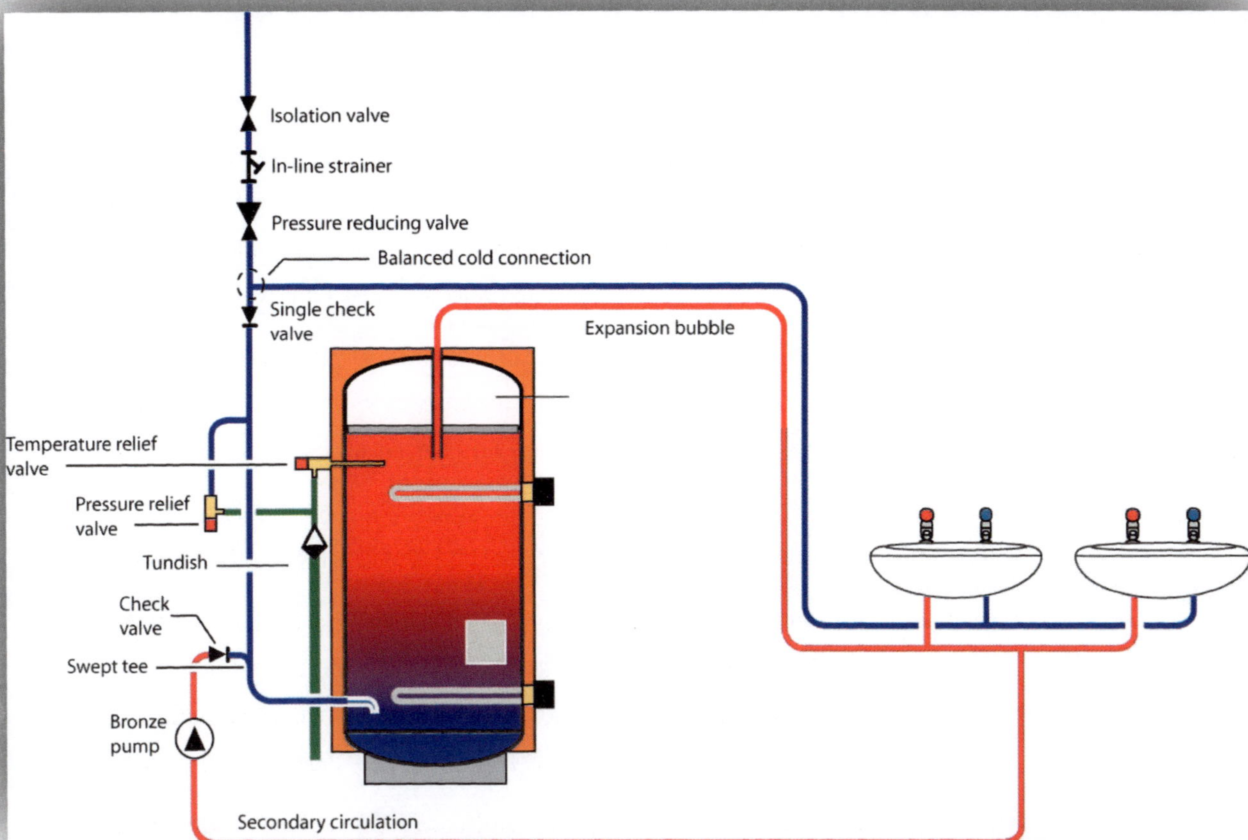

Decommission Hot water systems

Learning Outcome 2 - Decommission Hot water systems

In Learning Outcome 2, there are two Assessment Criteria:

AC2.1 Explain **procedures** for **decommissioning systems**

AC2.2 Carry out decommissioning **procedures**

There are often times when systems are required to be taken out of service, either for service and repair or extension or renewal. There are also times when systems are taken out of service permanently. Both of these procedures are called decommissioning.

In Learning Outcome 2, we will look at the decommissioning process and the tasks that must be performed to ensure that this is done safely.

AC2.1 Explain procedures for decommissioning systems

There are two types of decommissioning that we may be asked to perform:

- Permanent decommissioning – This is performed when a system is no longer required and involves completely draining down the system and the removal of all pipework, fittings, clips and appliances. The cold water supply should be turned off at the external (boundary) stop tap and the internal stop tap capped off as near to the stop tap as is possible. This is to avoid the stagnation of water.
- Temporary decommissioning – Temporary decommissioning is conducted when a system is either being serviced, repaired, renewed or extended. It involves isolating the water supply either to the whole property or to a particular section of pipework, while the work is performed. There are points to remember:

- **Notify relevant person** – Before the decommissioning process commences, notify the home owner or any other relevant person that the water will be isolated for a period of time. Advise them to save water, i.e. fill saucepans and kettle etc., to use while the system is off. If there are other trades on site that require water to work, such as plasterers, bricklayers etc. let them know in advance so they can make alternative water arrangements.
- **Isolate fuel/electricity supply to the system as appropriate** – isolating the hot water supply may also require the isolation the cold supply as well. You must ensure that any electricity using appliances or components, such as immersion heaters and boilers etc. are isolated also.
- **Isolate water supply** – if possible, only isolate the section of the system you are working on. If this is not possible, then the whole system may need isolating at the mains internal stop tap.
- **Apply warning notices and signs** – leave a warning notice that the system is turned off at the point where you have turned it off. This will prevent people from turning it on and causing problems. If possible, make sure the notice gives an approximate time that the supply will be reinstated and leave a phone number so that you may be contacted.
- **Drain system to a suitable location** – using the lowest possible drain point, drain the system water to a safe location. Avoid discharging the water onto the pathway of the property or the road, especially in cold weather. If it freezes it may cause an ice hazard for passers-by. If there is no drain point, then drain the system at the lowest draw-off point. While the system is drained, it may be a good idea to install a drain point for future use.

  When draining open vented cylinders, where there is no drain-off valve on the cold feed, will require that the water is syphoned out of the top of the hot water storage cylinder. This should be done with caution using the following procedure:

  - Isolate the cold water feed cistern and open the lowest hot water tap. This will drain the cistern and the system pipework, but it **WILL NOT** drain the hot water cylinder.
  - Disconnect the hot water draw-off after draining the cold water feed cistern at the lowest hot tap.

   - Attach a piece of copper tube at least the height of the cylinder to a hose pipe and carefully feed the pipe down the hot water draw-off connection until you can feel the bottom of the cylinder.
   - With the hose in a safe place for draining outside the property, fill the hose with water from the cold water main from an outside tap.
   - Once the hose is full, release the hose from the tap quickly. After a few seconds, the water in the cylinder will begin to syphon out. Once the cylinder is empty, the water flow from the hose will stop.

CAUTION! If there is no outside tap, the syphonic process can be started by sucking on the open end of the hose. However, this should be done very carefully and very quickly so that the water in the cylinder does not backflow into the mouth. This procedure should only be completed as a last resort.

- **Appropriately dispose of contents and any additives –** in most cases, water drained from a hot water system can be disposed of down the foul water drain. You should always be aware of top water drains as these may not be appropriate drain points.
- **Continuity bonding as required –** when cutting into an existing system, always remember to apply temporary continuity bonding clips first.
- **Temporary capping of pipework sections as required –** any pipework being left drained for a period of time will require capping to prevent the ingress of dirt and vermin/insects. This can be done with cap ends of the appropriate tube size.

AC2.2 Carry out decommissioning procedures

You will be asked to perform decommissioning procedures as part of your practical assessment. It is advisable to follow the procedure as stated in AC2.1.

Learning Outcome 4

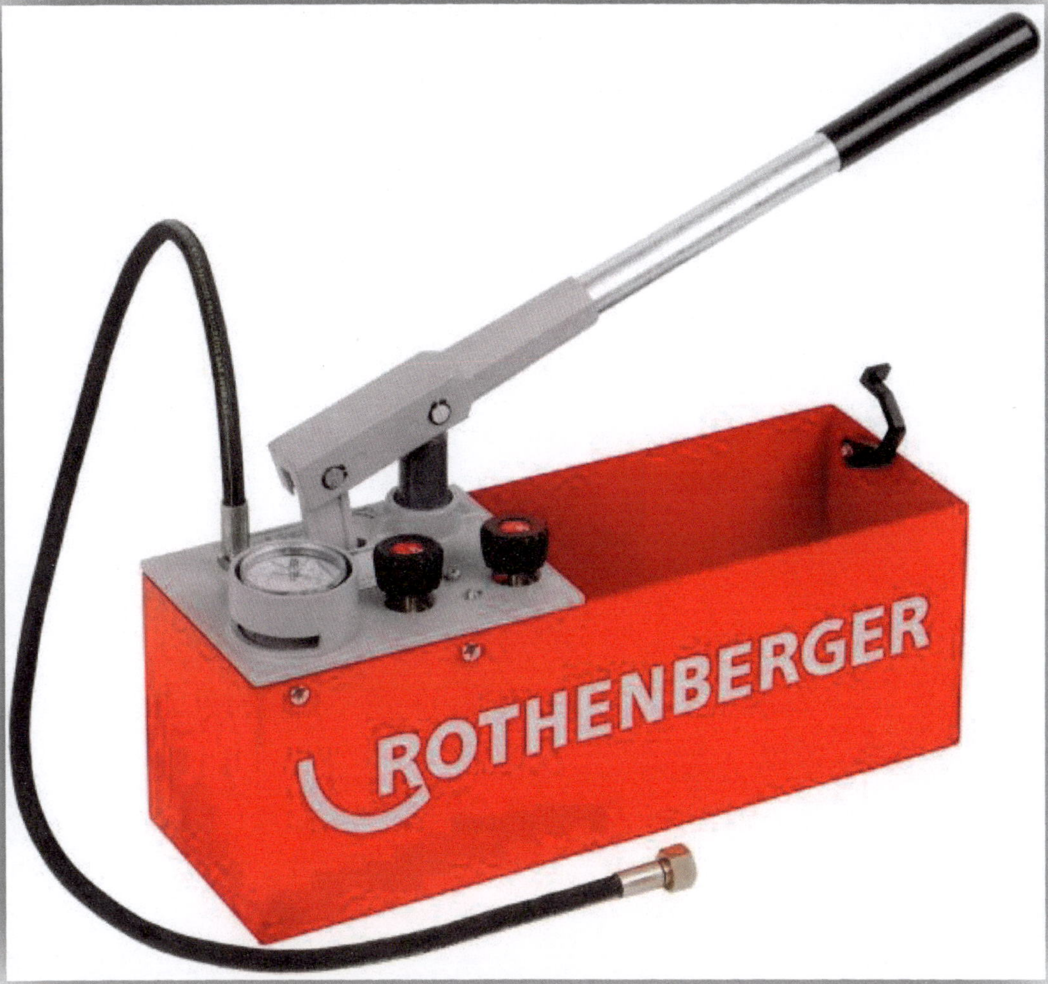

Perform A Soundness Test and Commission Hot Water Systems and Components

Learning Outcome 4 - Perform A Soundness Test and Commission Hot Water Systems and Components

There are six Assessment Criteria in this Learning Outcome:

AC4.1 Identify information sources required to complete testing and commissioning

AC4.2 Describe how to fill and vent hot water systems

AC4.3 Describe a visual inspection of a hot water system to confirm that it is ready to be soundness tested

AC4.4 Describe a **soundness test** to industry requirements on hot water systems **pipework** and components

AC4.5 State the **flushing requirements** including the use of **system additives** for new and existing hot water systems

AC4.6 Carry out a **soundness test**

AC4.1 Identify information sources required to complete testing and commissioning

The testing of hot and cold water installations is detailed in the following documents:

- The Water Supply (water fittings) Regulations 1999: Schedule 2: Paragraph 5 and Guidance G5.1
- British Standard BS EN 806 – 4: 2010 **Specifications for installations inside buildings conveying water for human consumption.**
- Manufacturer's instructions are required with regard to commissioning flow rates and pressures.

AC4.2 Describe how to fill and vent hot water systems, and

AC4.3 Describe a visual inspection of a hot water system to confirm that it is ready to be soundness tested

Visual inspections

Before the system is filled with water and a soundness test is performed, the system must first be visually inspected:

Point 1: walk the system. This is to ensure that all taps and isolation valves are in the 'off' position so we can conduct a controlled fill, testing the installation section by section. We do not want a situation to develop where ALL parts of the system are filling at the same time. If leaks begin to develop, it makes it extremely difficult to cope with.

Point 2: **walk the system part 2.** Whilst you are checking the isolation valves as described in point 1, take the time to check that you are happy with the installation in general:

- Is the pipework neatly installed?
- Is the storage vessel correctly sited and installed?
- Is the vent pipe (if fitted) terminated correctly?
- Is the discharge pipework (if fitted) terminated correctly?
- Are all the compression joints tight?
- Are the tap connectors tight?
- Has any residual flux been wiped off capillary joints?
- Have enough pipe clips been used to prevent excessive movement?
- Are the float operated valves set to a provisional water level?

Once these points have been satisfied, we may then proceed with filling the hot water system. Each system has its own particular method of filling. Both methods assume that the cold water system has been tested and is in operation.

Filling and venting the open vented hot water storage system

1) Ensure that the gate valve on the cold feed pipe and all hot outlets and taps are turned off. Also ensure that all drain valves are shut.
2) Ensure that any heat sources such as immersion heaters are switched off.
3) Turn on the cold water to the cold water storage system and allow the cistern to fill until the cistern is full and the float operated valve has shut off.
4) Check that the water line in the cistern is correct and well clear of the cistern overflow.
5) Open the gate valve fully (remembering to turn it back half a turn when fully open to prevent it seizing in the open position) and allow the hot water cylinder to fill with water. Stay with the cylinder whilst performing a cylinder fill so that any leaks can be detected immediately. Hold the draw-off pipe by the hand. When the water reaches it, the pipe temperature will change and you will feel it go cold.
6) Once the cylinder is full, allow it to stand for a couple of minutes. this will allow any air trapped in the cylinder to disperse up the vent pipe. Once you are sure there are no leaks, visit each of the hot water outlets in turn and turn on the tap/valve. Allow the water to run to clear any excess flux/swarf/dirt from the pipework.
7) Once all taps have been flushed through, turn off the isolation valve to the float operated valve in the cold feed cistern and completely drain the pipework and cistern.
8) Connect a hose pipe to the drain off on the cylinder and completely drain the cylinder.
9) Turn off all taps and outlets and refill the system.

Filling and venting the unvented hot water storage system

Once the system is completed, it can then be filled with water. It is a good idea at this stage to walk around the installation to make sure that you are happy that all pipework is installed correctly, that the pipework is neat and well clipped, that all joints are tight and jointed correctly and that the hot water storage vessel is positioned in accordance with the customer's wishes and manufacturer's instructions.

The storage vessel will contain a lot of air that will require purging from the system. This will mean that the water will splutter violently when it exits the taps for the first time. To prevent this, manufacturers recommend that unvented hot water storage systems are filled with at least one hot tap open. Where a bath is fitted, the hot bath tap is the best outlet to purge the air from because any spluttering is confined to the bath:

- Before turning the water on, check that the pressure in the external expansion vessel (if fitted) is correct to the manufacturer's recommendations. If the system has internal expansion, this step can be disregarded.

- Turn on the water to the hot water system. The hot water storage vessel will begin to fill immediately.
- Check the pipework for any initial leaks while the vessel is filling. The system, because of the high pressure and flow rate, should fill quickly.
- Fill the system until the water reaches the open outlet and allow the water to run until the spluttering ceases. When this is done, turn of the tap.
- Visit each hot tap in turn and run it until the spluttering stops.
- Let the system stand and check for leaks.
- Briefly turn on the pressure relief and temperature relief valves and check that the discharge pipework is operating correctly
- Once you are sure there are no leaks, turn off the isolation valve to the system, open all hot taps and drain the system of water via the system drain off valve.
- Once drained, refill the system and thoroughly flush with clean cold water.

AC4.4 Describe a soundness test to industry requirements on hot water systems pipework and components

Soundness testing was covered in much detail in **Unit 5 – Cold Water Systems** but it is of sufficient importance to be looked at again here.

The type of testing procedure will depend on the pipe work material used. Hot and cold water systems testing is detailed in BSEN806 pts 2 and 4, central heating systems testing is detailed in **BSEN14336:2004 Heating systems in buildings — Installation and commissioning of water based heating systems**. Above ground sanitation systems should be tested in accordance with Document H of the Building Regulations.

The recommendations of BSEN806 Pts 2 and 4

For many years the recommendations of BS6700 and the Water Supply (water fittings) Regulations 1999 stated that a hot/cold water system should be tested at 1.5 times the maximum operating pressure of the system. However, since the introduction of BSEN806, the test pressures for hot and cold water systems has been updated and is no longer as straightforward as it once was.

Calculating the test pressure

BSEN806 pt 1 gives a definition of the Maximum Design Pressure or MDP of a system. It states that the MDP is:

> *'the maximum hydrostatic pressure at which the potable water installation is designed to work.'*

BSEN806 pt 2, clause 3.4.2 states that:

> *'To ensure adequate strength, all components of the system shall be designed to meet the test pressure requirements of the local and national laws and regulations. The test pressure shall be **at least** 1.5 times the allowable Maximum Operating Pressure.'*

So, the maximum design pressure as stated in BSEN806 pt 1 is:

$$Max.\,Operating\ Pressure\ x\ 1.5 = Max.\,Design\ Pressure\ (MDP)$$

BSEN806 pt 4, which covers the installation and commissioning of hot and cold water systems, identifies in clauses 6.1.2 and 6.1.3 that the test pressure shall be:

$$MDP\ x\ 1.1\ or\ MOP\ x\ 1.65$$

How does this work in practice?

If a system is designed to operate at a Maximum Operating Pressure of 2 bar pressure, then?

$$2\ bar\ (MOP)\ x\ 1.5 = 3bar\ (MDP)$$

$$3\ bar\ (MDP)\ x\ 1.1 = 3.3\ bar\ (test\ pressure)$$

Or

$$2\ bar\ (MOP)\ x\ 1.65 = 3.3\ bar\ (test\ pressure)$$

This results in a test pressure that is slightly higher than that required by the Water Supply (water fittings) Regulations 1999 in Schedule 2, Paragraph 5 (1.5 x MOP).

The testing procedures

BSEN806 Part 4 states that there are 3 systems for testing pipework and these testing procedures are determined by the material from which the pipework is made. The testing procedures are classed as:

a) **Test A**
b) **Test B**
c) **Test C**

The table below illustrates which pipe material is associated with each of the tests:

Table taken from BSEN806 Part 4 page 20	
Type of material	**Hydrostatic test procedure**
Linear elastic materials (i e. metals: copper, LCS and stainless-steel)	**Test procedure A**
Elastic materials (PVC-U, PVC-C, etc.) and multi-layer materials	
Visco-elastic materials (i. e. PP, PE, PEX, PA, PB, etc.) with DN/OD ≤ 63mm	
Visco-elastic materials with DN/OD > 63mm (i.e. PP, PE, PEX, PA, PB, etc.)	**Test procedure B or C**
Combined system with DN/OD ≤ 63mm (metals and plastics)	**Test procedure A**
Combined systems with DN/OD > 63mm (metals and plastics)	**Test procedure B or C**

Test A requirements page 21, BSEN806 pt 4

The system should be filled with water and all air vented to atmosphere. All outlet valves and vents should be sealed once the air is removed. Apply the test pressure calculated (equal to 1.1 times MDP) for a period of 10 minutes. The test pressure must stay constant for 10 minutes. If there is any pressure loss, the system must be maintained at the test pressure until all leaks have been found. The test is shown in graph form right.

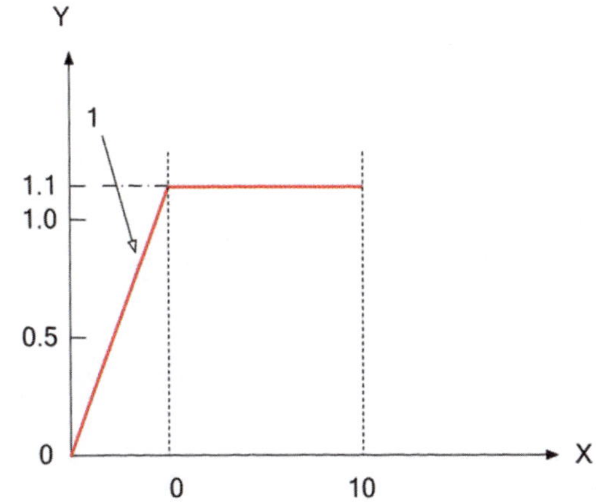

X = Time in minutes
Y = Test pressure
1 = Pumping

Test B requirements page 21, BSEN806 pt 4

The system should be filled with water and all air vented to atmosphere. All outlet valves and vents should be sealed once the air is removed. Apply the test pressure calculated (equal to 1.1 times MDP) for a period of 30 minutes. Inspect the system to identify any leaks.

Reduce the pressure by bleeding water from the system until the pressure reads 0.5 times the test pressure.

The pressure must remain constant for a period of 30 minutes after pressure reduction. Check the system visually for leaks. If, after the pressure has been reduced there is a further pressure drop, there is a leak on the system. The test pressure must be maintained while the leak is identified. Again, the test is shown in graph form.

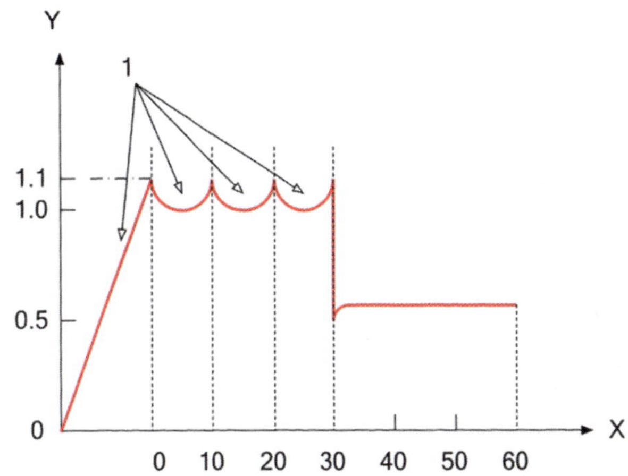

X = Time in minutes
Y = Test pressure
1 = Pumping

Test C requirements page 22, BSEN806 pt 4

The system should be filled with water and all air vented to atmosphere. All outlet valves and vents should be sealed once the air is removed. Apply the test pressure calculated (equal to 1.1 times MDP) for a period of 30 minutes. Inspect the system to identify any leaks.

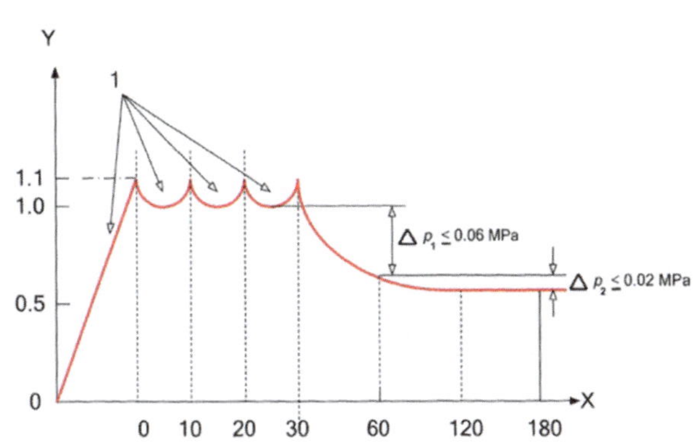

X = Time in minutes
Y = Test pressure
1 = Pumping
Δ P1 = Maximum pressure drop between 30 mins and 60 mins of test pressure
Δ P2 = Maximum pressure drop between 30 mins and 60 mins of test pressure

Note the pressure after a period of 30 mins and recheck for leaks.

Note the pressure after a further 30 mins. If the

pressure drop is less than 0.06 MPa or 0.6 bar, the system can be considered to be sound. Continue the test without further pumping.

Continue the test for a further 120 minutes (2 hours). If the pressure drops more than 0.02 MPa or 0.2 bar in this time, then this would indicate that the system has a leak. Maintain the pressure and identify the leak.

AC4.5 State the flushing requirements including the use of system additives for new and existing hot water systems

Again, flushing and disinfection of hot and cold water systems was covered in much detail in **Unit 5 – Cold Water Systems** but it is worth recapping some of the main points again here.

PD 855468:2015 Guide to the flushing and disinfection of services supplying water for domestic use within buildings and their curtilages

Flushing and disinfection of systems used to be part of BS8558:2011. However, the latest version of BS8558:2015 excludes flushing and disinfection. Instead, flushing and disinfection of systems is now a separate document PD 855468:2015.

PD 855468:2015 provides guidance on the cleaning, flushing and disinfection of cold water systems to control microbiological growth and the removal of debris. This includes guidance on:

- Deployment of the correct tools and personnel
- The use of the correct disinfectants
- How to respond if a microbiological problem is identified
- Keeping records of cleaning and disinfection

This new document applies to systems supplying water to domestic purposes within buildings and their curtilages, and includes water used in food preparation.

System Flushing

All systems must be flushed with cold water taken directly from the water undertaker's mains cold water supply before handing over to the end user. This is a requirement of the British Standards. It should be completed as soon as possible on completion of the installation and commissioning process. System flushing removes any dirt and debris, such as flux residues, swarf, jointing compounds etc. from the system that may be classed as contaminates.

The flushing process

Flushing hot water systems was dealt with on pages 178 and 179 as part of the system filling process. It is generally accepted that all systems must be flushed with cold water taken directly from the water undertaker's mains cold water supply before handing over to the end user. This is a requirement of the British Standards. It should be completed as soon as possible on completion of the installation and commissioning process. System flushing removes any dirt and debris, such as flux residues, swarf, jointing compounds etc. from the system that may be classed as contaminates.

It is accepted practice, that once a system has been flushed, it should not be left charged with water if the property is not going to be occupied straight away. Unused cold water systems can become stagnant if left for more than 4-6 weeks. However, draining a system can often be detrimental, since it's almost impossible to completely drain a system of water, especially if there are long, horizontal runs on the system. The water left in the system becomes stagnant, which immediately raises contamination issues. Also corrosion can set in, which often goes undetected. It is recommended that, in these instances, the system should be left fully charged with water and the system flushed through completely twice weekly. Again, this will involve opening all taps and outlets and should be completed until the build is permanently occupied.

Disinfection of Hot water systems

At the present time, there is no requirement that domestic hot water systems be disinfected. However, large systems should be disinfected using a special water system disinfecting agent that contains either sodium hypochlorite or calcium hypochlorite. The British Standard tells us that systems should be disinfected in the following circumstances:

- In all new installations, except those used as private dwellings
- Where extensions and alterations have been made to large cold water systems.
- Where it is suspected that the system has been compromised by contamination
- Where the system has not been in regular use.
- Where underground pipework has been installed (this does not include localised repairs or where a junction has been installed after the fittings have been immersed in a solution of sodium hypochlorite diluted to 200ppm) (1ppm = 1mg/l)

Systems should be cleaned and disinfected as part of a good maintenance regime. It should also be remembered that, where large amounts of water are stored, these systems must also be disinfected to prevent the onset of legionella pneumophila bacteria leading to Legionnaire's disease.

Disinfection types

There are two disinfectants approved by the Drinking Water Inspectorate:

⇒ EN900 – Calcium Hypochlorite

⇒ EN901 – Sodium Hypochlorite

Most disinfection processes for cold water installations use EN901 – Sodium Hypochlorite.

The disinfection process

Before disinfection can take place, the system must be first thoroughly flushed with fresh water from the water undertaker's cold water main and then completely drained. It should then be refilled adding the disinfecting agent as the system fills. As already stated, most systems are disinfected using a solution of EN901 – Sodium Hypochlorite and this should be added at a concentration of 50ppm. Once the system is full and all Float Operated Valves have shut off, the system must be allowed to stand for 1 hour. At the end of this period, the furthest outlets point should be tested to ensure that at least 30ppm of chlorine are present.

All person either working in or using the building MUST be informed that disinfection is going to be conducted BEFORE it takes place. This includes those persons who are not normally present during working hours. During disinfection, WARNING NOTICES must be posted to warn people that the system is being disinfected and that the water MUST NOT be used. These should be placed at ALL outlets.

Household bleaches, toilet cleaners and chemicals MUST NOT be allowed to come into contact with the disinfected water as this can produce highly reactive toxic fumes. All operatives must be made aware of the COSHH requirements of the disinfecting solution and PPE must be issued to all operatives beforehand. This should include goggles, face mask and protective gloves or gauntlets.

Calculating the disinfection solution required

To calculate the correct amount of solution to administer, we must first calculate the exact capacity of the system including the storage cylinder, the cistern and pipework capacities.

Sodium Hypochlorite contains around 5% chlorine, which is equivalent to 50,000ppm. So, to calculate the amount of solution to add to the system that is equal to 50ppm:

$$50 \div 1{,}000{,}000 \; x \; 100 \div 5 = $$
$$0.001 \; part \; of \; Sodium \; Hypochlorite \; solution \; to \; 1 \; litre \; of \; water$$

In this form, the calculation doesn't mean very much because we need to know how much Sodium Hypochlorite to add. So, it is generally accepted that a ration of 1 litre of Sodium Hypochlorite (containing 5% chlorine) for every 1000 litre of water in the system be used.

Calculating the capacity of the cold water system

To add the correct amount of solution, we must first calculate the total system capacity:

- ⇒ Calculate the capacity of the cistern up to the water line (in litres)
- ⇒ Calculate the capacity of the pipework

Calculate the capacity of the cistern up to the water line (in litres)

Cistern capacity is calculated by:

$$Length \; of \; the \; cistern \; x \; width \; of \; cistern \; x \; height \; of \; the \; cistern \; to \; the \; water \; line \; x \; 1000 = cistern \; capacity$$

Let's assume:

Length of cistern = 2m

Width of cistern = 1.25m

Height to water line = 1.25m

$$2 \; x \; 1.25 \; x \; 1.25 = 3.125m^3 \; x \; 1000 \; (to \; convert \; to \; litres) = 3125 \; litres$$

Calculate the capacity of the system pipework (in litres)

Again, let us assume the following amounts of copper tube are used in the installation. in reality these should be measured accurately. The table left gives the capacity of the various sizes of copper tube:

⇒ 50m of 42mm copper tube

⇒ 25m of 35mm copper tube

⇒ 25m of 28mm copper tube

⇒ 10m of 22m copper tube

⇒ 30m of 15mm copper tube

The capacity of copper tube per metre		
Standard/Spec	Tube Dimension (OD)	Carrying Capacity (L/m)
BS EN 1057 R250 half-hard straight lengths	6	0.018
	8	0.036
	10	0.061
	12	0.092
	15	0.145
	22	0.320
	28	0.539
	35	0.835
	42	1.232
	54	2.091

Calculating the capacity of the installation pipework				
Pipe	Size (mm)	Length (m)	Capacity per metre (l)	Capacity of the pipe (l)
1	**42**	**50**	**1.232**	**61.6**
2	**35**	**25**	**0.835**	**20.875**
3	**28**	**25**	**0.539**	**13.475**
4	**22**	**10**	**0.320**	**3.2**
5	**15**	**30**	**0.145**	**4.35**
Total pipework capacity (l)				**103.5 litres**

Now add to this the capacity of the hot water storage cylinder. In most cases, the manufacturer's data will give the exact cylinder capacity in litres:

Table showing sizes of open vented hot water cylinders						
Diameter	**300mm**	**350mm**	**375mm**	**400mm**	**450mm**	**500mm**
Height	**litres**	**litres**	**litres**	**litres**	**litres**	**litres**
900mm	55	72	86	95	122	150
975mm	59	81	95	105	130	163
1050mm	63	90	100	113	145	177
1125mm	68	95	110	122	154	190
1200mm	72	100	118	130	168	204
1350mm	81	113	130	150	190	231
1425mm	90	122	140	160	204	245
1500mm	95	130	150	168	213	263
1575mm	100	136	154	177	227	277
1650mm	105	145	163	188	236	290
1725mm	110	150	72	195	250	304
1800mm	113	160	181	204	263	322

For the purpose of this calculation, we will assume that a 900mm x 450mm hot water storage cylinder has been installed with a capacity of 122 litres:

$$103.5 + 3125 + 122 = 3350.5 \text{ litres total capacity}$$

So, to calculate the amount of disinfectant:

$$3350.5 \times 0.001 = 3.505 \text{ litres of sodium hypochlorite}$$

How to disinfect a large hot water system

This method presumes that capacity calculations have already been conducted and that the amount of disinfectant is known:

Stage 1 Thorough flush the system with clean cold water from the water undertaker's cold mains supply and drain down completely. This will remove any system installation debris.

Stage 2 Refill the cold water storage cistern adding the disinfectant to the water as it enters the cistern. This will ensure that the solution of disinfectant is completely and evenly diluted into the system water. Add the disinfectant slowly allowing the water to reach all parts of the system.

Stage 3 Working away from the cistern, open each hot water outlet in turn until the smell of chlorine is detected.

Stage 4 Once the system is full and the cistern Float Operated Valves are shut down, allow the system to stand for 1 hour. A free residual chlorine content of 30ppm must be detected at the furthest outlets point. If it is below this, then the system must be drained down, flushed and the process started again from the beginning.

Stage 5 If the disinfection process has been successful, then the system should be completely drained and thoroughly flushed with clean water until the free residual chlorine content of the water is no higher than that of the water undertaker's mains cold water supply. ***See page 124***

Water colour	Chlorine level (ppm or mg/l)
Clear	**None**
Faint pink/pink	**0.2 – 1**
Pink/red	**1 – 5**
Red/purple	**5 – 10**
Purple/blue	**10 – 20**
Blue/grey-green	**20 – 30**
Grey-green/yellow	**30 – 50**
Muddy brown	**Over 50**

Testing for free residual chlorine content

The test involves taking a sample of system water from the furthest outlet and testing it using a specialised kit. The kit contains a test tube and special chemicals in tablet form that are dropped into the test tube of water. As the tablet dissolves, the water changes colour, the colour indicating what the free residual content is. The colours and their representative free residual contents are shown in the table above:

Flushing the system after disinfection has been completed

It is extremely important to remember that Sodium Hypochlorite and Calcium Hypochlorite are extremely dangerous to wild life because of the chlorine content. Because of this, we cannot simply drain the contents to a foul water or surface water drain. Permission must be sought from the water undertaker so that they can expect water with a high chlorine content arriving at the water treatment plant when the system is flushed. Alternatively, a neutralising agent can be added to the water. The neutralising agent is known as **Sodium Thiosulphate** and must be added at the following strength:

$$System\ volume\ x\ ppm\ of\ chlorine\ in\ the\ water\ x\ 2\ =\ grams\ of\ sodium\ thiosulphate\ required$$

At the end of the process, the system should be completely drained and thoroughly flushed with clean water until the free residual chlorine content of the water is no higher than that of the water undertaker's mains cold water supply at all outlets.

General points to remember

- ⇒ Warn all persons before disinfection is conducted
- ⇒ Handle all chemicals with care and always where the appropriate PPE
- ⇒ Check the free residual chlorine content at the end of the disinfection test period
- ⇒ NEVER leave the disinfected water in the system overnight
- ⇒ NEVER discharge chlorinated water directly down the drainage system without seeking advice from the water undertaker or the Environment Agency first
- ⇒ Always complete the disinfection record paperwork as appropriate.

AC4.6 Carry out a soundness test

During your Phase 2 practical assessment, you will be asked to carry out a soundness test to industry standards.

bpec LEARNING FOR LIFE

Unit 7

Central Heating – Phase 2

Unit 7 – Central Heating – Phase 2

In Phase 2, there is 1 Learning Outcome:

LO1 - Install Central Heating Systems:

AC1.1 Identify types and layout features of heating systems (completed at Phase 1)

AC1.2 State advantages and disadvantages of types and layout features of heating systems

AC1.3 Identify typical pipe sizes used in central heating systems types within dwellings

AC1.4 Describe working principles of types of central heating systems, positioning, fixing, connection and operation of components

AC1.5 Explain the importance of pump positioning

AC1.6 Identify operating principles for system control

AC1.7 Define zoning and control requirements of central heating systems in accordance with statutory legislation

AC1.8 Describe insulation requirements and system frost protection

AC1.9 Identify the positioning and fixing of pipework within the building fabric

AC1.10 Explain expansion and contraction in central heating systems and negative effects

AC1.11 Identify sources of information required when undertaking work on central heating systems

AC1.12 Describe procedures for filling and venting system types

AC1.13 State the operating principles of heat-producing appliances

Learning Outcome 1

Install Central Heating Systems

Learning Outcome 1: Install Central Heating Systems

In Phase 2, you will be investigating the following Assessment Criteria from Learning Outcome 1:

AC1.2 State advantages and disadvantages of types and layout features of heating systems

AC1.3 Identify typical pipe sizes used in central heating systems types within dwellings

AC1.4 Describe working principles of types of central heating systems, positioning, fixing, connection and operation of components

AC1.5 Explain the importance of pump positioning

AC1.6 Identify operating principles for system control

AC1.7 Define zoning and control requirements of central heating systems in accordance with statutory legislation

AC1.8 Describe insulation requirements and system frost protection

AC1.9 Identify the positioning and fixing of pipework within the building fabric

AC1.10 Explain expansion and contraction in central heating systems and negative effects

AC1.11 Identify sources of information required when undertaking work on central heating systems

AC1.12 Describe procedures for filling and venting system types

AC1.13 State the operating principles of heat-producing appliances

AC1.2 State advantages and disadvantages of types and layout features of heating systems

Wet central heating systems

In all modern wet central heating systems, the space heating medium is water that is centrally heated in a boiler, which is then circulated via a system of pipes to heat emitters. The heat emitters distribute the heat from the water into the living space by either conduction, convection or radiation. In some instances, it may be a mixture of two or more of these methods. Heat emitters include:

- Radiators
- Convectors – both fan assisted and natural draft types
- Underfloor heating coils
- Skirting heating
- Radiant panels

There are basically four types of wet central heating systems:

- One pipe
- Two pipe
- Manifold (micro and minibore)
- Underfloor heating

Each of these can be further sub-divided. We will look at each system type in turn, assessing their advantages and disadvantages.

The One Pipe System

The one pipe system used a continuous loop of pipework that travelled to and from the boiler. In domestic dwellings, the main system used was the Pumped heating with gravity hot water (semi-gravity) system. This obsolete system is no longer installed as it is very inefficient, but there may still be instances when these systems have to worked on. The layout of the system is shown on page 191.

Each radiator is connected to the heating circuit that flows from the boiler and back to the boiler in a single unbroken loop of pipework. The radiators are usually connected via swept tees, as shown in the diagram, with the flow pipe connected to the top connection on the radiator. The reason for this is that although the water in the circuit is pumped, the water is not actually pumped through the radiators. When water is circulated mechanically, it takes the least line of resistance. Radiators offer too much resistance to flow and so the water in the radiator heats up through gravity. By installing the flow at the top of the radiator, the hot water is circulated directly to the top section of the radiator, where it immediately begins to cool by a few degrees. The cooler water begins to descend through the radiator a lot more quickly than the heated water

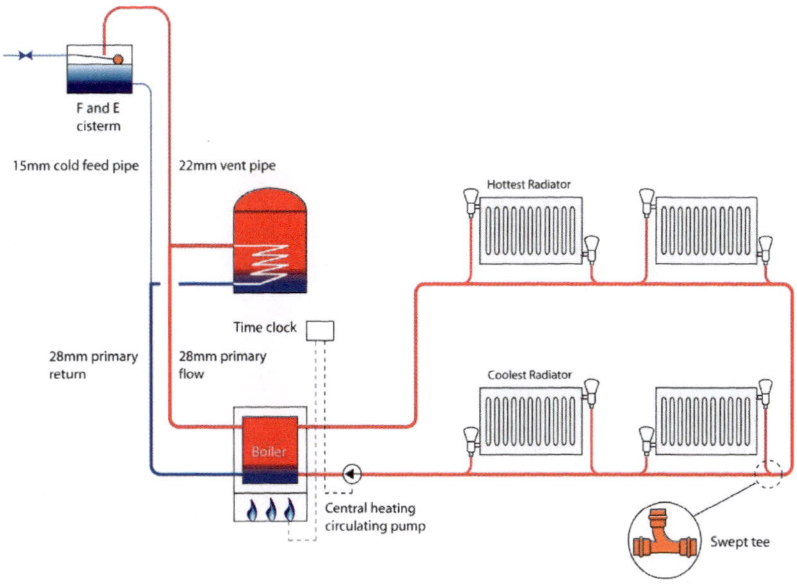

One pipe system

rises through it, so the radiator heats up more rapidly than if the hot water had entered at the bottom of the radiator. This creates a better gravity circulation.

A detrimental feature of this system was that each successive radiator is fed with water that is around 4°C cooler than the previous radiator. The first radiator on the system will always be the hottest, usually around 70°C, but the cooler water leaving the radiator mixes with the pipework loop, making the loop water somewhat cooler. This cooler water is then circulated through the next radiator, leaving the radiator cooler still and when mixed with the water in the loop, it cooled down further and so on, and so on. If we look at the four radiators on the diagram, the first radiator will be around 70°C but the last radiator, just before the loop enters the boiler, will be considerably cooler at around 56°C. The net effect of this is that a greater surface of heat emitter was required to heat the room. In other words, the radiators got bigger in proportion to the heat output.

Thermostatic control of both the hot water and the heating within this system was solely reliant on the boiler thermostat, which often meant that the hot water in the storage cylinder was excessively hot, especially in cold weather when the heating temperature was increased.

Advantages	Disadvantages
Most of the pipe work was surface mounted	The one pipe system is not Approved Document L compliant
Cheaper to install as there is less pipework in the installation when compared to more modern heating systems.	No thermostatic control of the primary circuit, which led to excessive hot water temperature in the hot water storage cylinder
	The boiler was only around 40 – 60% efficient
	Radiators were fed with cooler water the further away from the boiler they were. This leads to increased emitter sizes the further away they are from the boiler
	Often took a long time to reach temperature
	Larger pipe sizes were required
	The lack of proper control means excessive boiler cycling, which wastes energy
	Pipe runs were often difficult to install
	Condensing boilers cannot be used on one pipe systems

The Two Pipe System

Two pipe systems were a direct development of the one pipe system and gave an immediate improvement both in heat up times and temperature outputs of the heat emitters. Two pipe systems use two circuits; a flow circuit that goes from the boiler to every heat emitter and a return circuit that travels from every radiator back to the boiler. The benefit of this is that every radiator, in effect, has its own circuit to and from the boiler. This ensures that all radiators achieve similar temperatures because each radiator is supplied with water at the same temperature. There are two basic forms:

1. Pumped heating system with gravity hot water (known as the 'semi-gravity' system)
2. Fully pumped heating systems

1. Pumped heating with gravity hot water (semi-gravity)

With this system, the secondary water hot water storage cylinder is heated by the heat exchanger in the cylinder. The primary flow and return are heated by the boiler and the water circulated through the coil heat exchanger by gravity circulation.

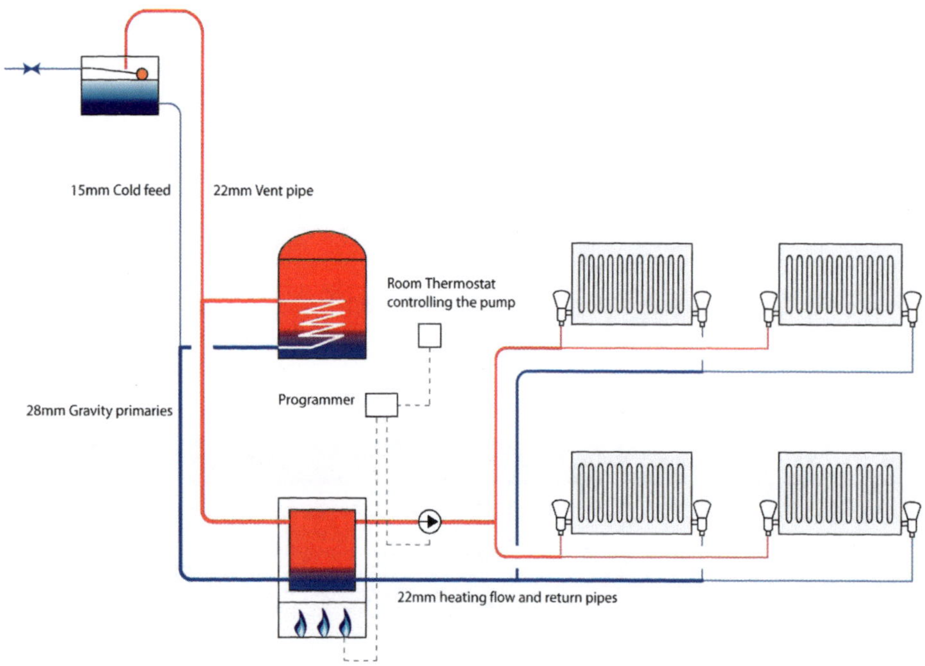

Two pipe system with gravity hot water (semi-gravity system)

The heating circuits are heated with water that is pump circulated from the boiler. The pump is installed on the flow pipe, as this gives better flow rates through the radiators, because the hotter, less dense water is easier to circulate through the system. A simple room thermostat helps to control the room temperatures by switching the pump off when the thermostat is satisfied. It is used in conjunction with thermostatic radiator valves. The main disadvantage with the two-pipe system is the lack of secondary hot water temperature control, as there is no thermostat fitted on to the primary flow and return. This means that the temperature of the secondary hot water in the cylinder will be dictated by the temperature of the heating circuit.

Advantages	Disadvantages
All heat emitters reach the same temperature at the same time	The two pipe system is not Approved Document L compliant
The two-pipe system has a far quicker heat up time, which means much improved system efficiency	No thermostatic control of the primary circuit, which led to excessive hot water temperature in the hot water storage cylinder

Smaller heat emitters can be used	The boiler was only around 78% efficient
	Constant boiler cycling even when both heating and hot water are at the correct temperatures. This leads to fuel wastage and is detrimental to overall system efficiency
	Condensing boilers cannot be used on these systems due to the requirement for gravity circulation of the primary circuit

Introduction of the C Plan semi-gravity system

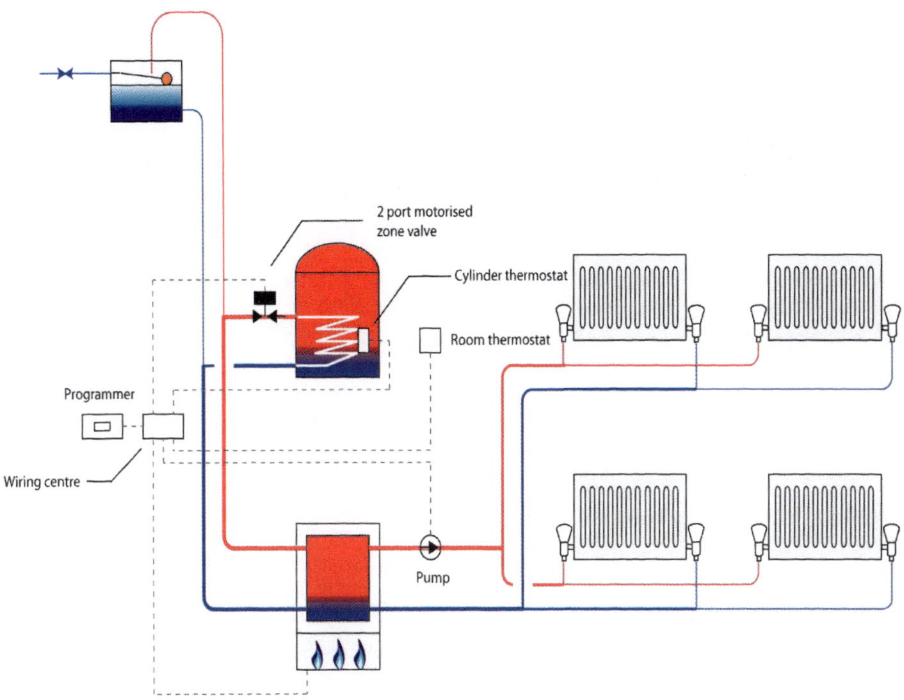

The C Plan semi-gravity heating system

Despite the obvious drawbacks of the semi-gravity system, it was further improved by the inclusion of a two-port motorized zone valve and a cylinder thermostat to control the temperature of the stored hot water in the hot water storage vessel. This system became known as the C Plan system and gave much improved overall efficiency.

The C Plan system is Approved Document L compliant and is the minimum standard for upgrading existing heating systems.

Advantages	Disadvantages
The C Plan system is Approved Document L compliant	The boiler is only around 78% efficient
The two-pipe system has a far quicker heat up time, which means much improved system efficiency	Still suffers from constant boiler cycling even when both heating and hot water thermostats are satisfied.
All heat emitters reach the same temperature at the same time	Not as controllable as modern heating systems
Full thermostatic control over all circuits	

2. Fully pumped heating systems

As Domestic Heating systems developed, the need to eliminate gravity circulation became apparent. Gravity circulation is very wasteful in terms of heat energy and fuel and so, by the inclusion of a motorised valve or a series of motorised valves, heating systems became fully pumped. In other words, the circulating pump now pumped the water around the heat exchanger/coil in the storage cylinder, as well as the heating system itself. Temperature control was maintained by using a room thermostat to control the heating and a cylinder thermostat to control the secondary hot water. Three initial systems were developed:

a) The Honeywell Y Plan using a three-port motorised mid-position valve,
b) The Honeywell W Plan using a three-port motorised diverter valve, and;
c) The Honeywell S Plan using two, two-port motorised zone valves

Fully pumped systems give a greater choice of system design, especially when positioning components such as the hot water storage cylinder. No longer was it necessary to have the cylinder higher than the boiler because gravity circulation had been eliminated. Other benefits are greater choice of boiler position and quicker hot water heating times. These are just a selection of the benefits of the fully pumped system.

a) The Honeywell Y Plan using a three-port motorised mid-position valve

3-port mid-position valve

The Y Plan uses a three-port mid-position motorised valve to control the flow of water from the boiler. The valve reacts to the demands of the system. If both heating and hot water are needed, the valve enters a mid-position allowing water to flow around both the heating and hot water circuits. If heating only is required, the valve swings to close off the hot water port (Port B on the valve) to allow water to flow around the heating system only. If hot water only is required, then the valve swings to close off the heating port (Port A on the valve). When both hot water and heating are satisfied, the system will shut down completely until heat is required to either of the circuits and the valve returns to the mid-position. Water temperatures are controlled by the room thermostat and the cylinder thermostat, with individual room temperatures controlled by thermostatic radiator valves (TRVs). A system by-pass is fitted to allow water circulation through the boiler to keep the boiler below its high energy cut-out temperature (around 85°C), as this would result in the boiler overheating and the boiler 'locking out' preventing the system from operating

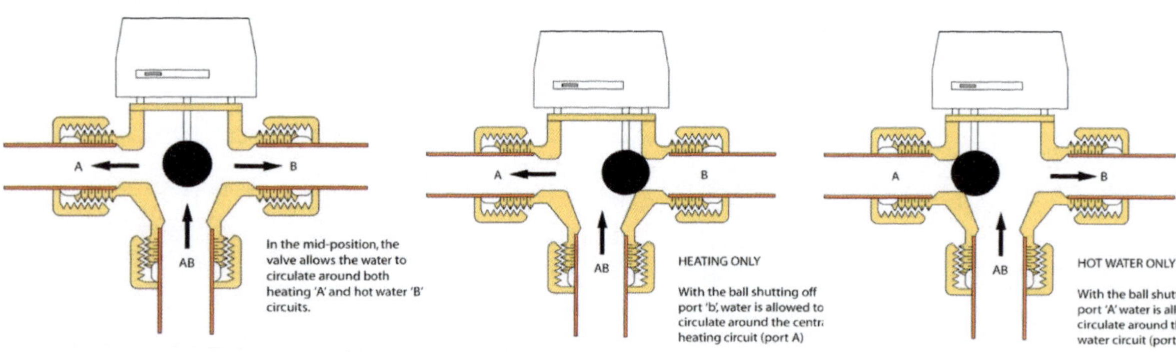

The various positions of the three-port mid-position valve

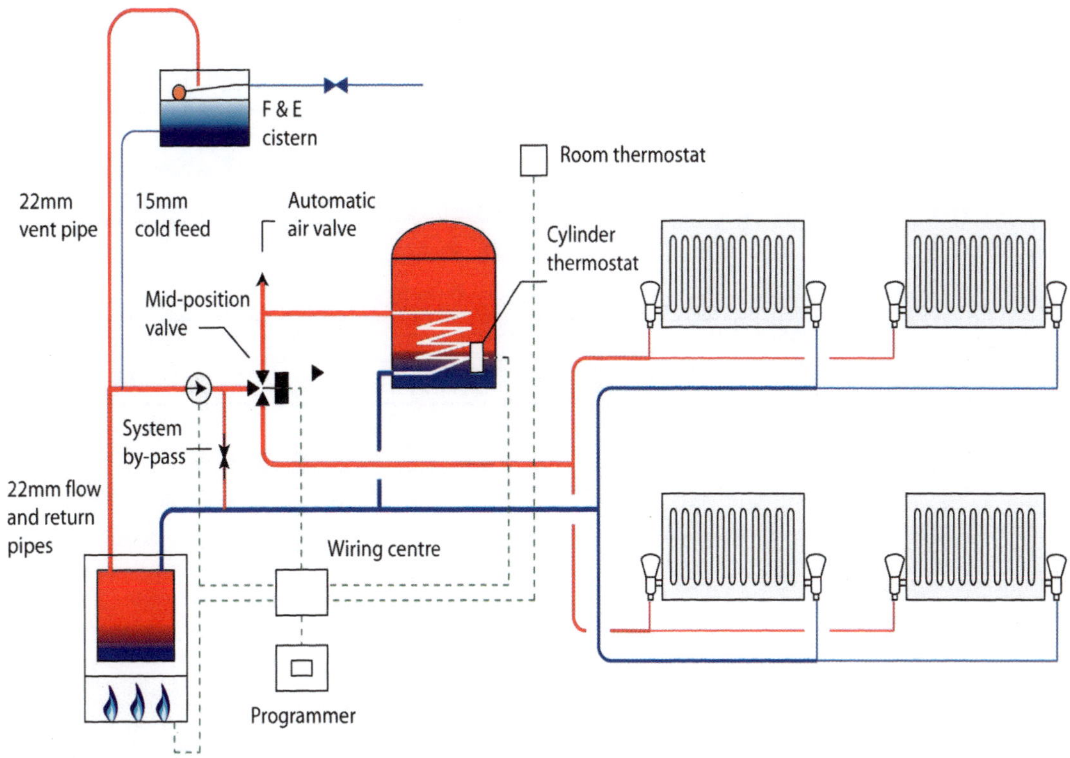

Y Plan system

Advantages	Disadvantages
The Y Plan system is Approved Document L compliant	The three-port mid-position valve can be restrictive of flow rate with larger systems.
The Y Plan system has a far quicker heat up time, which means much improved system efficiency	Zoning the system can be difficult
All heat emitters reach the same temperature at the same time	
Full thermostatic control over all circuits	
High efficiency Condensing boilers can be used	
Does not suffer from boiler cycling	

b) The Honeywell W Plan using a three-port motorised diverter valve

The W Plan system uses a three-port diverter valve. The three-port valve can either open the heating circuit or the hot water circuit but not both at the same time. One circuit is closed when the other circuit is open. It is designed to give priority to the hot water circuit, which means if both hot water and heating circuits are calling for heat, the heating circuit will not be opened until the hot water circuit is satisfied and up to the required temperature. This system is not recommended where high hot water demand is likely as it could lead to the space temperature dropping below the normal comfort levels. The system layout is identical to the Y Plan system.

Advantages	Disadvantages
The W Plan system is Approved Document L compliant	The three-port mid-position valve can be restrictive of flow rate with larger systems.
The W Plan system has a far quicker heat up time, which means much improved system efficiency	Zoning the system can be difficult
All heat emitters reach the same temperature at the same time	The heating circuit will not work until the hot water is satisfied and up to temperature. This is known as a hot water priority system
Full thermostatic control over all circuits	This system is not recommended where a high hot water demand is likely
High efficiency Condensing boilers can be used	
Does not suffer from boiler cycling	

c) The Honeywell S Plan system using <u>two</u>, two-port motorised zone valves

The S Plan system uses two, two-port motorised zone valves to independently control both the heating and hot water circuits. The room thermostat operates the heating zone valve and the cylinder thermostat operates the hot water zone valve. A system by-pass is installed in case the system needs to circulate water through the boiler to maintain the system temperature below the high energy cut-out temperature. TRVs maintain individual room temperatures.

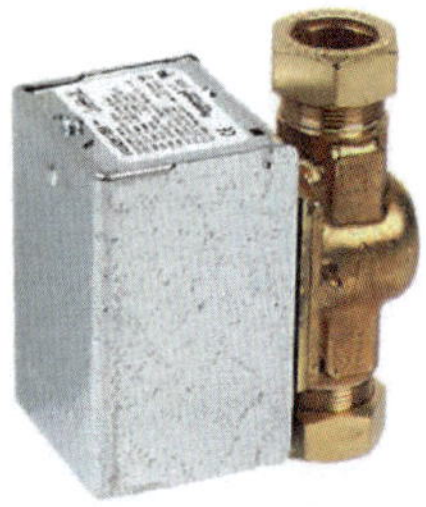

2-port motorised zone valve

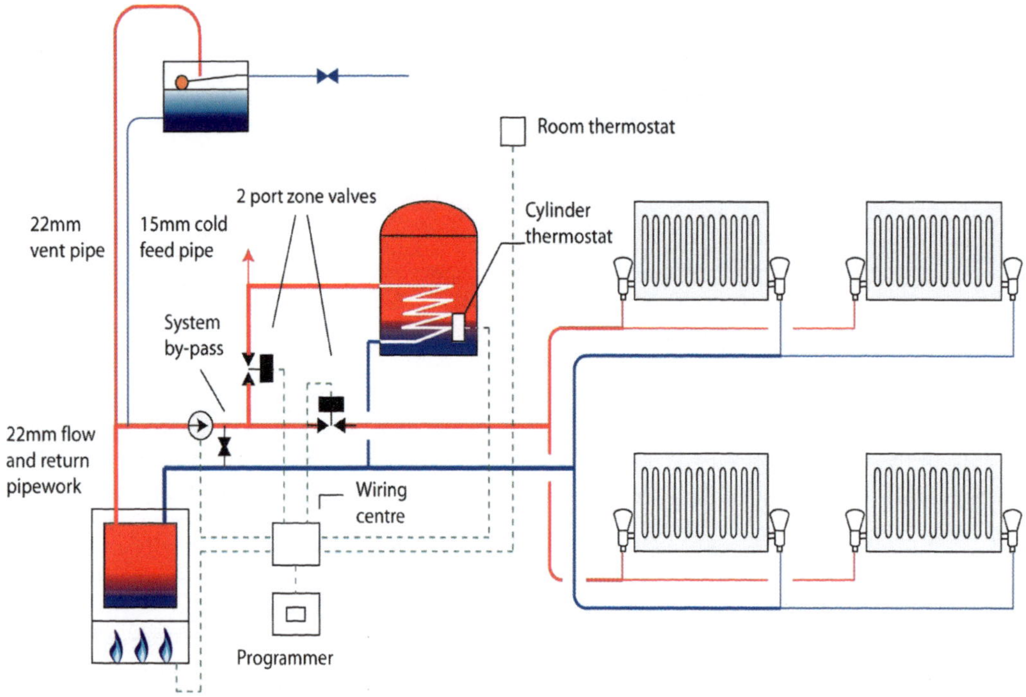

S Plan system

Advantages	Disadvantages
The S Plan system is Approved Document L compliant	Can be a complicated system to install, especially where multiple zones are required
The S Plan system has a far quicker heat up time, which means much improved system efficiency	It's an expensive system to install
All heat emitters reach the same temperature at the same time	
Full thermostatic control over all circuits	
High efficiency Condensing boilers can be used	
Does not suffer from boiler cycling	
Is suited to large domestic systems	
Zoning the system is possible (see S Plan Plus)	
Fully compatible with modern underfloor heating systems and hybrid systems	

The next development – sealed, fully pumped systems

In the late 1990s, Domestic Heating systems underwent another design change but still based on the two-pipe system. Boilers became smaller and more efficient, and feed and expansion cisterns, with the introduction of the expansion vessel, became obsolete overnight. No longer was a second (feed and expansion) cistern needed in the roof space. Systems could be filled direct from the mains cold water supply through a temporary filling loop. Gone, too, was the open vent pipe that kept the system at atmospheric pressure and safeguarded against the boiler heating the water to excessive, potentially boiling, temperatures. These new systems operated at pressures above atmospheric pressures.

They became known as 'Sealed Heating Systems'.

What is a Sealed Heating System?

A sealed heating system is a system that does not contain an open vent pipe and is, in effect, sealed off from the atmosphere. It does not have a Feed and Expansion cistern to fill the system or take up the expansion of the water when it is heated. Instead, it contains an expansion vessel to accommodate the expansion of water, a pressure/expansion relief valve to relieve excessive pressure created by expansion vessel failure and temporary filling loop to fill the system directly from the mains cold water supply. It is called 'temporary' because once the heating system is full, the filling loop MUST be disconnected to create an air gap between the water main and the heating system. This is because the water in a domestic heating system could contain chemicals to prevent corrosion, called inhibitors. The Water Supply (water fittings) Regulations 1999 classify the water contained in a central heating as fluid category 3. Water from the water main is classified as fluid category 1. The filling loop, in effect, creates a cross connection between water at fluid category 1 and water at fluid category 3, which, under the Water Regulations is not allowed without protection against back pressure or back syphonage. Because of this, the filling loop arrangement must also include an EC/ED back flow prevention device, better known as a double check valve.

Heating the water that is around 1 bar pressure gives benefits that were not present in older systems:

1. Air infiltration through the open vent pipe is eliminated.
2. Higher water pressure means better flow rate through the system.
3. Higher temperatures can be used if required.
4. Less pipework required
5. No need for pipework in the roof space
6. Systems can be fitted in flats and apartments where installing an F&E cistern at high level is problematic.

Initially, the sealed system components were installed externally to the boiler, the expansion vessel and filling loop being fitted to the return pipework, close to the boiler. Later, system boilers were designed where all the necessary components were installed within the boiler casing.

3 Two pipe sealed systems (using system boilers or sealed system components)

a) Fully pumped, 2 x two port valves (S Plan) system using external sealed system components

The S Plan sealed system layout is almost the same as its low-pressure version. The main difference is the absence of a Feed and Expansion cistern, cold feed and open vent pipes. These are replaced by a temporary filling loop, expansion vessel, pressure relief/expansion valve and a pressure gauge to monitor the system pressure. The system cold pressure is 1 bar. This will rise to around 1.5 bar when the system is hot.

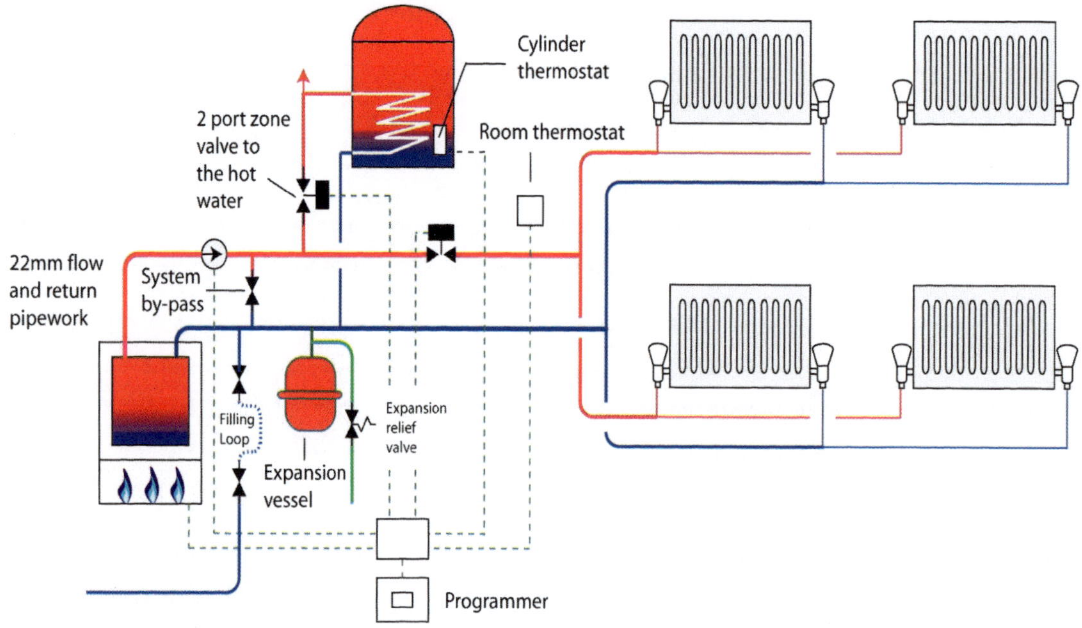

Sealed S Plan system with external sealed heating components

Advantages	Disadvantages
The S Plan system is Approved Document L compliant	Can be a complicated system to install, especially where multiple zones are required
The S Plan system has a far quicker heat up time, which means much improved system efficiency	It's an expensive system to install
All heat emitters reach the same temperature at the same time	
Full thermostatic control over all circuits	
High efficiency Condensing boilers can be used	
Does not suffer from boiler cycling	
Is suited to large domestic systems	
Zoning the system is possible (see S Plan Plus)	
Fully compatible with modern underfloor heating systems and hybrid systems	
Less pipework to install	
Sealed heating system means less chance of air infiltration, reducing corrosion	

b) Fully pumped, 3 x two port valves (S Plan Plus) system using a sealed system boiler

The S Plan Plus system was designed as a result of a change in the Building Regulations Approved Document L1, which states that all dwellings must be zoned between the living space and the sleeping space in a dwelling to give individual time and temperature control across these zones.

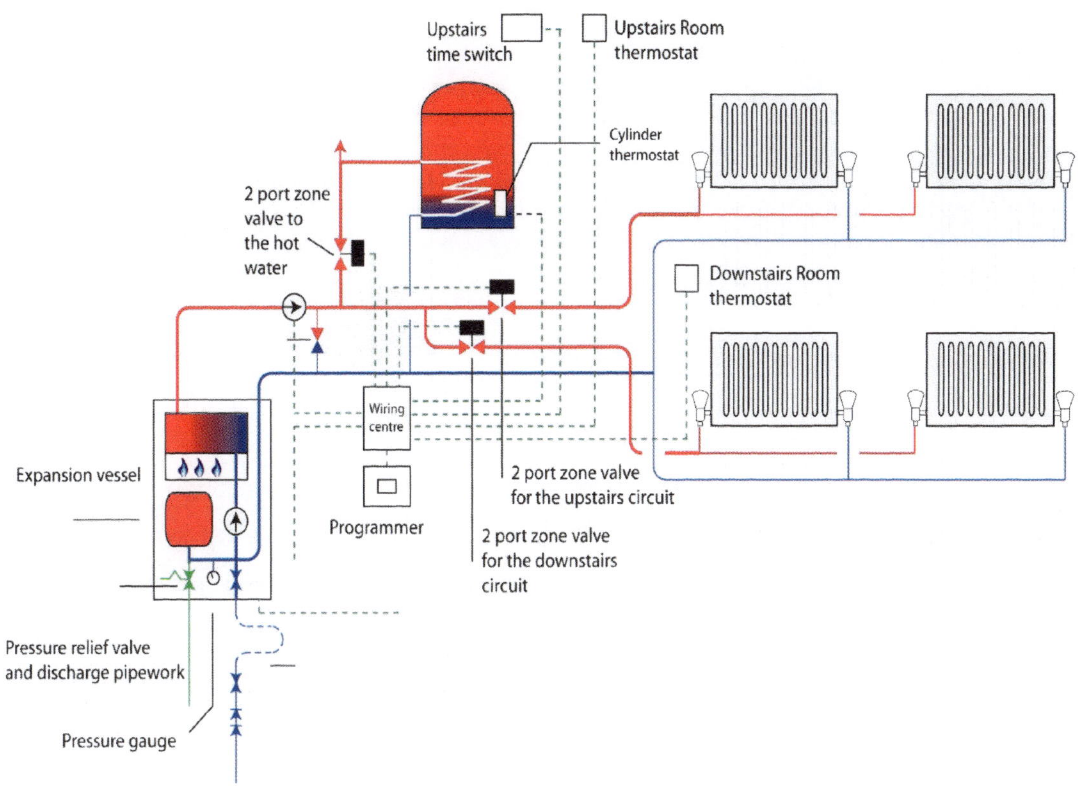

The S Plan Plus with sealed system boiler

Here, the upstairs heating circuit can be individually controlled from the downstairs living space by the inclusion of a third 2-port zone valve on the upstairs circuit. The third motorised valve is controlled by its own room thermostat and time control/programmer. The temperature of the individual rooms are controlled by programmable thermostatic radiator valves.

The drawing shows an S Plan Plus system with a sealed heating system boiler, where all components such as the expansion vessel, filling loop and pump are factory installed inside the boiler casing. This has the effect of simplifying the installation and uses less pipework.

Modern boilers must be the energy saving condensing type boiler, which has been designed to extract as much of the generated heat out of the combustion of gas and oil as possible.

Advantages	Disadvantages
The S Plan Plus system is Approved Document L compliant	Can be a complicated system to install, especially where multiple zones are required
The S Plan system has a far quicker heat up time, which means much improved system efficiency	It's an expensive system to install
All heat emitters reach the same temperature at the same time	
Full thermostatic control over all circuits and zones	
High efficiency Condensing boilers can be used	
Does not suffer from boiler cycling	
Is suited to large domestic systems	
Zoning the system means greater temperature control over all spaces	
Fully compatible with modern underfloor heating systems and hybrid systems	
Less pipework to install	
Sealed heating system means less chance of air infiltration, reducing corrosion	
Wide choice of boilers/heat sources available	

c) Heating systems incorporating a combination (combi) boiler

These are probably the easiest of all heating systems in terms of design and installation. Heating systems that incorporate a combination boiler simply require flow and return pipework to and from the boiler. The flow pipe from the boiler must be separated into two separate systems using two, two-port motorised zone valves, one for the living space circuit and one for the sleeping space circuit. These should be separately controlled by individual time clocks/programmers and room thermostats. A circuit to the hot water coil/heat exchanger is not required in this case, since combination boilers use an instantaneous method of hot water supply.

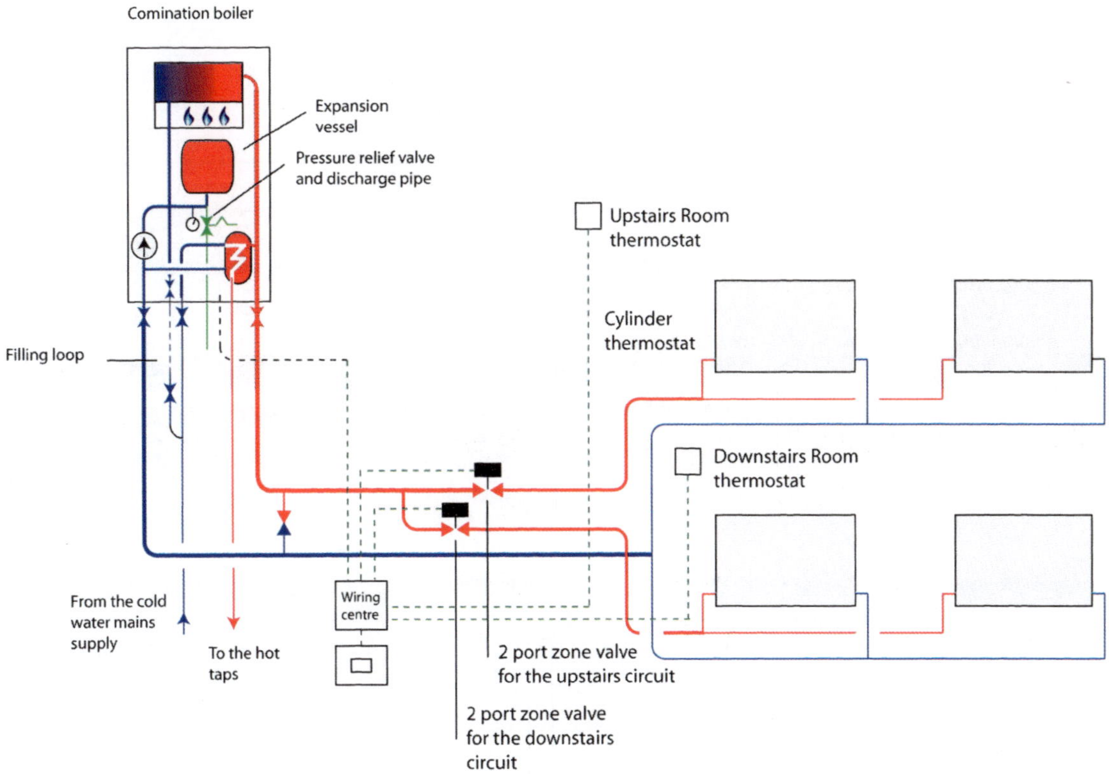

Heating system with combination boiler

Microbore systems using a manifold

Microbore systems are derivatives of the two-pipe system, but the heat emitters are fed using pipework that is very much smaller than conventional systems.

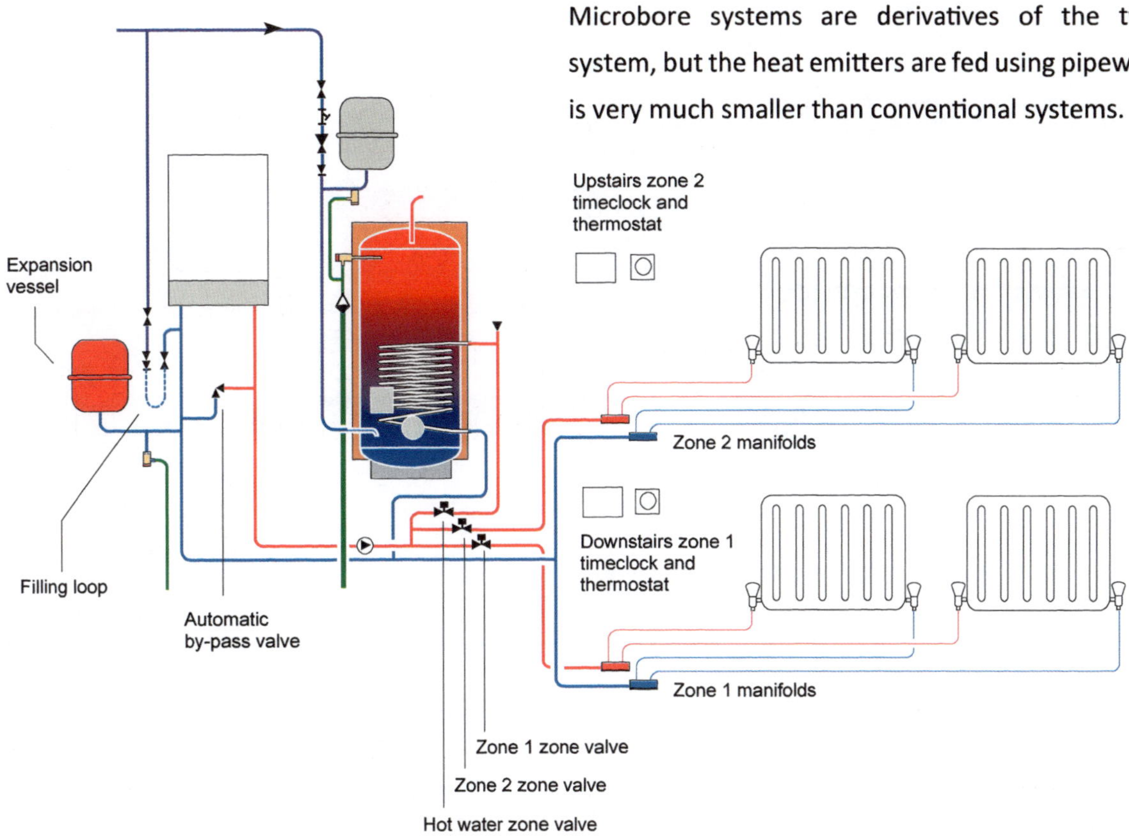

Typical microbore system

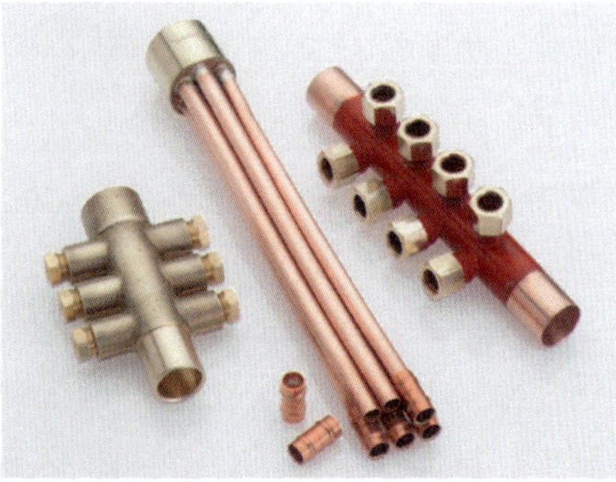

Typical microbore manifold

Because of the pipework size, each radiator is fed individually from centrally placed manifolds – one on the flow and one on the return. The simple rule is one heat emitter to one pipe run. Two radiators from a single microbore run would not work because of the restrictive flow rate of the microbore pipe. The heating main runs are usually either 22mm or 28mm pipe depending on the size of the system. Microbore pipe is typically either 8mm or 10mm in diameter.

In small domestic dwellings, one pair of manifolds will accommodate up to 8 heat emitters. Generally, though, it is accepted practice to fit two pairs of manifolds, one pair for the upstairs circuit and one for the downstairs circuit. This is shown on the diagram on page 201.

Advantages	Disadvantages
The system is Approved Document L compliant	Microbore piping is very easily damaged
The microbore system has a far quicker heat up time because the pipe work is smaller and contains less water	Can be susceptible to sludging if the system is installed poorly or the system water not treated
All heat emitters reach the same temperature at the same time	
Full thermostatic control over all circuits and zones	
High efficiency Condensing boilers can be used	
Does not suffer from boiler cycling	
Is suited to large domestic systems	
Zoning the system means greater temperature control over all spaces	
It can often be a cheaper form of heating installation	
Long pipework pipe work lengths mean fewer joints	
Sealed heating system types means less chance of air infiltration, reducing corrosion	

Underfloor heating

Underfloor heating systems have been around for over 2000 years. They were first designed and installed by the Romans who used air warmed by a furnace drawn under floors to heat their villas and homes through a system of hypocaust tiles.

Modern underfloor heating systems use a pattern of water filled pipes buried under a concrete screed laid to a specific thickness, that are heated by a gas or oil-fired boiler, or an air source or ground source heat pump. They create a uniform heat because the whole floor acts as a heat emitter, warming the room through radiated heat.

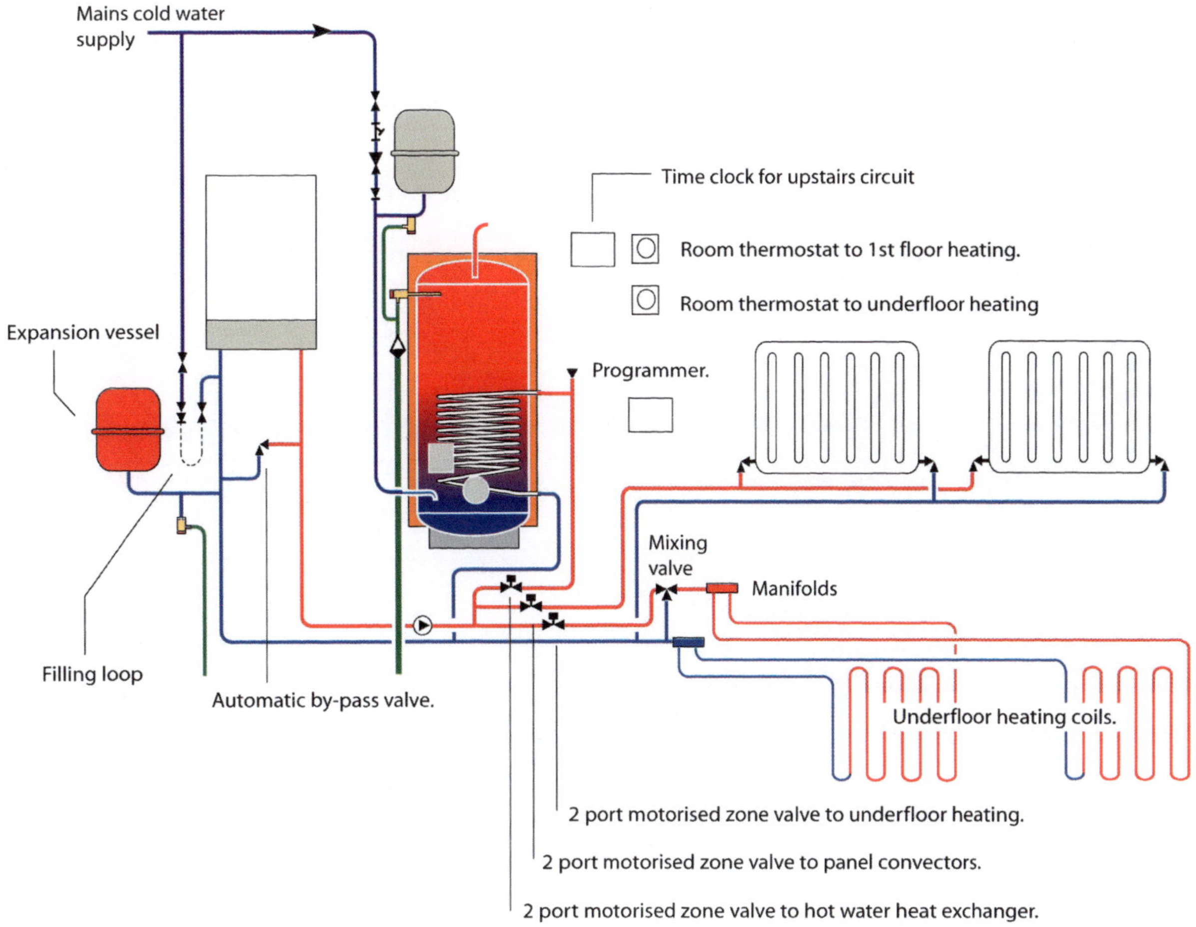

Underfloor heating with wet radiator upstairs.

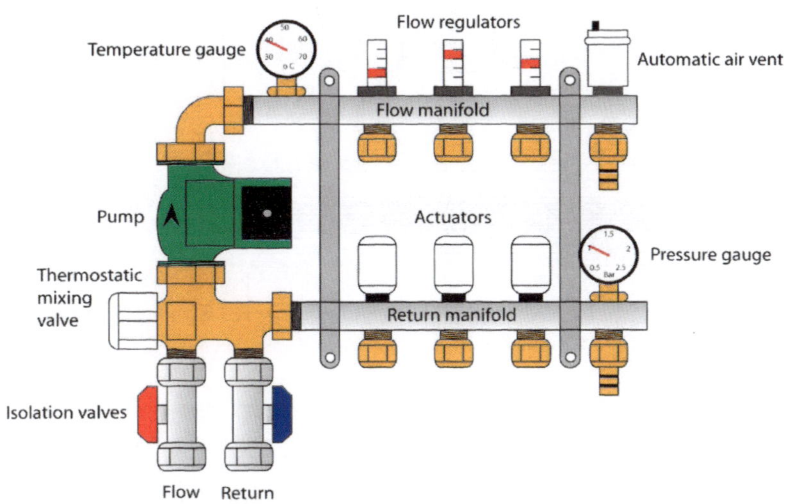

An underfloor heating manifold

Underfloor heating systems work on low temperatures. The screeded floor should be at around 27°C for optimum heating performance. This will give a room temperature of around 21 °C in most cases. To achieve this, the water in the system should not be any higher than:

- 40 – 45°C for concrete screeded floors, and;
- 50 – 60°C for timber constructed floors.

The underfloor pipe runs are laid to specific centre to centre measurements depending on the heat requirements of the rooms. Each room is fed by a single loop of pipework that travels from a thermostatically controlled manifold sited centrally in the dwelling. Each room is individually temperature controlled, with its own room thermostat that is linked to a temperature-controlled actuator on the manifold. The flow pipe from the boiler is connected to a thermostatic blending or mixing valve, which mixes the hot flow water with the cooler return water to control the temperature of the water to the individual room pipe coils.

Because resistance through the system is high, underfloor heating systems have their own circulating pump positioned at the mixing valve, to circulate the water through the system. The maximum floor area to be heated by a single coil is $100m^2$. Larger rooms may require two or more coils from the manifold.

The drawing above shows an underfloor heating system on the ground floor with wet radiators on the upper floor. This kind of system is often used in barn conversions and other older buildings where underfloor heating upstairs is not practical.

Advantages	Disadvantages
The system is Approved Document L compliant	It's a very expensive system
Underfloor heating creates a better overall heat and is much better for those people who suffer from hay fever and asthma, since there are no convection currents to spread dust and pollen	The floor often takes a long time both to heat up and cool down
Full thermostatic control over all circuits and zones with accurate room temperatures	When used with heat pumps, a secondary heat source will be needed to heat the hot water in the storage cylinder to full temperature.
The system is suited to alternative forms of heat source such as air source heat pumps, as well as condensing boilers	Not suitable for existing properties due to the difficulty in installing the underfloor heating coils.
Does not suffer from boiler cycling	Longer installation time
Is suited to large domestic systems including new build houses and barn conversions	
Greater safety with no very hot, unsightly heat emitters on walls	

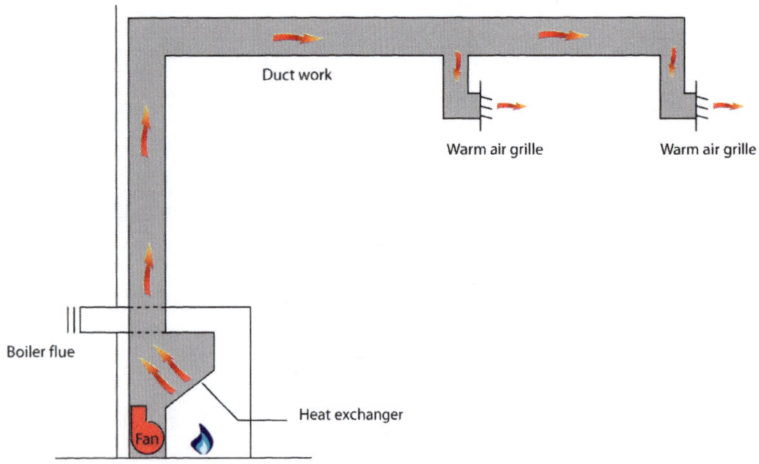

Typical warm air system

Warm air systems

Warm air systems in domestic dwellings use a gas boiler to warm the air through a heat exchanger. The warmed air is then forced by a fan through a series of ducts to the individual rooms. The warm air enters the room through a grille that can be installed in the floor, wall or ceiling. Often the boiler will also contain a wet heat exchanger to heat the water in a domestic hot water storage cylinder.

Advantages	Disadvantages
The system is Approved Document L compliant	It's a very expensive system to install
Reduced fuel use as gas fired warm air systems are particularly fuel efficient	Siting the ductwork in domestic properties can be troublesome
Quick warm-up of the living space	Requires high level of maintenance on the warm air heat exchanger to ensure that the heat exchanger integrity is maintained.
Helps to reduce damp by moving hot air around the building, preventing damp formation	
Quiet operation. Warm air systems are relatively quiet compared to other systems	
No unsightly heat emitters on walls	

Electric Storage heaters

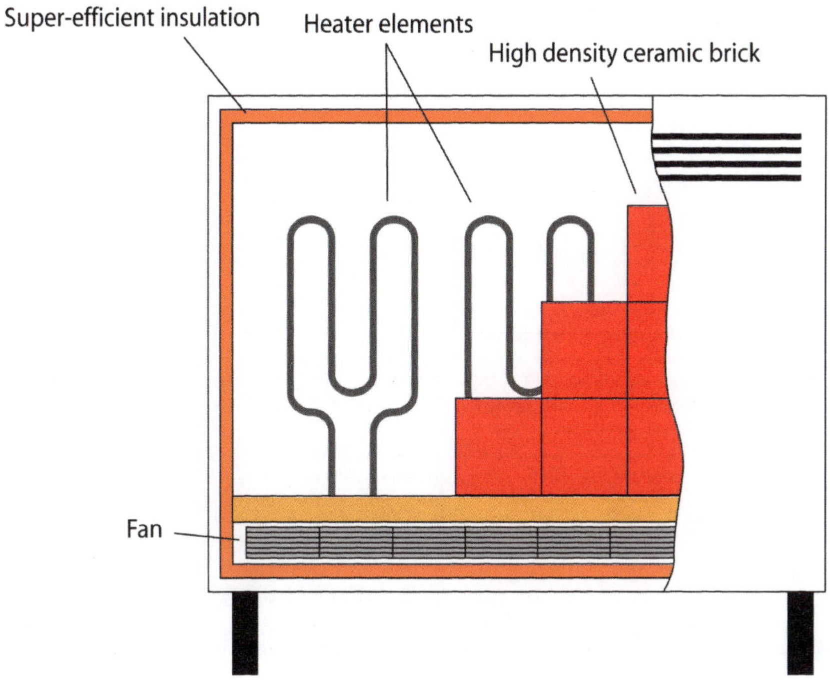

An electrical storage heater

A storage heater is an electrical heater that stores thermal energy overnight and then releases it throughout the day. They are usually used where a dual tariff electricity meter is installed. Dual tariff electricity meters record separately the amount of electricity used during the day and electricity used at night when the price of electricity is usually only 33% of daytime electricity.

They work by storing heat generated by a series of heating elements inside ceramic bricks. The heating elements heat the bricks, where the heat is stored until it is released when the heaters are operated. Some heaters contain a daytime top-up element, but these use standard tariff electricity and are very expensive to run.

Storage heaters generally only have two controls, the charge which, controls the amount of heat stored and the draught, which controls the rate at which the heat is released. The controls can either be operated manually or automatically using a room thermostat.

Installing storage heaters usually requires two separate circuits. One for night time tariff and one for during the day. These are switched off during the summer to save electricity when the heat is not required. During the winter months, the off-peak night time circuit is left on continuously. The normal tariff switch should be used when required.

Fan assisted models, which blow the heat out, are available but these use more electricity because of the electrically operated fan. Storage heaters have a poor efficiency rating.

Advantages	Disadvantages
The system is Approved Document L compliant	It's a very expensive system to run
No pipework to install	Requires dedicated circuits and a separate consumer unit to allow running on cheap rate electricity
Quiet operation	Some storage heaters are quite large in size
Storage heaters are fairly cheap to buy	

District Heating

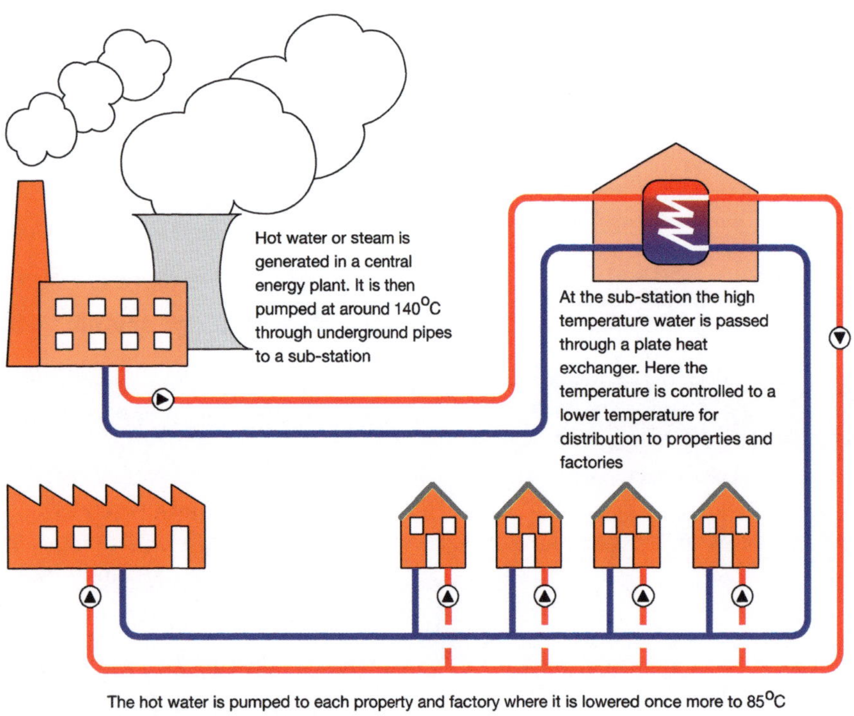

The principle of District Heating

District heating is the supply of hot water heating to many properties from a central boiler plant. The pipework is laid beneath the ground and enters each building at a district heating interface. From here, the building's heating system is installed. The heating interface replaces the boiler as the heat source and can be used to supply a standard wet central heating system or underfloor heating systems. Often, instantaneous hot water heat exchangers can be used to supply the hot water demand for the property because of the high temperature at the building's district heating interface.

In many cases the heat source is a combined heat and power plant (CHP) providing electricity and heating from a central source located outside of the district. The hot water often travels many hundreds of metres to reach each property before returning to the plant to be re-heated.

There are two heating mediums used for district heating systems. These are:

- High temperature hot water
- Steam

Steam district heating is used where high temperature is needed for industrial processes.

Advantages	Disadvantages
The system is Approved Document L compliant	The initial installation of the street mains is very disruptive.
Lower energy costs. Heat is not generated by a domestic boiler. Gas/oil/solid fuel usage in the property is reduced	Water temperature is extremely hot and requires caution at the district heating interface
Security of supply. Once the heating mains are installed in the street, the heat source is always available.	
Environmentally friendly. District heating systems utilise heat from other processes, that would otherwise be wasted	

AC1.3 Identify typical pipe sizes used in central heating systems types within dwellings

The pipe sizing of central heating systems should be carried out to ascertain the correct size of pipes required by the system. Not all systems are the same and larger systems will, naturally, require a larger size of pipe to deliver the correct amount of heat.

System	Main heating runs	Radiator runs
One pipe	Up to 35mm depending on dwelling size	Generally, 22mm or 15mm radiator connections depending on heat emitter size
Two pipe semi-gravity	28mm for the gravity primary circuit. 28mm or 22mm heating circuits depending on dwelling size	15mm heat emitter connections
Two pipe fully pumped systems (all types)	28mm or 22mm heating circuits depending on dwelling size	15mm heat emitter connections
Microbore systems	28mm or 22mm heating circuits to the manifolds depending on dwelling size	Microbore heating systems use 8mm and 10mm size piping depending on the system design and manifold size
Underfloor heating	22mm connections to the underfloor heating manifold	14mm to 16mm underfloor heating pipe, depending on manufacturer

LEARNING FOR LIFE

AC1.4 Describe working principles of types of central heating systems, positioning, fixing, connection and operation of components

1. Radiator valves – thermostatic and manual valves

Thermostatic Radiator valves

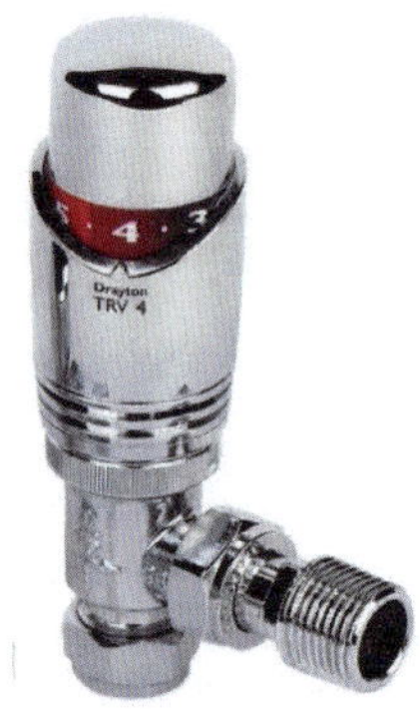

A TRV

These control the temperature of the room by controlling water flow to the radiator. TRVs contain heat sensitive head workings that reacts to air temperature (not water temperature!). The head is usually filled with a heat sensitive wax that expands and contracts with heat. When the room heats up, the wax in the head of the valve expands. This expansion causes a central pin in the valve to be pushed downwards closing the valve. The pin opens/closes the valve in response to the expansion/contraction of the wax. Each TRV has a range of temperature settings to allow different temperatures as required by the end user. TRVs can be installed on either the flow or the return, depending on the manufacturer.

Building Regulations Approved Document L1A/B requires that TRVs be fitted in all rooms except where a room thermostat is sited.

Manual Radiator valves

There are two types of manual radiator valve:

- Wheel head valves – Wheel head valves can be controlled by the end user by turning clockwise to turn off and anti-clockwise to turn on.
- Lockshield valves – Lockshield type valves do not have an operating wheel head as they are designed to be set by the plumber during the balancing operation at the commissioning stage of the heating installation.

A wheel head radiator valve

The water flow is shut off by a metal-to-metal seating that is carefully designed stop the flow of water through the radiator.

2. Automatic air vents

Automatic air valves are used to automatically release the air from the system. They are usually placed at high points in the system pipework but can be placed in any position where air is likely to collect.

How do AAVs work?

When the system is filling, the water entering the system pushes the air into high spots on the system. The air, being less dense than water, allows the float to drop away from the valve and the air is released. When the water reaches the AAV, the float literally floats on the rising water until the valve is closed. As more air arrives at the valve, the float will drop, releasing the air from the system. They are often used in conjunction with a single check valve or non-return valve so that air is prevented from entering the system through the AAV.

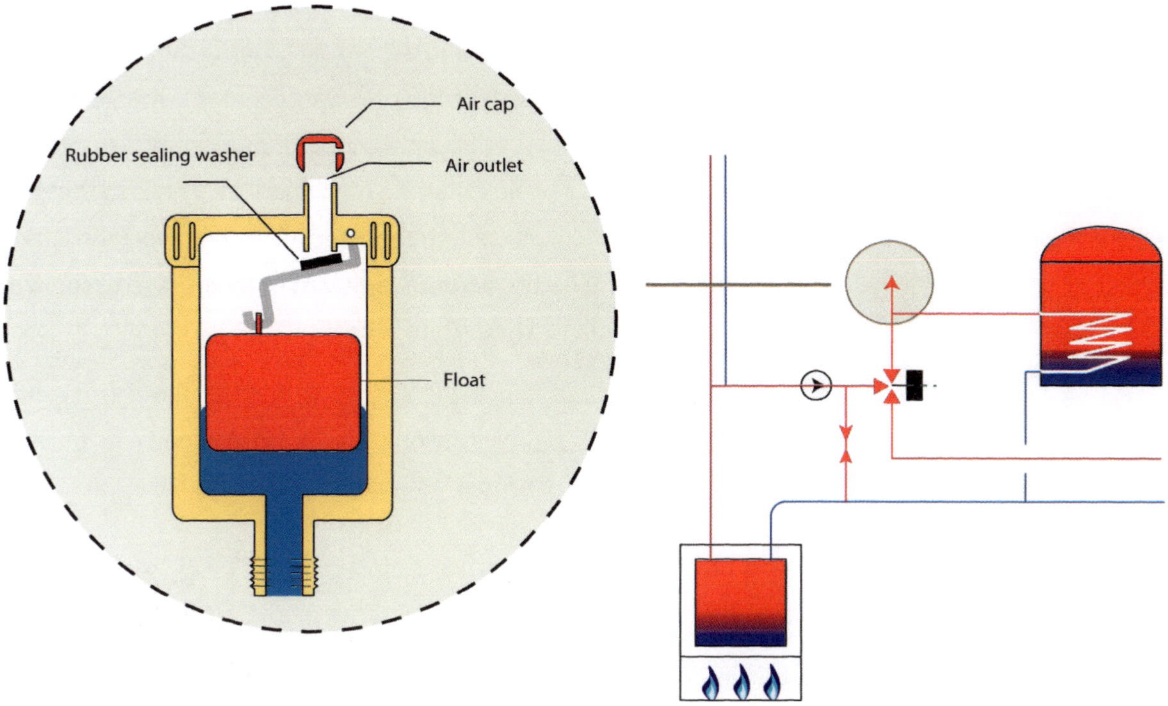

The operation of an Automatic Air Valve

3. Temporary Filling loops

This is a temporary connection between the mains cold water supply and the heating system for the purposes of filling and pressurisation. To comply with the Water Supply (water fittings) Regulations, filling loops must be removed after the filling, venting and pressurization tasks are completed to prevent a cross connection and possible contamination of the water undertaker's mains cold water supply in the event of a fault developing.

4. Pressure gauges

Water pressure gauges, often called 'Bourdon Pressure Gauges' are fitted to sealed heating systems so that a check can be made on the system pressure. Sealed heating system boilers and combination boilers have pressure gauges fitted as standard by the boiler manufacturer.

5. Feed and expansion cisterns

These small cisterns are designed for use with open vented hot water heating systems. They supply cold feed water to an open vented heating system, and they accommodate any expansion of the water due to the water being heated. The cistern must be large enough to accommodate any expanded water and must be sized accordingly. For most domestic heating systems, an 18 litre feed and expansion cistern will be large enough to accommodate any expansion of water.

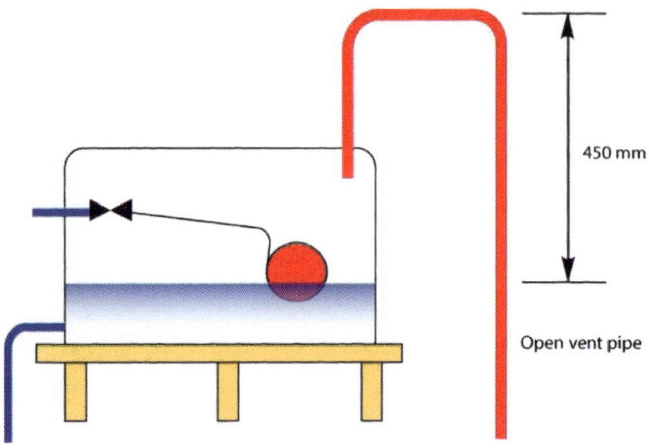

A F&E cistern

Feed and expansion cisterns should be installed so that the cold water content when the system is full is only about 100mm from the bottom of the cistern to the water line. This is so that there is enough room for the water to expand without it causing the cistern overflow pipe to run when the system is hot.

Since the cistern is manufactured from either polyethylene or polypropylene, it must be properly supported across its base.

6. Circulating pumps

Domestic heating circulating pump

The circulating pump operates by circulating the water around the system. It is not a pump in the true sense of the word, since it needs water both sides of the impeller to operate correctly.

Domestic circulating pumps have either a 6m or 10m head of pressure capability meaning that they will lift the water vertically either 6m or 10m. This is sufficient for most domestic systems.

7. Thermo-mechanical cylinder control valves

Mainly used with gravity circulation on the primary circuit of semi-gravity heating systems, these control the temperature of the hot water stored in the hot water storage vessel/cylinder. They work in a similar manner to thermostatic radiator valves. A phial containing a heat sensitive liquid/gas is fastened to the side of the hot water storage cylinder about $^{1}/_{3}$ up from the bottom. As the cylinder heats up, the liquid/gas expands activating a valve installed on the pipework. When the valve is closed, circulation cannot take place. According to the Domestic Heating Compliance Guide they are:

A thermo-mechanical cylinder control valve

*'For **replacement systems** where only the hot water cylinder is being replaced and where hot water is on a gravity circulation system, a thermo-mechanical cylinder thermostat should be installed as a minimum provision.'*

These valves must always be installed on the return pipework so that incorrect placement on the flow, potentially closing off the vent pipe, is avoided.

8. Anti-gravity valves

Anti-gravity valves are designed to prevent unwanted gravity hot water circulation in semi-gravity heating systems. These are very similar in design to single check valves. They only allow water to flow in one direction and only when the pump is running. When the hot water circuit only is in operation, they remain closed, therefore preventing circulation to the upstairs radiators.

9. Drain off valves

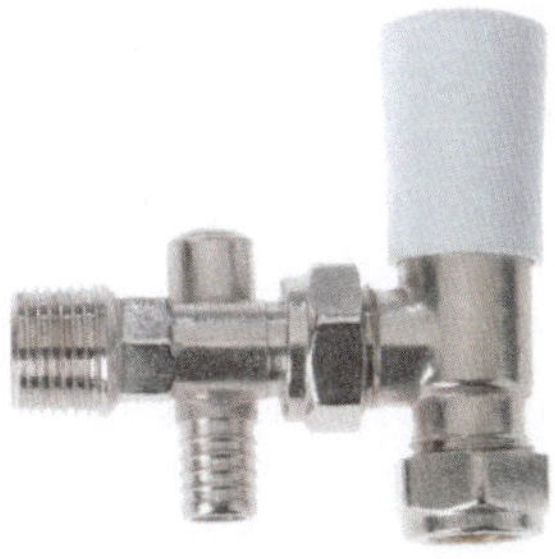

Radiator valve with drain-off

Drain valves should be positioned at all low points in the heating system to allow total draining of the system water. If the radiators are connected via pipe drops from above, then every radiator should have its own drain-off valve to facilitate draining. Radiator valves with built-in drain-off valves are available.

Drain off valves are manufactured to BS 2879 for Type A drain off valves and BS 864 for plain inlet types.

10. Additives

Additives are usually liquid chemicals that are added to the central heating water. They are used in two ways:

- To prevent problems, such as corrosion and sludge formation, from occurring.
- To remedy a problem, such as scale build up and sludge formation

There are four types of additive that can be used in central heating systems:

- Inhibitor – added to new systems to prevent electrolytic corrosion from dissimilar metals or air infiltration
- De-scaler – used to alleviate noise problems from hard water scale build up
- De-sludger – mainly used in older systems to remove sludge build up that has occurred due to system corrosion
- Neutraliser – primarily used to counter act the effects of strong, aggressive chemicals such as de-sludger

Inhibitor

De-sludger

De-scaler

System cleanser

11. Low loss headers and multiple boiler installations

A low loss header provides a hydraulic break between the primary circuit of a heating system and the secondary circuits. Low loss headers work at their most efficient in heating systems where boilers are connected directly to the loads, as they allow the installation of circuits with different heat emitter types and flow rates. They allow for independently pumped secondary heating circuits. The low loss header ensures that the boiler operates on a constant flow rate that is optimum for the boiler heat exchanger. This greatly improves the efficiency and lifespan of the boiler.

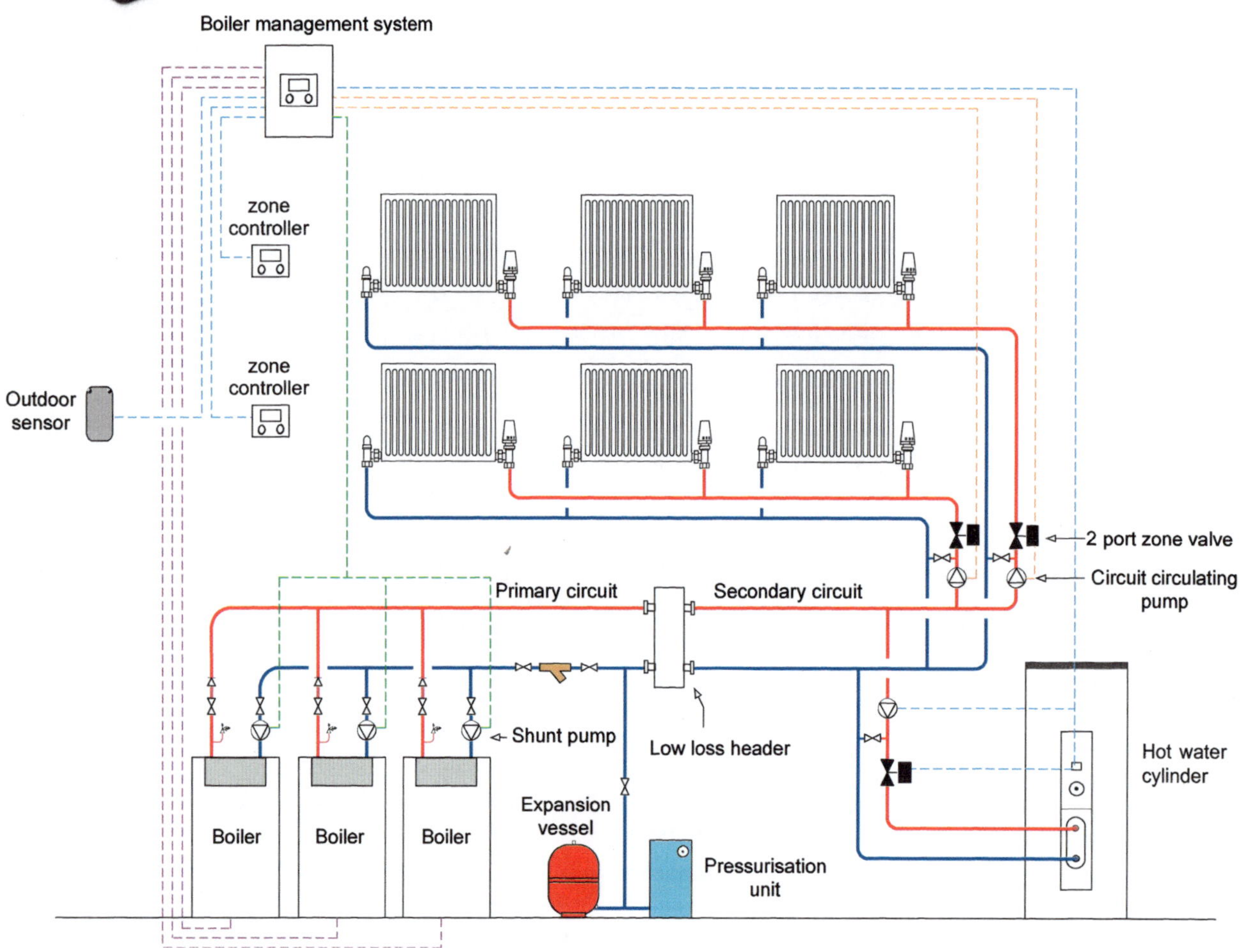

Low loss header and multiple boiler installation

For a boiler to work at maximum efficiency, the flow rate through the boiler needs to be in line with the boiler manufacturer's design limits. Condensing boilers only work in full condensing mode when there is a certain temperature drop across the flow and return. Installing a low loss header allows two distinct circuits to be created

The Primary circuit – This circuit connects the boilers with the low loss header. It is here where the efficiency of the boilers can be maximised by maintaining the manufacturer's recommended flow rate through the boilers, regardless of the flow rate demands of the secondary circuit. Each boiler has its own pump (the shunt pump) maintaining equal velocity through the heat exchangers.

The Secondary circuit – On systems where different heat emitters are used for different circuits, the low loss header allows for individual zones to be created with different flow rates. These can be balanced individually as required by the circuit demands without affecting the flow rate of the primary system, thus system efficiency is maintained. Each circuit has its own pump, motorized zone valve and thermostat for independent control, linked to an outside temperature sensor.

Another important factor is water temperature. For condensing boilers to work in full condensing mode, a temperature of 55°C to 60°C is recommended. If the difference in temperature between the flow and return is excessive, the boiler risks going into thermal shock because of the rapid expansion and contraction of the heat exchanger. Because of this, some systems allow for the installation of temperature sensors to be fitted to the low loss header to allow temperature control over the primary circuit.

Many low loss headers are fitted with automatic air valves for removing system air and drain valves to allow the removal of any sediment.

12. Buffer tanks

Buffer tanks are used as a secondary heat source for the heating system. By maintaining a buffer tank at a constant temperature, the need to constantly heat the system up from cold is eliminated. This not only provides a 'thermal store' of hot water for the heating system to draw from but also saves energy, reducing the carbon footprint of the building. They are recommended for use with renewable energy sources, such as solar thermal systems, as well as gas and oil heat sources.

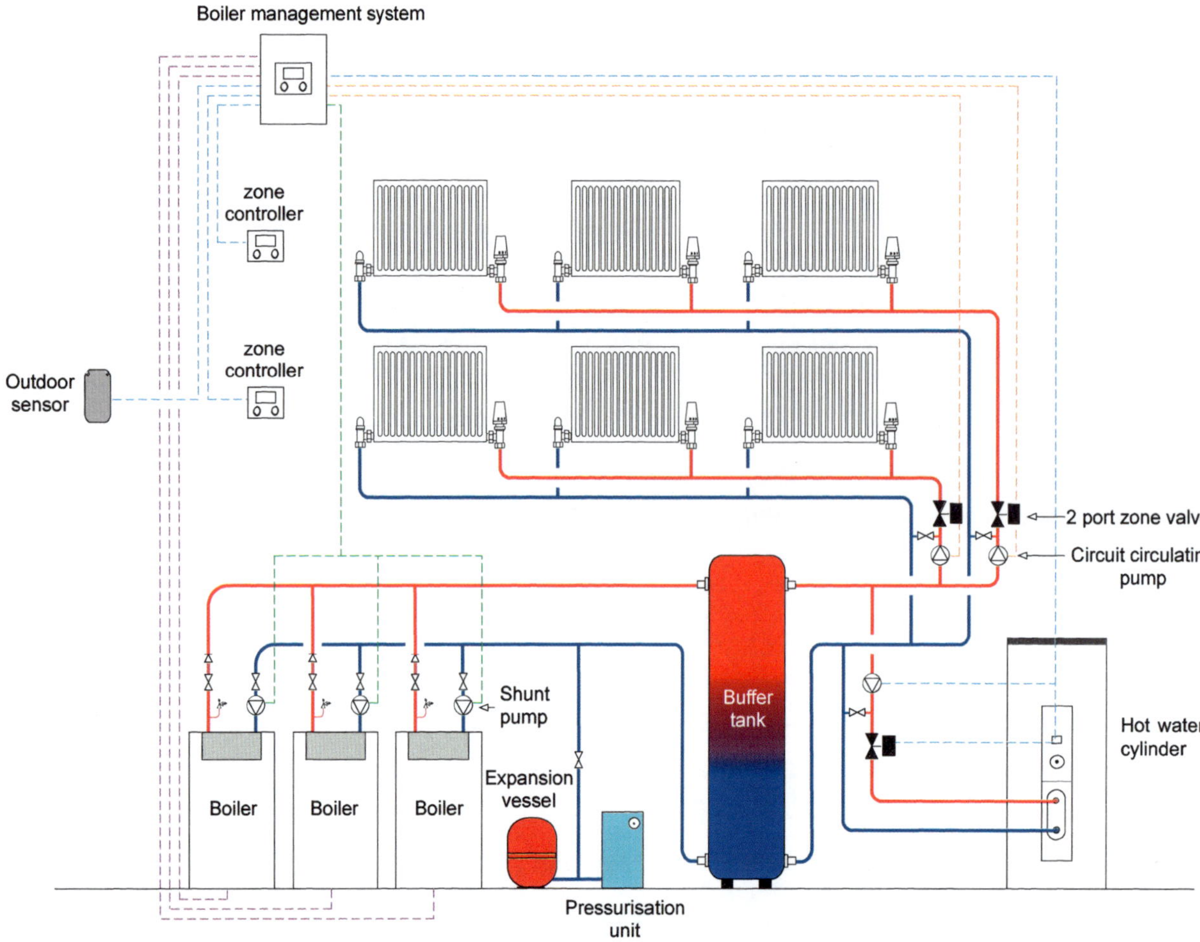

The use of a buffer tank

One of the most popular uses for buffer tanks is in conjunction with biomass boilers. The boiler heats the water in the buffer tank, which, because of a high standard of thermal insulation, maintains the water at the required temperature. The heated water in the buffer tank is then used as required to heat the heat emitters on the system. When sizing a buffer tank, it is advisable to install one large enough to provide a full day's heating requirement. Temperature monitoring equipment should be used to control the buffer tank temperature.

13. Pressure relief valves

Designed to discharge water from the heating system at a pre-determined pressure to prevent high pressure damage to the boiler and the system. It will discharge water in the event of failure of the expansion vessel.

Pressure relief valve

14. Expansion joints

Expansion joints in pipe runs accommodate the expansion and contraction of metallic pipes due to the pipes being heated and cooled by water. Typical types of expansion joint include expansion bellows and flexible joints. Metallic expansion joints are designed to absorb dimensional and length changes as well as vibration and other forms of movement in pipework and movement due to building settlement.

15. Corrosion filters

Magnetic filters

Magnetic corrosion central heating filters use a very powerful magnet to remove metallic sediment as it circulates through the central heating system. The filter should be installed on the return pipe, close to the boiler. It should be regularly cleaned to prevent black sludge and sediment build up.

CAUTION! It is recommended that people with heart pacemakers should not handle the magnet that the filter contains. Whilst the magnet is enclosed within the casing, the magnet has little or no effect with a strength of around $^{1}/_{10}$ of that of the average fridge magnet. Manufacturer's instructions should always be followed.

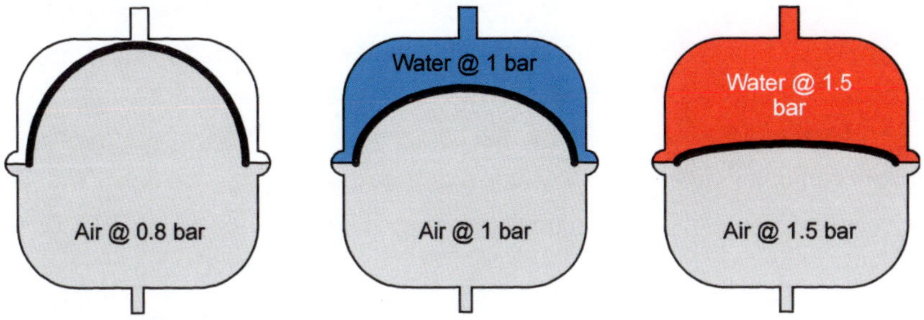

The central heating expansion vessel

16. Expansion vessel

This is required to accommodate the expansion of the central heating water when heated. Some systems will have a separate external expansion vessel, some systems utilise an in-built expansion vessel, built in to the boiler casing. Central Heating expansion vessels utilise a neoprene rubber membrane with air on one side and water accommodation on the other.

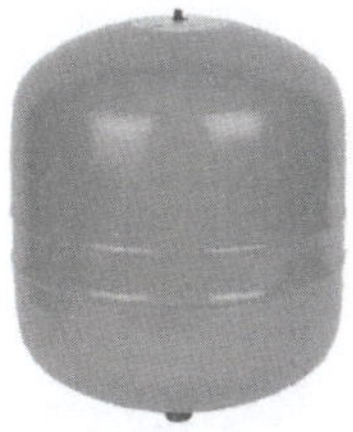

Expansion vessel

The expansion vessel should be positioned on the return pipework as this is around 20°C cooler than the flow pipework. If this is not possible, expansion vessels can be positioned on the flow pipe but they must be connected on the suction side of the pump so that the pump does not artificially raise the pressure causing activation of the pressure relief valve.

17. Automatic by-pass valve

The automatic by-pass valve

This is a spring-operated valve that is designed to open when all zone valves and thermostatic radiator valves are closed. This usually occurs when the thermostats have been satisfied and the system is up to the correct temperature. The valve is opened by the flow of water created by the pump and prevents the boiler from overheating by allowing circulation of water through the heat exchanger, preventing activation of the high limit thermostat on the boiler.

Automatic by-pass valves can be set to respond to the flow rate of the system. The top of the valve is marked in litres/second flow rate and is fully adjustable.

Central Heating Controls

Modern central heating systems require modern, efficient controls if they are to operate at their maximum efficiency. It is a requirement that all DOMESTIC heating systems comply with Building Regulations Approved Document L1: Conservation of fuel and power. App. Doc. L1A for new installations and L1B for existing installations. These documents are retrospective. In other words, ALL heating systems, regardless of age, must comply with the regulations laid down in L1A and L1B. It is worth looking at these requirements in detail:

The following information is quoted verbatim from the TACMA-guide-to-Part-L1A/B.

CONTROLS FOR NEW DWELLINGS **Part L1A, England and Wales:**	**CONTROLS FOR EXISTING DWELLINGS** **Part L 1B, England and Wales.**
1. Provide installations that are fully pumped. 2. Provide independent time & temperature control to both the heating and hot water circuits. Part J, Scotland requires the installation of 7-day timing. 3. Provide control systems with a boiler interlock. **Note 1** 4. Install an automatic by-pass valve if a bypass is fitted. **Note 2** 5. Split the heating circuit into zones using either: a. room thermostats or programmable room thermostats in all zones. b. a room thermostat or programmable room thermostat in the main zone and Thermostatic Radiator Valves c. Thermostatic Radiator Valves (TRVs) on all radiators in the other zones. 6. Ensure installation of a cylinder thermostat and a zone valve to control stored hot water. **Note 3** 7. Provide time control by the use of either: a. a full programmer with separate timing to each circuit. b. two or more separate timers providing timing control to each circuit c. programmable room thermostat(s) to the heating circuit(s), with separate timing of the hot water. 8. For dwellings with a total usable floor space greater than 150m^2, then: a. the heating circuit should be split into a minimum of two zones plus a hot water service zone and: b. each zone should be separately timed by the use of a multi-channel programmer or multiple heating programmers or programmable room thermostats	When a new installation is fitted, the controls should be as for a new dwelling. When replacing the boiler and/or the hot water vessel, the opportunity to improve the controls should be considered. To be confident that the requirements of the Building Regulations are met this would entail the following: 1) Provide fully pumped installations with a boiler interlock. **Note 1** 2) Install an automatic by-pass valve, if a bypass is fitted. **Note 2** 3) Provide time and temperature control to both the heating & hot water circuits, for fully pumped installations. 4.1) Separate the space heating system into zones and: a. if a room thermostat is already fitted fit TRVs on at least all radiators in the sleeping areas and check if a new room thermostat or programmable room thermostat should be fitted. b. if there is no room thermostat, install either a room thermostat or programmable room thermostat and fit TRVs on at least all radiators in the sleeping areas. 4.2) Provide time control by the use of either: **a)** a full, standard or mini programmer, or **b)** one or more separate time switches, or **c)** programmable room thermostat(s). 5) For semi-gravity installations, the recommended option is to convert it to fully pumped. The controls should then be as detailed above. 5.1) If either the new boiler or hot water storage vessel can only be used on a pumped circuit, the installation must be converted to fully pumped. Install controls as detailed above.

9. Any Boiler Management Control System that meets the specified zoning, timing & temperature requirements is a wholly acceptable alternative.	5.2) If the new boiler or storage vessel is designed for gravity hot water, or it is impractical to convert to a fully pumped installation, provide the following controls: a. a cylinder thermostat & zone valve to control the hot water and provide a boiler interlock. **Note 3** b. a room or programmable room thermostat. c. a programmer or time switch. d. TRVs on at least all radiators in the bedrooms. 6) Any Boiler Management Control System that meets the specified zoning, timing & temperature requirements, is a wholly acceptable alternative.

Note 1: Boiler interlock is achieved by the correct use of the room thermostat(s) or programmable room thermostat(s), the cylinder thermostat and zone valve(s) in conjunction with the timing device(s). These should be wired such that when there is no demand from the heating or hot water both the pump and boiler are switched off. The use of TRVs alone does not provide interlock.

Note 2: Although the regulations may not always require a bypass to be fitted, the performance of any installation with multi-zoning or TRVs is improved by the use of an automatic by-pass valve.

Note 3: The use of non-electric hot water controllers does not meet this requirement.

1. Zone valves (2 port, 3 port, mid position and diverter)

The two port motorised zone valve

These are used primarily on S Plan and S Plan Plus heating systems but may also be used on C Plan Plus systems. Each valve is wired to control either a room thermostat or a hot water cylinder thermostat. They are mostly used to 'zone' parts of heating systems for better thermostatic and energy efficiency control of heating systems.

The three port motorised mid-position valve

These are used exclusively on Y Plan domestic heating systems. The valve is linked to both room and cylinder thermostats and operates by opening or closing the heating or hot water circuits. With the mid position valve, at least one circuit is open all of the time. It is not possible to close both hot water and heating circuits simultaneously. When both circuits are required, the valve swings to the mid-position allowing hot water to flow to both circuits.

The three-port motorised diverter valve

These are very similar in appearance to the three port mid-position valve but they will only open one circuit at a time. There is no mid-position function. They are specifically designed for the W Plan fully pumped system.

2. Programmers and time clocks

A typical programmer

Programmer

Programmers are two-way timeclocks, able to switch two circuits at multiple times throughout the day. There are three types:

- **Mini programmers –** these allow the heating and hot water circuits to be on together, hot water independently but not heating independently.
- **Standard programmers** – these use the same time settings for all circuits.
- **Full programmers** – these allow multiple independent settings for both circuits. Each circuit can be used on its own or together.

Time clocks

Time clocks provide a simple way of activating a central heating system. They are only suitable for switching on or off one circuit, such as the heating circuit. This makes them ideal for systems that incorporate a combination boiler. Mechanical and digital types are available

3. Thermostats

A hive room thermostat

Programmable room thermostat

A room thermostat controls the air temperature of the room. It is basically a thermostatically controlled switch that either operates an electrical connection controlling the heating circuit.

Some room thermostats are programmable, allowing different temperatures at different times of the day and on different days of the week and many provide a 'night setback' facility where a minimum temperature can be maintained throughout the night. This is particularly useful during the winter months.

A cylinder thermostat

Cylinder thermostat

These control the temperature of the hot water that is stored in the hot water storage cylinder. They are strapped to the cylinder about ¼ of the way up from the bottom of the storage vessel. They are used in close conjunction with a motorised zone valve or diverter/mid position valve to provide accurate temperature control.

Frost thermostat

Frost thermostat

These are wired to vulnerable parts of the heating system to provide protection against sub-zero temperatures. They are designed to override all other controls, timeclocks and programmers to activate the heating when the temperature reaches 3 – 5°C. They are used in conjunction with pipe thermostats.

Pipe thermostat

Pipe thermostat

These are wired in series with frost thermostats and provide an accurate temperature of the water in the heating system. They are strapped to the pipework that is installed in a vulnerable position. For example, a boiler installed in an unheated garage. They are designed to activate when the system water reaches temperatures close to 0°C.

Wiring centre

Fully pumped central heating system contain many electrical controls that are linked. A wiring centre enables these controls to be wired to a central box. The wiring centre contains numbered electrical connections for easy, trouble-free wiring.

4. Heat emitters

Bespoke heat emitters

There are many designs for heat emitters available to suit all building types and environments from domestic properties through to commercial and industrial applications. The heat output from bespoke radiators should be considered cautiously, as many are not manufactured to the British Standards.

Panel radiators

Most modern panel radiators are designed to emit heat by two methods:

- Convection
- Radiation

Almost all panel radiators now have fins on the back, and it is these fins that create effective convection currents to warm the room. 70% of the heat emitted is via convection. Cool air is drawn from the bottom of the radiator and passes through the fins, where it is heated. The heated air rises upwards to circulate through the room, eventually cooling and sinking once again to floor level, where the process starts all over again.

There are still some radiators without fins. These rely on radiated heat only to heat the room.

Panel radiators should be installed in accordance with manufacturer's instructions, paying particular attention to the distance from the bottom of the heat emitter to the finished floor level, as it is here where the process of convection begins. They should be installed in a suitable location on a clear wall without furniture obstruction, as this can inhibit heat flow into the room.

Domestic panel radiators have ½ inch BSP female threads on all ends for installation of the radiator valves. All radiators have an air release valves and plugs. These are usually supplied by the radiator manufacturer. There are several different types of panel radiator.

Compact

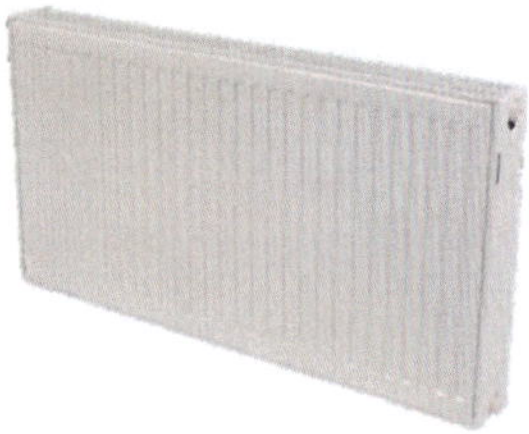

Compact radiators have factory fitted top grill and side panels and are more aesthetically pleasing than other radiator types. The top grill and side panels can be removed to match seamed top radiators if desired.

Rolled top

Not as popular as compact radiators, these do not have any top grills or side panels. The top of the radiator has a rolled seamless appearance with welds along the sides and across the bottom. Generally, these are more expensive that compact radiators.

Seamed top

The seamed top radiators are made by welding two flat plates of steel together. At one time, these were the most popular radiator installed but have fallen out of popularity due to the introduction of the compact radiator.

Flat panel

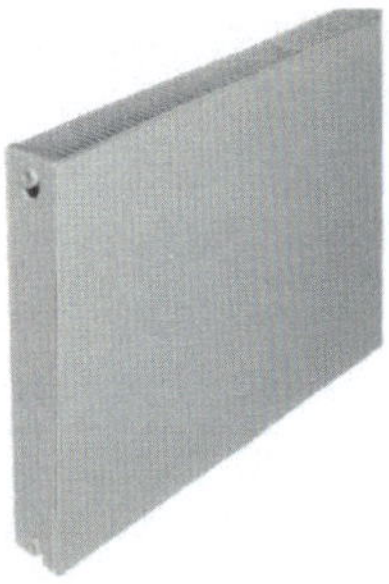

These are becoming increasingly popular due to the smooth finish and stylish design that compliments modern living environments. They are very easy to keep clean and have a low profile, maximising the wall space. They can be mounted horizontally and vertically.

The types of panel radiator

There are many types of panel radiator from single panel to double panel with fins. These are shown in the drawing on the following page:

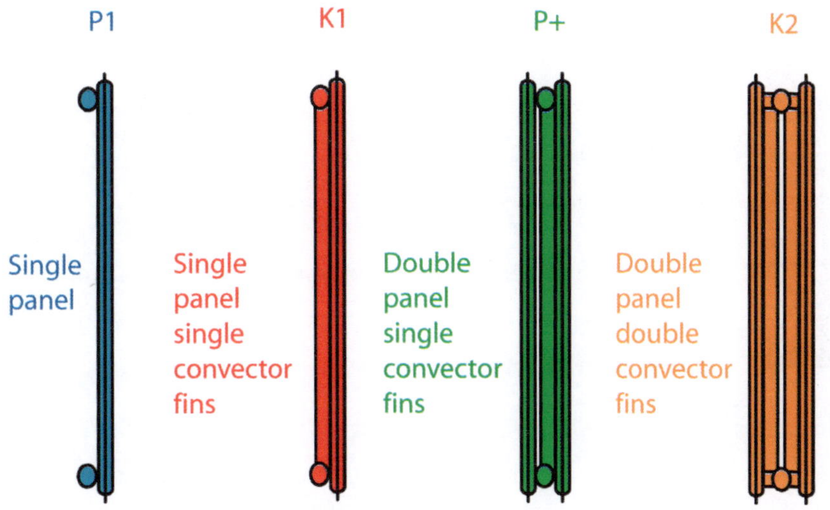

Connections to radiators

Radiator connections are classified by their abbreviations. For example:

- **TBOE**: Top-Bottom-Opposite-End (used on solid fuel systems for heat sink radiators and one pipe systems)
- **BBOE**: Bottom-Bottom-Opposite-End (the usual radiator connection method)
- **TBSE:** Top-Bottom-Same-End (used occassionally on one pipe systems)

Column radiators

Column radiators are, as the name suggests, made from columns historically made from cast iron. The more columns a radiator has, the greater the output is. Modern column radiators may be made from cast aluminium or steel. They are often called church or hospital radiators because they were designed to withstand being installed in large public spaces. They are increasingly used on refurbishments where an historic 'period-feel' is important.

There are two disadvantages with column radiators:

1. They contain a lot of water, which means that they may take longer to heat up than convention panel radiators.
2. The heat is predominantly radiated heat with very little in the way of convection. This means that the radiators may be on the large side.

Low surface temperature radiators

Designed to comply to an HSE Health Services Information Sheet no.6, which states that:

> *Contact with surfaces above 43 °C can lead to serious injury. Prolonged contact often occurs because people have fallen and are unable to move or are trapped by furniture. Incidents often occur in areas*

where there are low levels of supervision, for example in bedrooms, bathrooms and some communal areas.

This standard has been accepted by many local authorities and NHS establishments to include any place where the general public have access, including residential homes, care homes and schools. Low surface temperature radiators are also becoming popular in many domestic situations too, especially in children's bedrooms and nurseries, and areas where the elderly or infirm may come into contact with radiators.

Inside each LST is a normal panel radiator. The outer shell is a specific distance away from the direct heat so that it never exceeds the 43°C limit. Bottom grill and top grill allow the passage of cooler air in and heated air out, so that the room is warmed sufficiently.

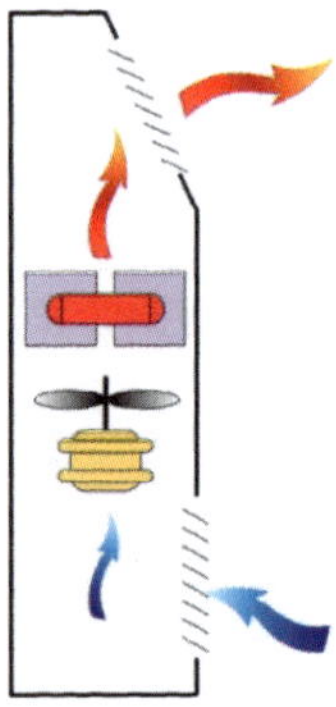

A fan heater

Fan convectors

Fan convectors work by using an electrically operated fan blowing warm air into the room. The heat exchanger is usually a coil of copper pipe with aluminium fins attached that is connected to the flow and return pipes of the heating system.

The fins trap the air and heat it quickly before the fan forces it out. A thermostat operates the fan by switching off the electricity once the temperature is reached. They are connected to the electricity supply via a switched fuse spur. There are two common types of fan convector:

Wall mounted

These are usually large in size. They can be successfully used on sealed heating systems where the temperatures can be slightly higher than open vented systems. Heat outputs vary with heater size and manufacturer's data should be checked for specific installation requirements.

Plinth heaters

These are specifically designed to fit below kitchen cupboards for use in kitchen areas with limited space for a radiator. They work in the same way as fan convectors.

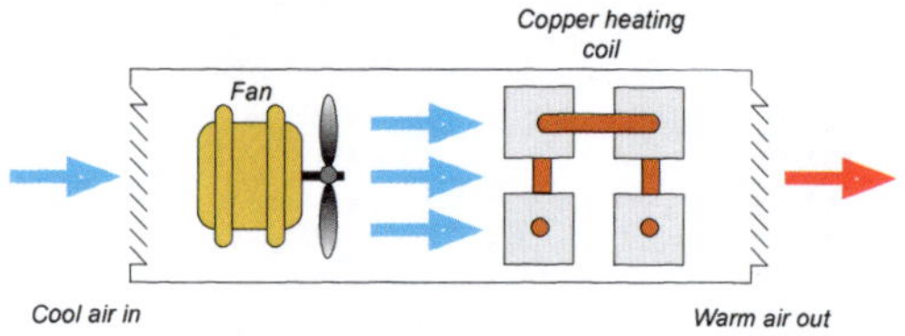

A kick space heater

Typical tubular towel rail

Towel warmers

Tubular towel warmers

Bathroom towel warmers are available in many sizes, styles and finishes and are often referred to as designer towel rails. Many have dual fuel capability, being used with wet central heating in the winter and using a small electrical element to heat them during the summer when the heating system is not used. They are usually vertically mounted.

Towel warmers with integral panel radiators

Not as popular as tubular towel warmers, these combine a towel warmer with a radiator. The idea is to warm the towel but not affect the convection currents warming the bathroom. They are usually used in period bathrooms, where modern heating efficiency meets classic styling.

Towel rail with column radiator

5. Underfloor heating components

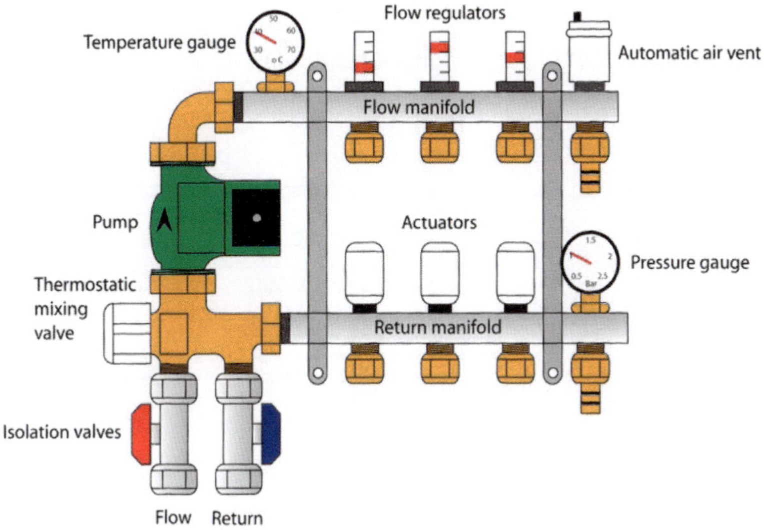

An underfloor heating manifold

Manifolds

The manifold of an underfloor heating system is the central point where all underfloor heating coils begin and end. There are two manifolds – a flow manifold and a return manifold. The flow manifold is fitted with flow regulators so that the correct flow rate through the system can be achieved. The return manifold contains the thermostatic actuators for thermostatic controls of the system. These are linked to individual room thermostats in each space. Manifolds also contain a pressure gauge to monitor system pressure and a temperature gauge. Each manifold also has a system drain point.

Pump control unit

The underfloor heating pump control unit is made up of several key underfloor heating components:

- The circulating pump – like all heating systems, the circulating pump circulates water through the underfloor circuits. It works independently of any other pump on the system.
- The thermostatic mixing valve – this valve mixes the flow and return water to the correct temperature for the underfloor heating coils. It can be set to the temperature required by the circuits and the floor surface, since each floor surface will require a different heat output.
- The manifold isolation valves – each manifold has an isolation valve fitted so that the pump control unit can be isolated for maintenance and repair. The valves are spherical ball-type valves and are similar in construction to isolation valves.

Insulation

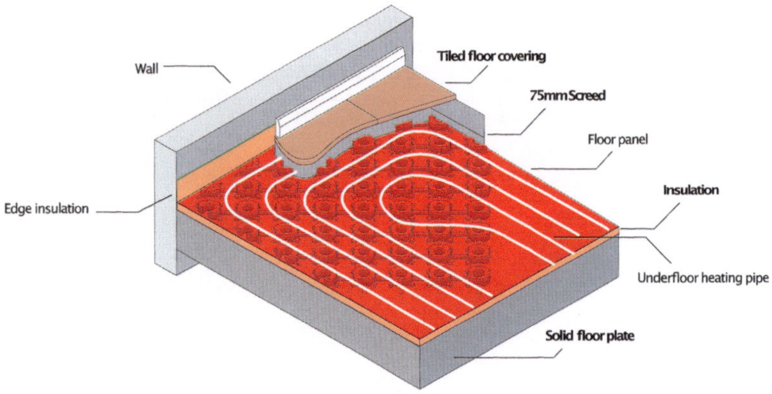

Solid floor

Here, the underfloor heating pipework is laid into pre-insulated sheets that have castellated grooves for the pipework to clip into. This allows for a more precise installation with regard to centre-to-centre measurements. The insulation material is either polystyrene or Kingspan kooltherm insulation board

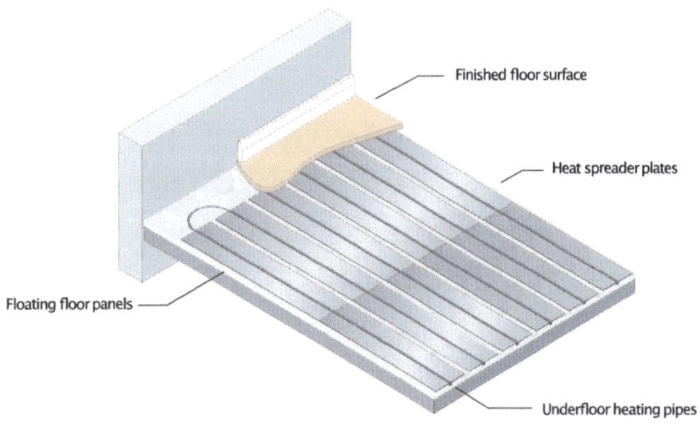

Floating floor

Here, the underfloor heating pipework is laid on to 50mm thick polystyrene panels that have pre-formed groves for the pipework to clip into. The panels are not fixed and 'float' on the top of the concrete sub-floor. The finished floor can be laid directly on to the pipework.

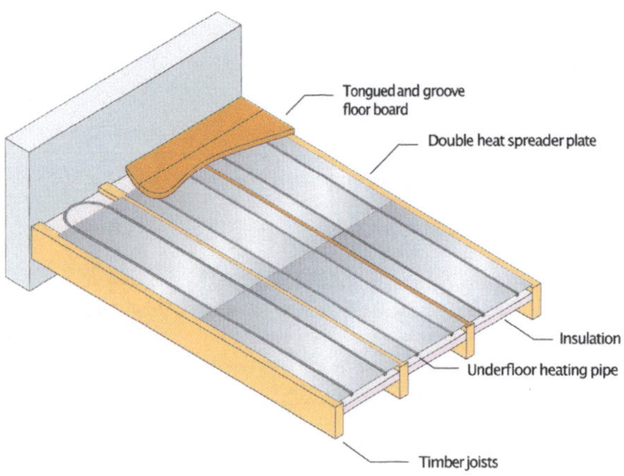

Suspended floor

This system uses aluminium double heat spreader plates to radiate the heat evenly across the timber floor. It is suitable for suspended timber floors with joist centres of 450mm. the aluminium plates are fixed to the joists using staples or small flat-head clout nails. Insulation or insulation board is placed under the plates to prevent downward penetration of heat.

Pipework

There are two common types of pipework systems for underfloor heating systems:

- The series pattern
- The snail pattern

Series pattern

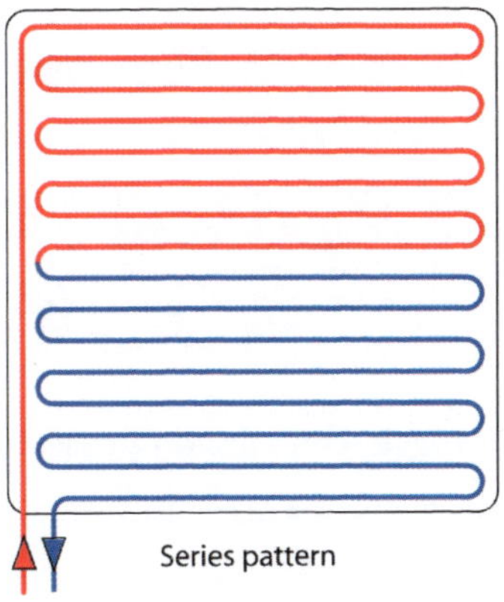

Series pattern

Also known as the 'meander' pattern, the series pattern is designed for long pipework runs that require an even distribution of heat. It is especially useful in areas of high heatloss. The flow part of the heating circuit must be directed where the heat requirement is greatest, such as those areas containing windows and areas that are expected to be cold. The heating pipework can be concentrated in this area before returning in a uniform manner back towards the manifold.

Pipework should be installed at defined centres to give the required heat input into the room.

Snail pattern

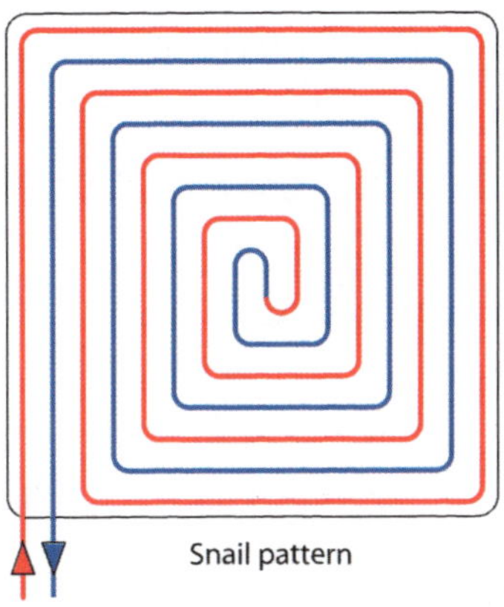

Snail pattern

Also known as the 'bifilar' pattern, the snail pattern is installed in rooms that require an even heat spread, such as those rooms containing hardwood flooring.

The pipework is installed to run in decreasing circuits to the centre of the room. The return pipework then parallels the flow pipe back to the underfloor heating manifold.

AC1.5 Explain the importance of pump positioning

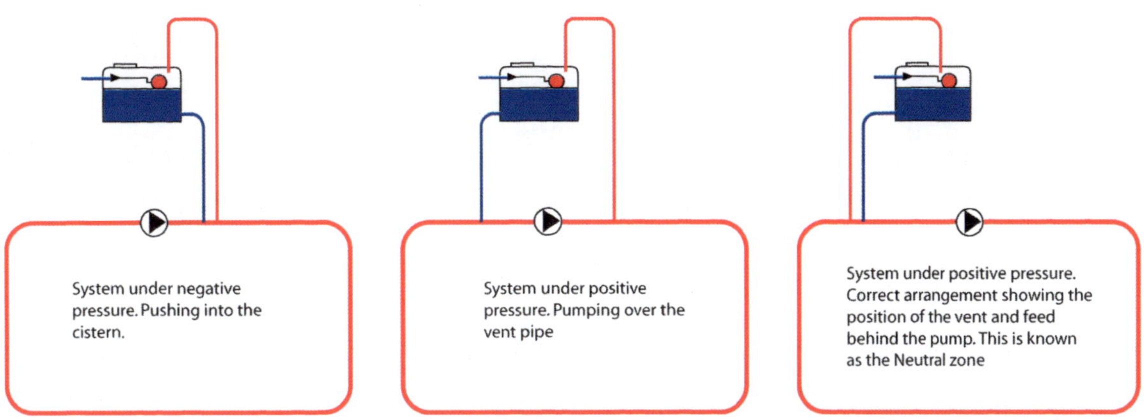

In fully pumped, open vented systems, the pump must be positioned on the flow pipe from the boiler with the cold feed and the open vent pipe installed on the suction (-) side of the pump. Take a look at the drawing above.

Left hand drawing – this shows the system under negative pressure. Activating the pump in this configuration would literally push water back up the cold feed and over the vent pipe. Circulation around the heat emitters would be almost impossible.

Middle drawing – here the heating feed and the heating vent pipe are installed either side of the pump. This would circulate water to and from the F&E cistern causing aeration of the water, which could lead to system corrosion. Circulation through the heat emitters would not occur.

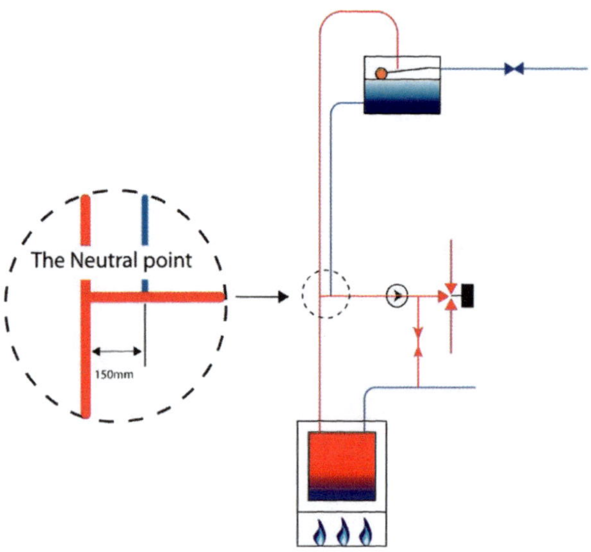

The neutral zone

Right hand drawing – in this drawing, the cold feed and the open vent are positioned correctly on the suction side of the pump. Because both cold feed and open vent are positioned in the same place, the suction pressure on the cold feed and the open vent is equal. This means that a state of equilibrium occurs and the connections create a neutral point (see the drawing left). For the neutral point to remain in balance, the distance between the open vent and the cold feed should be no more than 150mm.

An easy way to remember the order is:

Vent, **C**old feed, **P**ump = VCP = **V**ery **C**orrect **P**osition.

AC1.6 Identify operating principles for system control

Time control

Time control via programmers and time clocks was discussed on page 218 of the book.

Weather compensation control

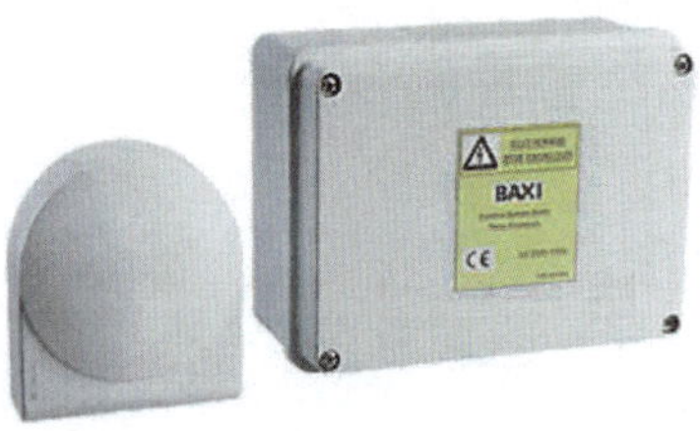

Weather compensation kit

Weather compensation control optimises the temperature of the flow pipe from the boiler, taking advantage of warmer spells of weather to reduce the water temperature to the radiators. It senses the temperature via an outside sensor (normally fitted on a north facing wall). When the weather becomes colder, the control is reversed and the temperature to the radiators is increased.

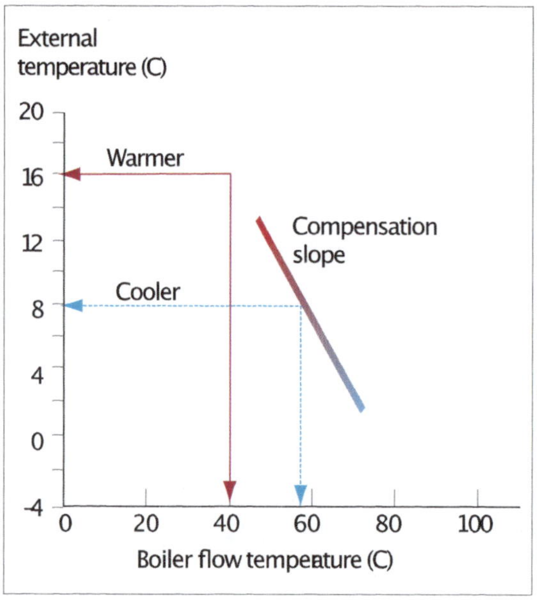

Weather compensation

Delayed start control

Delayed start control delays the time that the heating system is due to switch on based upon the time it takes to heat the property up to the required temperature. Delayed start analyses the current dwelling temperature and adjusts the startup time to compensate for heat already present in the property. The benefits of this system are the savings in fuel made during milder winter weather.

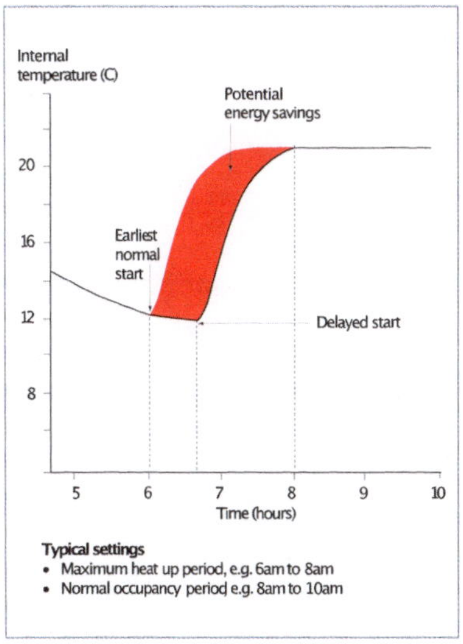

Delayed/optimum start

Optimum start control

With optimum start, the controller calculates the required heat up time from data inputted by the end user. The optimum start control requires a temperature setting and the times of occupancy to be set by the occupier and the controller then calculates the necessary required heat up time. The system is based around thermal comfort rather than energy saving features.

Home Automation and the Internet of Things

Home Automation Systems – Often called 'smart home', home automation systems will control, not only the home central heating system, but also air conditioning, lighting, household appliances and entertainment systems, through the home Wi-Fi system. Many systems also include automated intruder and fire alarm systems. When connected to the home internet, it forms an important part of the Internet of Things.

Most systems connect through a central hub, often called a 'gateway' and include a user interface, with wall mounted terminals, tablet connection and desk top computers to monitor the system. Many also include control from a mobile cell phone, via an app.

In larger buildings, these are known as Building Management Systems (BMEs).

The Internet of Things – This is an extension of home automation whereby everyday appliances connect to the internet to relay information such as diagnostics, analytics, software updates and user preferences. Modern voice activated systems of personal assistant such as 'Alexa', 'Siri' and 'Cortana' also fall into this category.

Multiple Boiler Sequence controls

In installations with multiple boilers, MBSC control the operation and sequence of boiler operation and use. Most modern boilers are very efficient at low heating loads. By controlling the output of multiple boilers, and using lower outputs, saves energy and makes the system more efficient. By running all boilers at lower heat input/output rates, the risk of the lead boiler firing at full capacity is eliminated.

AC1.7 Define zoning and control requirements of central heating systems in accordance with statutory legislation

What is 'zoning'?

Zoning a central heating system means dividing the heating system into areas that require separate time and temperature control with regards to heating that space or zone. It prevents the unnecessary heating of spaces that are not occupied all of the time. It is a mandatory requirement of Building Regulations Approved Document L1A/B, which states:

> *In order for a system to be specified with time and temperature zone control, it must be possible to programme the heating times of at least two* ***heating zones***[1] *(see note at bottom of the page) independently, as well as having independent temperature controls. These two heating zones must be space heating zones. For single storey dwellings in which the living area is greater than 70% of the total floor area, sub-zoning of temperature controls is not appropriate, and the recipe will default to a programmer and room thermostat.*

All central heating systems installed within domestic dwellings must be separated into at least two heating zones. If stored hot water is installed in the dwelling, then the minimum requirement would actually be 3 independently controlled zones. These are usually:

- Zone 1 – living space
- Zone 2 – sleeping space
- Zone 3 – hot water

Each of these zones would be independent in terms of temperature control and operating times.

Zoning in practice

The main aim of zoning is to avoid over heating the areas of the property that require less heat to maintain a comfortable living environment or because the temperature required is less than in other areas of the property, such as bedrooms. Although the minimum requirement is for two space heating zones, there are also benefits to be had from adding additional zones in certain spaces within the property. The choice of control is, to a large extent, dictated by the activity in those zones. However, there are any number of controls that, either used individually or collectively, achieve good, energy saving system control, including:

⇒ Installing every zone with individual time and temperature control

⇒ Multi-channel programming to allow timed activation of individual rooms or multiple zones set from a single point. This is often a more sensible approach than many programmers fitted at locations all over the property.

[1] The reference to 'at least two heating zones' simply means that the hot water zone is independent of the heating zones and MUST NOT be taken as 1 of the 2 zones. This would mean that the minimum requirement where stored hot water is installed would be 3 zones – 2 heating and 1 hot water.

⇒ The installation of digital programmable thermostatic radiator valves in rooms where solar gain adds free radiated heat into the room. These are usually south, south-east and south-west facing rooms. TRVs are fast acting and react quickly when the room heats up.

The benefits of separate time and temperature control of zones

Separate temperature control – this provides a more comfortable living environment because the building can be maintained at different temperatures in different areas without relying on a single room/room thermostat to dictate the temperature of the whole house. Lower temperatures can be set for those unoccupied areas, allowing the property to take advantage of the solar gains available mentioned earlier. Solar gain can be quite substantial, even during the winter months and this can lead to significant energy and fuel cost savings. Outside sensors linked to weather compensation and delayed/optimum start applications reduce energy costs further still.

Separate time control – this allows the zones to be activated at different times of the day. The benefits here are that, not only can the heat be focussed in those rooms that are occupied through the day, but that separate zone timing saves energy by allowing the zones that are not occupied during the day to activate early morning and evening only. This reduces energy usage whilst maintaining a comfortable living environment.

AC1.8 Describe insulation requirements and system frost protection

Frost protection and insulation is covered in detail in **Unit 5 – Cold water systems** on page 85.

AC1.9 Identify the positioning and fixing of pipework within the building fabric

This subject is covered in detail in **Unit 5 – Cold water systems** on page 87.

AC1.10 Explain expansion and contraction in central heating systems and negative effects

This subject is covered in detail in **Unit 6 – Hot water systems** on page 144.

AC1.11 Identify sources of information required when undertaking work on central heating systems

The British Standards

- **BS EN 12828: 2012+A1:2014:** Heating systems in buildings. Design for water-based heating systems.
- **BS EN 12831-1:2017**: Heating systems in buildings. Method for calculation of the design heat load. Space heating load, Module M3-3
- **BS EN 12831-3:2017**: Heating systems in buildings. Method for calculation of the design heat load. Domestic hot water systems heat load and characterisation of needs, Module M8-2, M8-3
- **BS EN 14336:2004**: Heating systems in buildings. Installation and commissioning of water-based heating systems
- **BS EN 1264-1:2011**: Water based surface embedded heating and cooling systems. Definitions and symbols
- **BS EN 1264-2:2008+A1:2012**: Water based surface embedded heating and cooling systems. Floor heating. Prove methods for the determination of the thermal output using calculation and test methods.
- **BS EN 1264-3:2009**: Water based surface embedded heating and cooling systems. Dimensioning.
- **BS EN 1264-4:2009**: Water based surface embedded heating and cooling systems. Installation.
- **BS EN 442-1 :2014**: Radiators and convectors. Technical specifications and requirements
- **BS EN 442-2 :2014**: Radiators and convectors. Test methods and rating

The Regulations

- The Building Regulations:
    - **Approved Document L:2010: 2013 edition with 2016 amendments:** Conservation of fuel and power.
        - **L1A: 2010: 2013 edition with 2016 amendments:** Conservation of fuel and power in new dwellings.
        - **L1B: 2010: 2013 edition with 2016 amendments**: Conservation of fuel and power in existing dwellings.
    - **Approved Document F: With 2013 amendments:** Ventilation
    - **Approved Document J: With 2013 amendments**: Combustion appliances and fuel storage systems.
- The Water Supply (water fittings) Regulations 1999
- The Gas (Installation and Use) Regulations
- The IET (18th edition) (BS 7671:2018) Requirements for Electrical Installations. IET Wiring Regulations

The Recommendations

- **The Domestic Building Services Compliance Guide 2013 edition** – can be down loaded free from: https://assets.publishing.service.gov.uk/government/uploads/system/uploads/attachment_data/file/697525/DBSCG_secure.pdf This document offers advice of how to comply with Approved Document L1A/B as well as hot water systems, mechanical ventilation, comfort cooling, fixed internal and external lighting, and renewable energy systems
- **The Central Heating System Specifications (CHeSS) 2008** – practical advice for compliance with good and best practice for the design and installation of domestic central heating systems.
- **HVDH Domestic Heating Design Guide 2017** - The Domestic heating guide was produced to help heating engineers to specify and design wet domestic central heating systems. It provides a simple means to design and understand central heating systems.

The Manufacturer's instructions

The manufacturer's instructions are probably the most important document to read and consult when installing, servicing and maintaining appliances, components and equipment, because they instruct us on the best methods to use whilst keeping to current legislation and regulations. In some cases, it may appear that these instructions contradict the regulations. This occurs because regulations and codes of practice are only updated periodically, whereas manufacturers are constantly reviewing and updating their literature in line with modifications and current good practice. Where a conflict exists, manufacturer's literature should always be followed. If not:

- The warranty of the equipment may be void.
- Regulations may be inadvertently broken
- The installation may be dangerous.

AC1.12 Describe procedures for filling and venting system types

The filling and venting requirements for open vented and sealed heating systems are very different. It is important that we know how to fill these systems correctly and how to avoid common filling problems, such as air locks, which can be difficult to clear.

Trapped air is one of the biggest problems in central heating systems. It can cause pump cavitation, leading to pump failure, excessive system noise and system corrosion.

Filling Open Vented Systems

a. Close all radiator valves and radiator air release valves before filling begins.
b. Check the F&E cistern. Make sure that the FOV shut off level has been provisionally set and that all joints are tight.
c. Replace the pump with a short piece of tubing. This will prevent debris from entering the pump.
d. Open manually all motorised valves and zone valves.
e. Check that all radiator valve unions are tight.

f. Turn on the service valve to the F&E cistern and allow the system to fill.
g. Filling the system should begin at the furthest radiator on the downstairs circuit. Open the radiator valves and then the air release valve until water is detected. Work backwards towards the boiler filling each downstairs radiator in turn. Filling the downstairs circuit first helps to eliminate pockets of air, which can prevent efficient circulation of the heating water. When the downstairs circuit is full, proceed to the upstairs circuit and follow the same procedure, starting with the furthest upstairs radiator.
h. Once the system is full, allow it to stand for a few minutes.
i. Visually check for leaks across the system. Visit each radiator and all exposed pipework.
j. Check the water level in the F&E cistern and adjust if necessary.
k. Drain down the entire system at each downstairs radiator. This is called the cold flush and will remove any flux residues, steel wool and swarf that entered the system during installation.
l. Refit the pump and turn on the pump valves.
m. Refill the system.

Filling sealed systems

a. Close all radiator valves and radiator air release valves before filling begins.
b. Connect the temporary filling loop to the system pipework.
c. Check the pressure in the expansion vessel at the Schrader valve to ensure it matches the manufacturer's data plate.
d. Replace the pump with a short piece of tubing. This will prevent debris from entering the pump.
e. Open manually all motorised valves and zone valves.
f. Turn on the service valve to the Filling loop and allow the system to fill to 1bar pressure and then turn off the service valve.
g. Filling the system should begin at the furthest radiator on the downstairs circuit. Open the radiator valves and then the air release valve until water is detected. Work backwards towards the boiler filling each downstairs radiator in turn. Filling the downstairs circuit first helps to eliminate pockets of air, which can prevent efficient circulation of the heating water. When the downstairs circuit is full, proceed to the upstairs circuit and follow the same procedure, starting with the furthest upstairs radiator. As filling progesses, the pressure in the system will fall towards zero pressure. When this happens, go back to the filling loop each time the pressure drops and fill back up to 1 bar pressure.
h. Visually check for leaks across the system. Visit each radiator and all exposed pipework.
i. Check the system water pressure at the pressure gauge.
j. Open the pressure relief valve to ensure that water discharges from the valve.
k. Drain down the entire system at each downstairs radiator. This is called the cold flush and will remove any flux residues, steel wool and swarf that entered the system during installation.
l. Refit the pump and turn on the pump valves.
m. Refill the system.

AC1.13 State the operating principles of heat-producing appliances

Traditional boilers

Also known as 'heat only boilers', traditional boilers do not provide instant hot water in the way that combination boilers do. They also require a hot water storage cylinder if hot water is also to be provided by the system. These boilers can be filled via a feed and expansion cistern positioned in the roof space in an open vented type system or filled by a filling loop. However, it must be remembered that if a filling loop is to be used, the system will also require an expansion vessel and an expansion relief valve. These are normally used as replacement boilers for older heating systems.

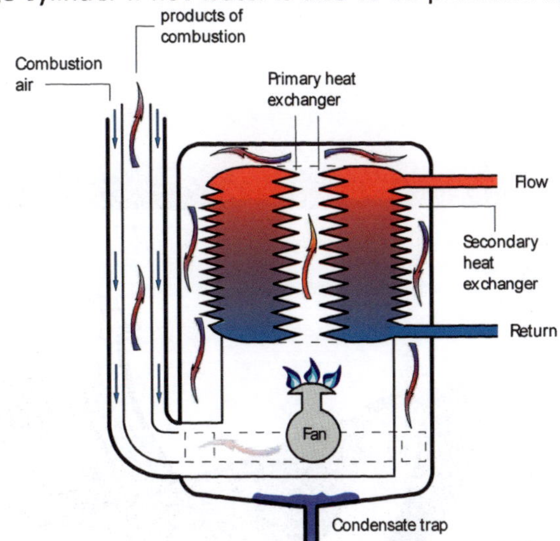

Condensing boiler

Condensing boilers

Condensing boilers have 2 heat exchangers. The first or primary heat exchanger heats the water when the boiler is fired. This produces hot exhaust gases which are made up of CO_2 and water vapour. These are removed via the flue to outside. However, during the flue exhaust gas removal, the exhaust gas, which has a temperature around 230°C, is directed over a second heat exchanger to remove even more heat from these hot, exhaust gases. During this process, the exhaust gas cools to around 90°C and the water vapour condenses back into water or condensate. The condensate is removed from the boiler by the condensate pipe to a safe position or drain. Condensing boilers are around 93% efficient with only 7% of the heat produced wasted through the remaining exhaust flue gases.

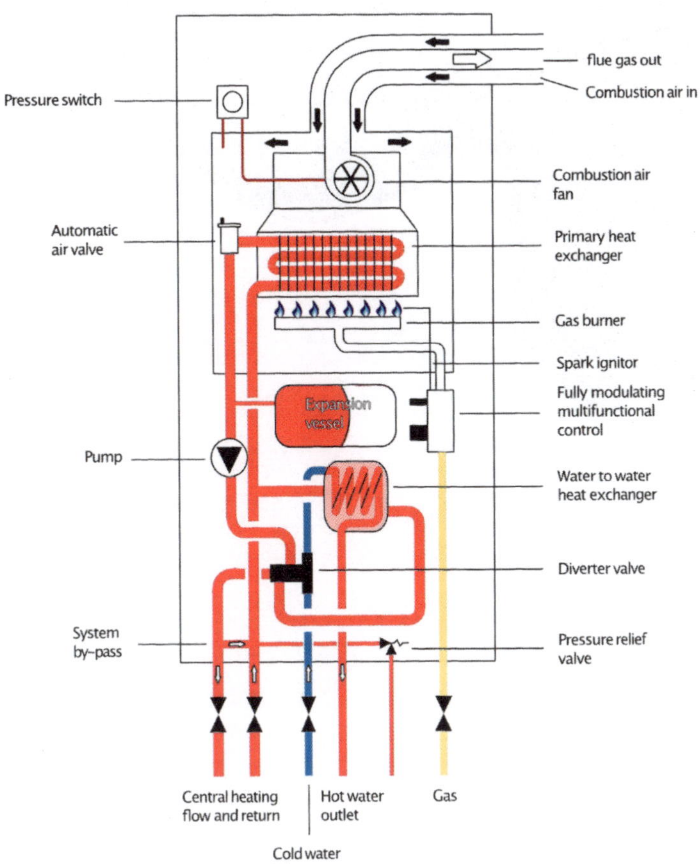

Combination boiler

Combination boilers

Combination or 'Combi' boilers as they are often referred to as, not only provide central heating, but also provide instantaneous hot water via a plate heat exchanger situated in the boiler casing.

The plate heat exchanger plays a key role in the operation of the boiler. It is linked to a diverter valve that diverts the water either through the heating system or through the plate heat exchanger. It is this diversion of the water in the boiler through the plate heat exchanger which heats the mains cold water passing through it to around 40°C for hot water delivery.

Combination boilers contain all of the necessary controls and components for a full central heating system within the boiler casing, such as heating circulation pump, expansion vessel, and expansion relief valve.

Freestanding boilers

These are floor mounted boilers for use where the boiler required is too large to mount on a wall

Wall mounted boilers

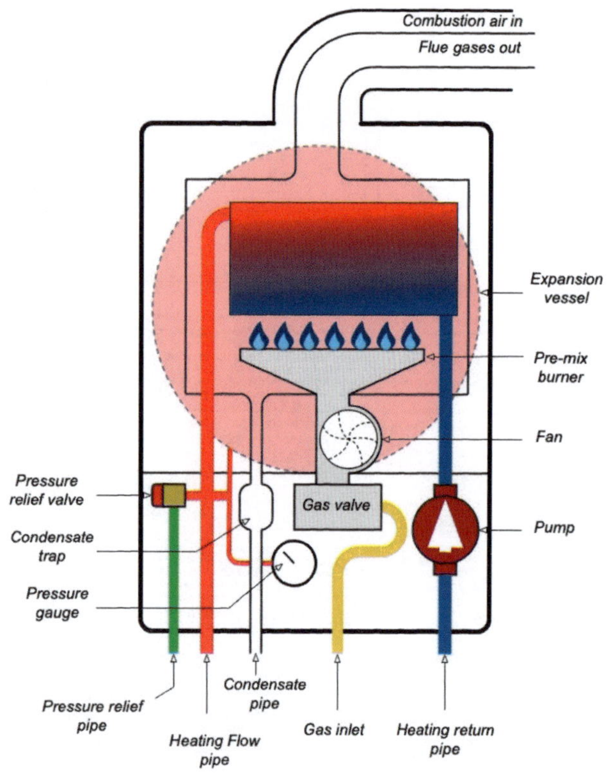

A system boiler

Most boiler types are now available as wall mounted types, including sealed system boilers, combination boilers and tradition boilers.

System boilers

System boilers are heat only boilers for use only on sealed heating systems. They do not provide instantaneous hot water. They can be installed on systems with hot water storage vessels in the same way as traditional boilers, in the form of an S plan Plus, W plan or Y plan type heating systems. System boilers contain all of the necessary controls and components for a full central heating system within the boiler casing, such as heating circulation pump, expansion vessel, and expansion relief valve.

Unit 08

Rainwater Systems – Phase 2

Unit 08 – Rainwater Systems – Phase 2

In Phase 2, there are 3 Learning Outcomes:

LO1 - Install Rainwater Systems:

AC1.1 Identify types and layout features of **rainwater systems**: **pipe (RWP)** and **gutter (completed at Phase 1)**

AC1.2 State advantages and disadvantages of rainwater systems: pipe (RWP) and gutter

AC1.3 Identify typical sizes and materials used in rainwater systems: pipe (RWP) and gutter Rainwater systems

AC1.4 Describe working principles of rainwater systems (positioning, fixing, connection and operation of components)

AC1.5 Explain expansion and contraction in rainwater systems and negative effects

AC1.6 Explain factors affecting gutter bracket selection and fixing for buildings

AC1.7 Identify sources of information required when undertaking work on rainwater systems

AC1.8 Explain working principles of rainwater recycling systems

AC1.9 Install rainwater systems

LO2 – Decommission rainwater systems

AC2.1 Explain **procedures** for decommissioning systems

AC2.2 Carry out decommissioning

LO4 – Perform a soundness test and commission rainwater systems

AC4.1 Identify information sources required to complete testing and commissioning

AC4.2 Describe a **visual inspection** of a rainwater system to confirm that it is ready to be soundness tested

AC4.3 Describe a **soundness test** to industry requirements on rainwater, gutter systems pipework and components

AC4.4 Describe **operational checks** required during commissioning

AC4.5 Identify actions that must be taken when commissioning reveals defects

AC4.6 Describe the procedure for handing over to the end user

AC4.7 Carry out soundness testing and **commissioning procedures**

Learning Outcome 1

Install Rainwater Systems

Learning Outcome 1: Install Rainwater Systems

In Phase 2, you will be investigating the following Assessment Criteria from Learning Outcome 1:

AC1.1 Identify types and layout features of **rainwater systems: pipe (RWP)** and **gutter (completed at Phase 1)**

AC1.2 State advantages and disadvantages of rainwater systems: pipe (RWP) and gutter

AC1.3 Identify typical sizes and materials used in rainwater systems: pipe (RWP) and gutter Rainwater systems

AC1.4 Describe working principles of rainwater systems (positioning, fixing, connection and operation of components)

AC1.5 Explain expansion and contraction in rainwater systems and negative effects

AC1.6 Explain factors affecting gutter bracket selection and fixing for buildings

AC1.7 Identify sources of information required when undertaking work on rainwater systems

AC1.8 Explain working principles of rainwater recycling systems

AC1.9 Install rainwater systems

AC1.2 State advantages and disadvantages of rainwater systems: pipe (RWP) and gutter

Gutter Profiles

In the last 200 years, rainwater and gutter systems have been manufactured in a variety of styles and shapes, from the simple lead lined box gutter to the more elaborate ornamental gutters or OGEE gutters, used to good effect on Victorian buildings from around 1850. Today, modern gutter profiles add much to the look of a building. Choosing the right type of gutter for the building is aesthetically important, especially when replacing old, worn gutters for new. Maintaining the 'period' feel of the building can enhance the overall look.

There are four main gutter profiles in use today. These are shown below:

Half round

This is probably the most common and widely used gutter profile used in the UK.

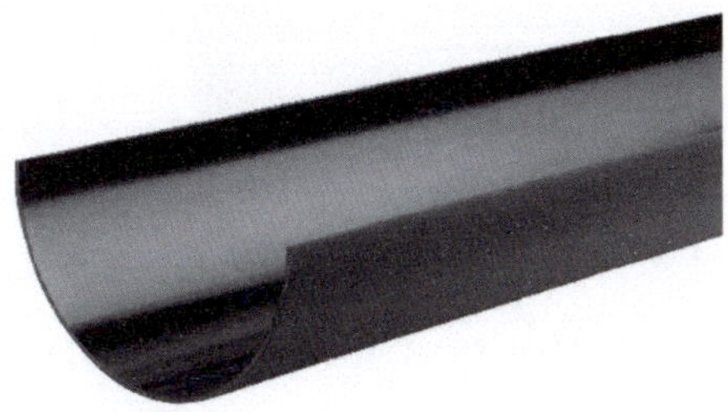

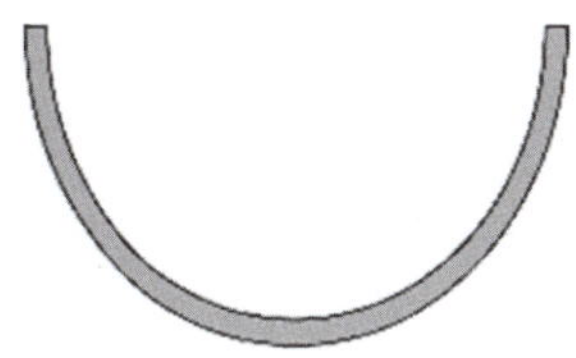

Advantages	Disadvantages
• It gives good flow when laid to the correct fall, with little sediment build up. • All manufacturers use a common size. • Fittings are interchangeable between manufacturers • Converters to other profiles and materials readily available	• Often incorrectly used on properties with large roof areas, causing the gutter to overflow • Must be installed to the correct fall

High capacity/Deep half round/storm flow

Similar in appearance to half round gutter but with a much deeper profile. It is slightly elliptical in shape and generally used where the roof area is large or steep angled. Specifically designed for high volume flow rate.

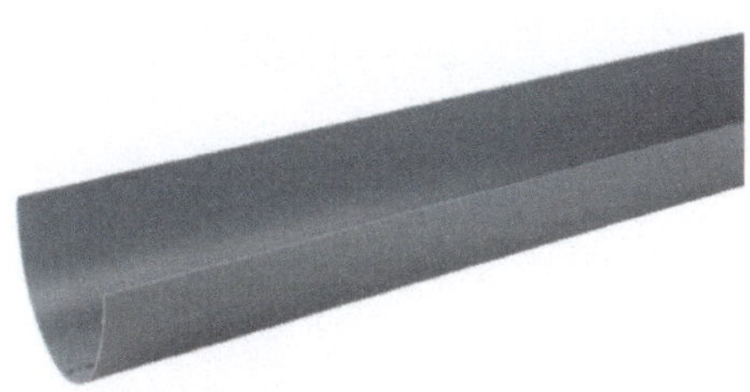

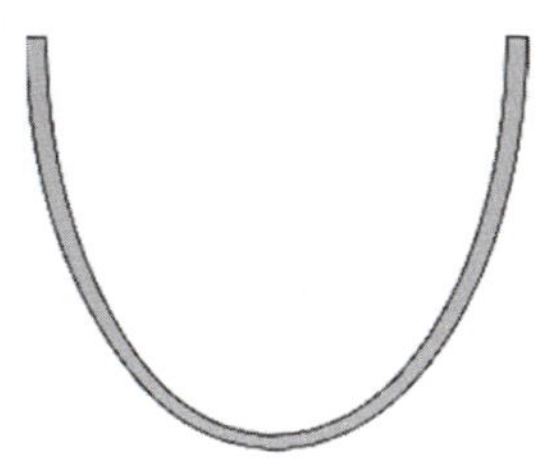

Advantages	Disadvantages
• Deeper profile means greater flow rate • Ideal for use on properties with large roof areas • More rigidity than other PVCu gutter types	• It's expensive when compared to other gutter profiles • Has a tendency to silt up due to its elliptical shape • Must be installed to the correct fall

Square Section

This type of gutter became very popular in the 1980s and 1990s and it is still used today on new build dwellings as well as replacement gutters.

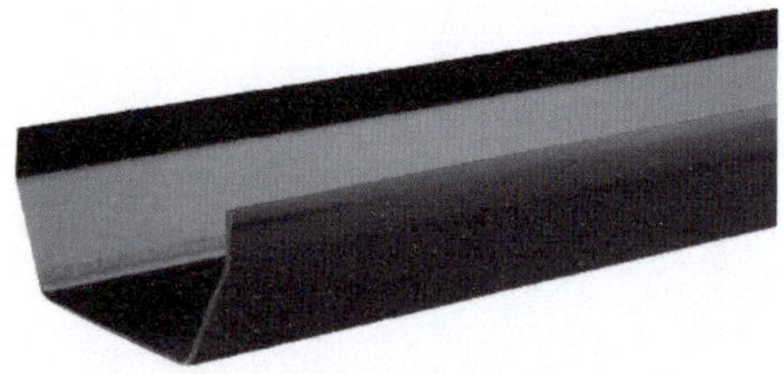

Advantages	Disadvantages
• Square section has almost twice the capacity of half round section and flow rates are acceptable when laid to the correct manufacturer's fall	• Some makes are no longer available and will not fit modern equivalent shapes • It can silt up very quickly because of its flat bottom. It requires regular cleaning. • Not all manufacturer's gutters and fittings are interchangeable. • Must be installed to the correct fall

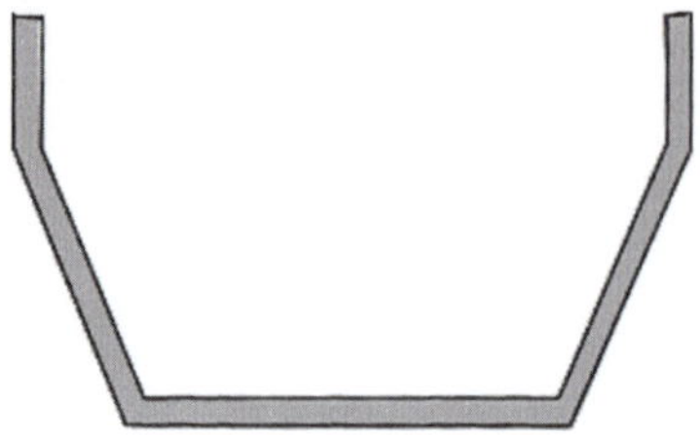

Ogee (O. G. or ornamental gutter)

A modern adaptation of a Victorian design, ogee guttering is used to very good effect on both old and new buildings alike.

Advantages	Disadvantages
• Offers a lightweight and long-lasting alternative to cast iron gutters when used on period buildings or renovations • More rigid than half-round type gutter profiles • Good capacity and flow rate	• Caution should be exercised when adding to existing gutter systems as each manufacturer tends to have their own designs and many are not interchangeable

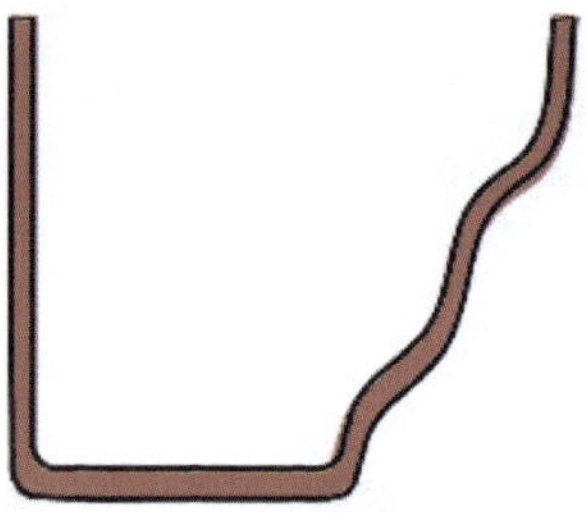

Rainwater pipe Profiles

There are 2 rainwater pipe profiles. These are:

- Round section
- Square section

Round rainwater pipes

Round rainwater pipes can be used with all forms of PVCu gutter systems. Most manufacturers use 68mm diameter pipe.

Advantages	Disadvantages
• Round pipe has a good flow rate • Fittings can be spun in any direction • Common pipe size across all manufacturers	• Not as rigid as square section pipe

Square rainwater pipes

Usually used with square section guttering systems.

Advantages	Disadvantages
• Most manufacturers use the same size (65mm square) and so are interchangeable. • Aesthetically pleasing when used with square section gutter • More rigidity than round profile pipe	• Tends to block easily • Fittings are not multi-directional. • Because it is square it has a poorer flow rate than round pipe.

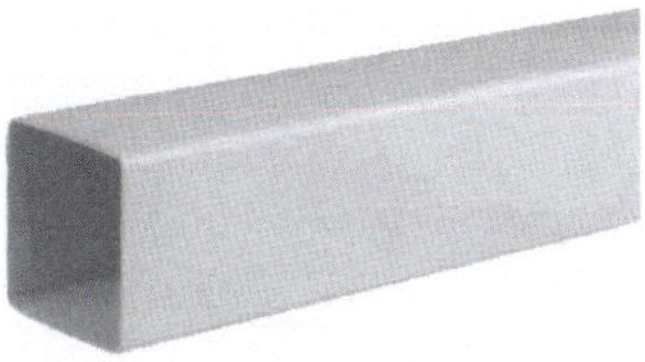

AC1.3 Identify typical sizes and materials used in rainwater systems PVCu

BS EN 607:2004 Eaves gutters and fittings made of PVCu,

BS EN 122001:2000 Plastic rainwater piping systems for above ground external use.

Most of the guttering used today is made from PVCu because it is light, flexible and relatively inexpensive. It does, however, suffer from light degradation when exposed to direct sunlight and it expands more than other guttering materials. Typical gutter size is 112mm. Rainwater pipe size is 68mm

Advantages of PVCu	Disadvantages of PVCu
• Smooth internal bore gives better flow rates • Light in weight making it easier to handle • Very little maintenance required when compared to other materials such as cast iron • Does not suffer from corrosion like other gutter and rainwater materials • Easy to joint and install • Various colours available means little or no painting • An expected lifespan of up to 50 years	• Greater co-efficient of linear thermal expansion when compared to other guttering and rainwater materials. • Tends to go brittle in very cold temperatures • Softens at fairly low temperature. • Can be affected by wood preservatives such as creosote because of the solvent content • Colour can fade in direct sunlight

Extruded Aluminium

BS 8530:2010: Traditional-style half round, beaded half round, Victorian ogee and moulded ogee **aluminium** rainwater systems.

BS 8530 specifies requirements for cast and **extruded aluminium** rainwater systems

Extruded seamless aluminium guttering systems are not readily available and are usually installed companies specialising in extruded guttering systems. It is very lightweight and corrosion resistant. Most gutters of this type are formed on site by a special machine that presses the gutter into the required profile from a roll of aluminium sheet. Because it is light and flexible, strengtheners are fitted to the gutter to give added rigidity before it is installed.

Gutter sizes	Rainwater pipe sizes
100mm x 75mm	100mm x 100mm
125mm x 100mm	75mm x 75mm
150mm x 100mm	
200mm x 150mm	

Advantages of extruded aluminium	Disadvantages of extruded aluminium
• It has strength and flexibility and is very light in weight • Long continuous lengths means fewer joints • Less leakage than other gutter systems • It can be formed into many profiles and many colours are available • Very little linear thermal expansion	• The initial cost is expensive • Not suited for terraced or semi-detached properties unless all buildings are having extruded aluminium installation

Cast Iron

BS 460:1964 Cast iron rainwater goods

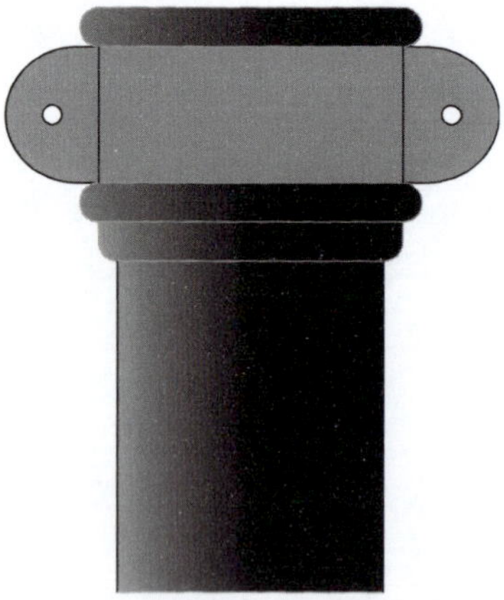

Cast iron gutters can be seen on many older properties and may still be fitted to buildings that want to retain the 'period' feel. It is still used on many of the UK's listed and historic buildings and even in local authority conservation areas.

Cast iron is strong but can be very high maintenance, requiring regular painting to prevent corrosion. Internally, cast iron has a fairly rough surface and this often means regular cleaning out of blockages and silt build up.

Advantages of extruded aluminium	Disadvantages of extruded aluminium
• It has strength • Maintains the character of older historic buildings	• Cast iron is expensive to purchase and has greater installation time. • It is very heavy • Requires regular maintenance and painting • Jointing is very messy and work intensive • Cast iron is brittle and cracks easily if mishandled

Cast iron gutter sizes	Cast iron Rainwater pipe sizes	Cast iron gutter profiles
100mm	65mm	Half round
115mm	75mm	Ogee
125mm	125mm	
150mm	150mm	

Lead

Lead has been used for gutters and rainwater pipes for hundreds of years as a method of taking rainwater water away from buildings. Nowadays, its use is restricted to church and cathedral use and buildings of historic and architectural importance. Many church restorations employ specialist lead working companies to fabricate exact replicas of exiting lead pipework. This can be achieved by casting, lead welding or bossing lead sheet into the required shape.

AC1.4 Describe working principles of rainwater systems (positioning, fixing, connection and operation of components)

In this Assessment Criteria, we will take a brief look at some of the more common components used with eaves gutter systems. The fittings shown below are available for all PVCu gutter profiles but not necessarily for all materials. The manufacturer should be contacted should there be any special requirements with regard to a particular installation.

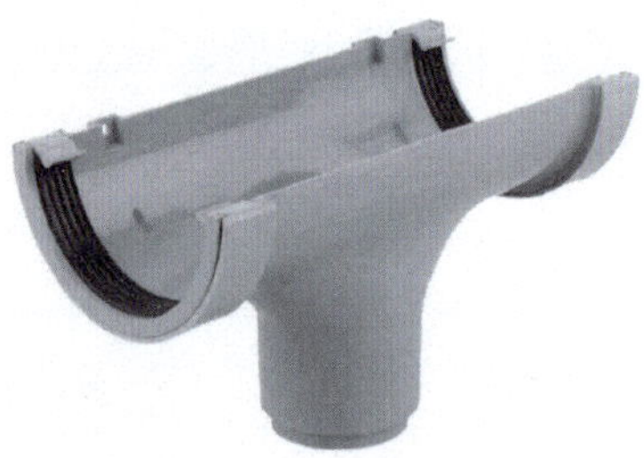

Running outlet – this is used to convey rainwater from the gutter system to the drain via the rainwater pipe

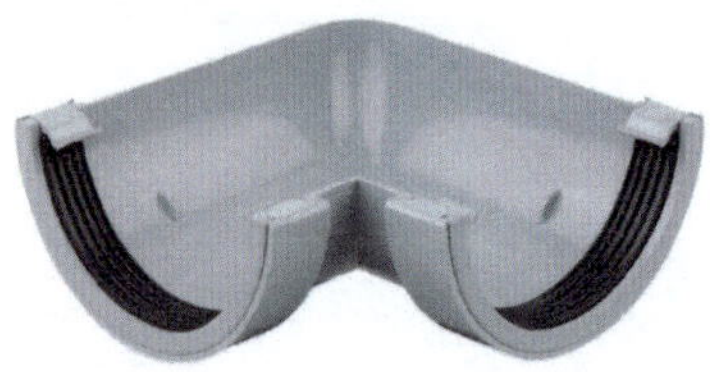

Gutter angles – changes direction of the gutter to suit the roof and gutter designs. There are many angles available including:

- 90°
- 135°
- 110°

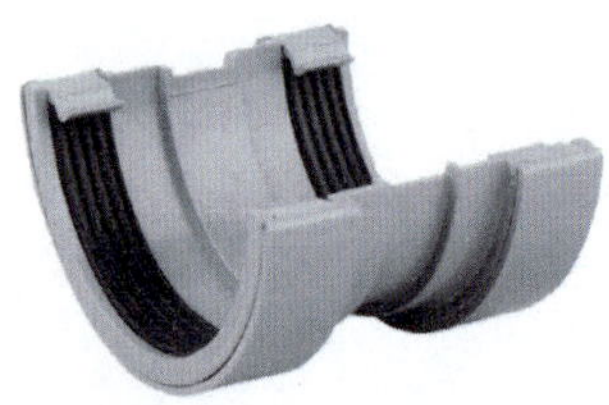

Gutter union – used to join lengths of gutter together. They also act as a gutter bracket as they are fixed to the fascia board.

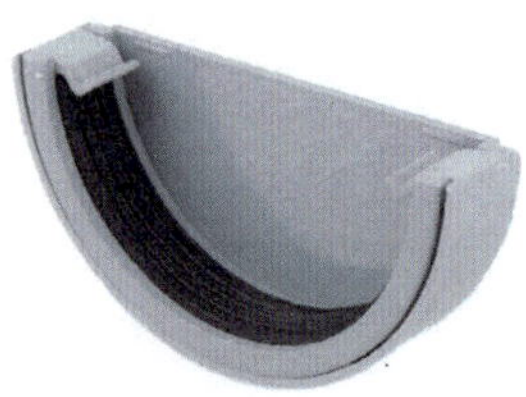

External gutter cap end – these are used to stop the rainwater from spilling from the gutter system at the end of a gutter run.

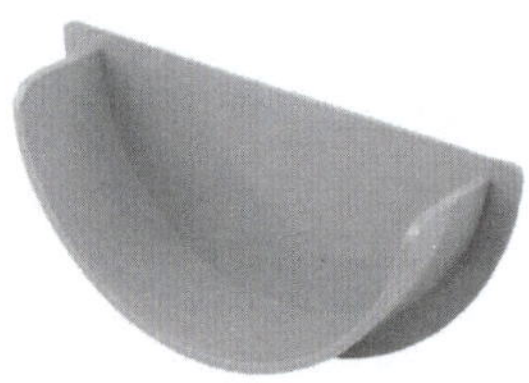

Internal gutter fitting cap end – these are used to cap off fittings. They are ideal for converting running outlets to stop-end outlets.

Specialist gutter adapters – these are used when converting from one gutter profile to another. The image shows a converter from half round PVCu to cast iron OG gutters.

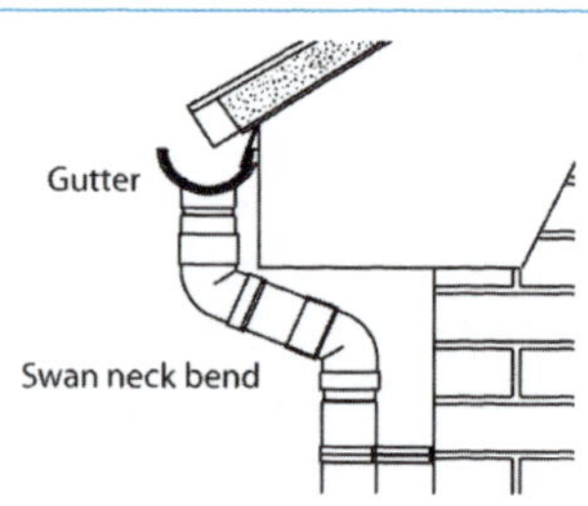

Offsets – often referred to as the 'swan neck', this is used to connect the running outlet to the rainwater pipe at eaves height. They are usually made from two 135° or 112.5° bends.

Angles – can be used to make offsets. Generally used for changes of direction of rainwater pipes. There several angles available, including 90°, 135° and 112.5°.

Branches – used for connecting two rainwater pipes into one.

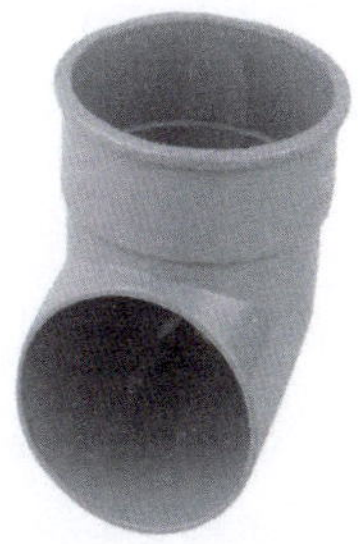

Shoes – these are fitted at the base of Rainwater pipes where the water is discharging into a gulley.

Hopper heads – these are often used to collect rainwater from flat roofs or where two rainwater pipes converge at the same point. They are connected to a rainwater pipe

Specialist connectors – these are used to connect the rainwater pipe to the drainage system. There several different types available.

Syphonic roof drainage systems for flat roofs

Conventional gravity rainwater systems rely on the rainwater spreading across the roof surface to be collected in gutters where the water runs to the outlet at the lowest point before flowing down through the rainwater pipe to the drainage system. As the water flows down the rainwater pipe, air is drawn in creating a vortex as the water spirals down the pipe. This significantly reduces the capacity of the rainwater pipe making the system less efficient. In addition, leaves and debris from the roof are pulled down the pipe, which then block the rainwater pipes. This often leads to significant damage to the property. On a flat roof, the problem is compounded by lying water, which often causes material damage to the roof and its substrate.

By contrast, a syphonic roof drainage system contains an anti-vortex baffle plate which prevents air and debris from entering the rainwater pipe, allowing the pipe to fill completely with water. The roof water is then sucked from the roof at high speed using syphonic action. The gravitational pull of the water and the fact that the pipe contains no air, creates a vacuum. The system is fast and efficient and creates less problems with lying water or blockages. The advantages of a syphonic roof drainage system are:

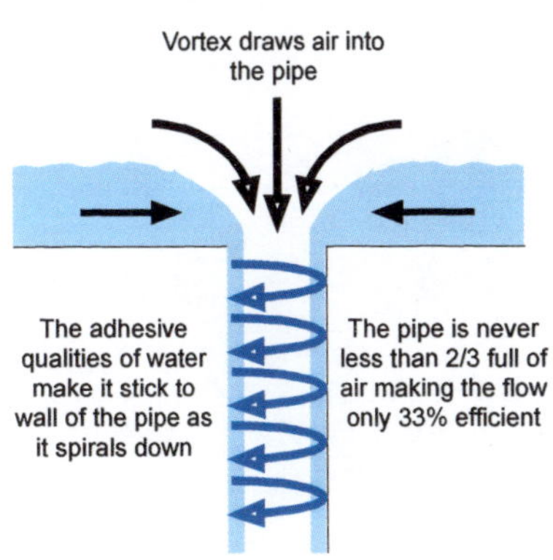

Gravity Rainwater System

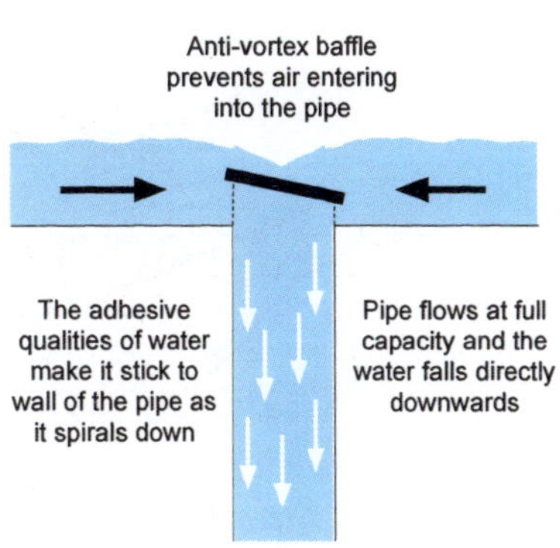

Syphonic Rainwater System

- Fewer outlets are required
- Pipe can be run level
- The lack of air in the pipe maximises flow rates and gives greater efficiency
- No gradient on the pipe is required as the system uses the height of the building and the pull of gravity

AC1.5 Explain expansion and contraction in rainwater systems and negative effects

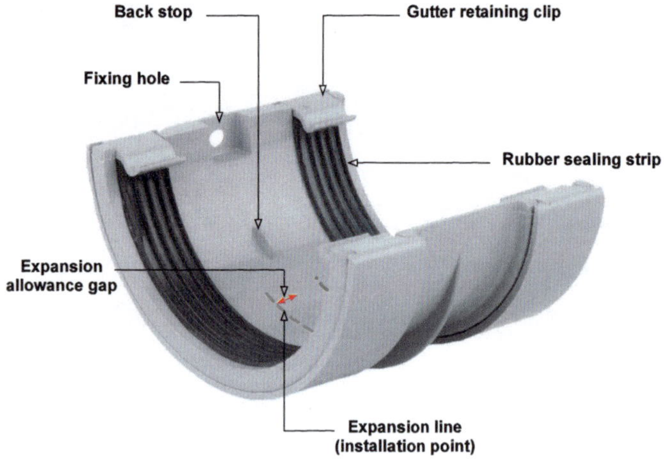

The thermal expansion mark on gutter fittings

PVCu has a large expansion rate when compared to other gutter materials. Gutters and rainwater pipes expand by 0.06mm/m/°C. This can lead to the gutter movement and joint leakage if the expansion is not accommodated for.

Gutter fittings have a thermal expansion mark on the joint. This is the point where the gutter is positioned. It allows for around 10mm of expansion up to the gutter stop, preventing joint leakage.

AC1.6 Explain factors affecting gutter bracket selection and fixing for buildings

New systems – on new buildings the gutters and rainwater pipes can be installed from a scaffold. Facia boards should have been previously painted and the roof finished before installation begins.

Replacing existing systems – Before installation of gutter and rainwater pipes can commence, the existing system must be de-commissioned and removed. This must be done with care and in such a way that the customer's building structure and possessions are not damaged.

Gutter brackets – There are several types of gutter bracket, depending on what the gutter is being fixed to:

- **Fascia brackets –** these should be installed at every 800mm centres for PVCu gutter systems and 150mm before gutter angle (internal and external types). Facia brackets should be screwed to the fascia with a 40mm x 10 zinc plated screw. Fascia brackets for cast iron gutters are also available.

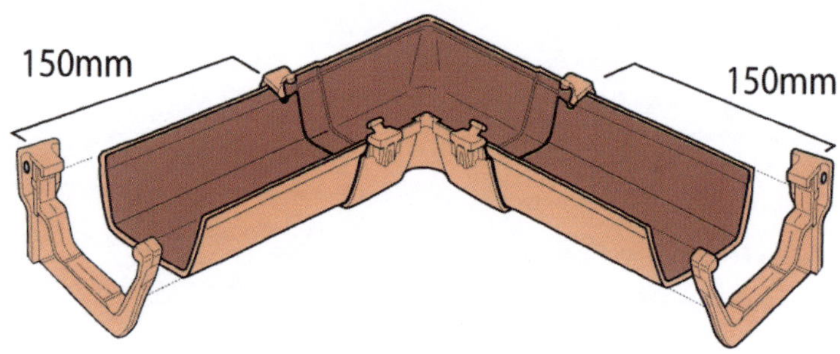

- **Rafter brackets –** since most rafters are installed at 600mm centres, these should be fixed at every rafter for PVCu gutter systems. For cast iron gutters, there should be at least 2 brackets per 1.8m length of gutter.

- **drive-in types (rise and fall) brackets** – rise and fall brackets should be positioned to suit the material that the gutter is made from. For PVCu gutters, they should be positioned in line with the manufacturer's installation instructions. For cast iron gutter, at least two brackets per 1.8m length of gutter.

Different buildings require different gutter fixing and bracketing methods. Below is a selection of gutter brackets and clips and their uses on buildings:

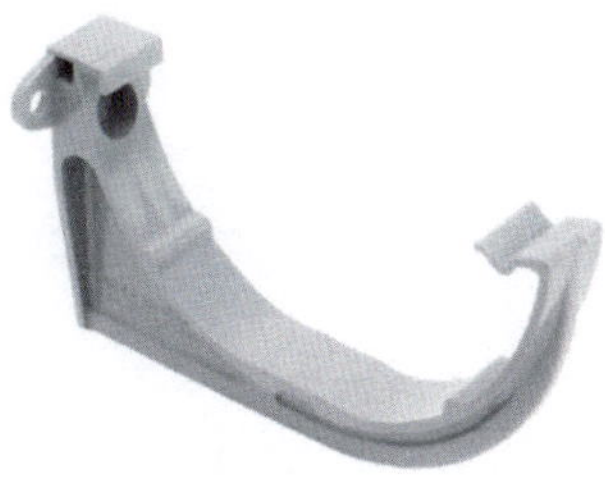

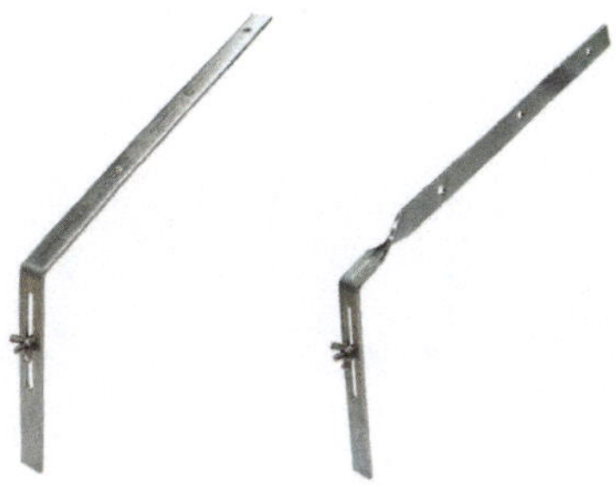

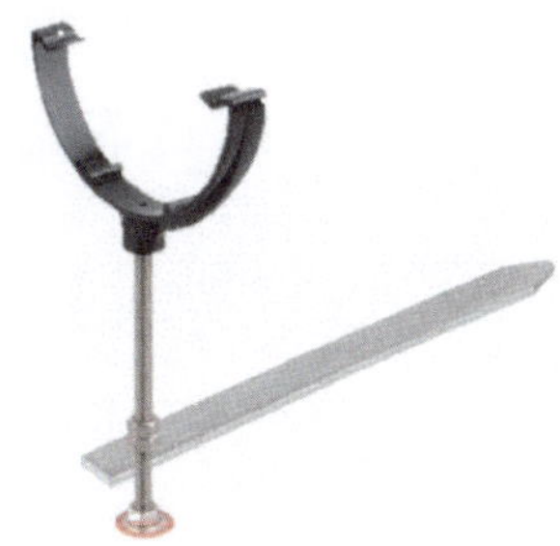

Fascia brackets – these fix directly to the facia board of a building. The gutter is simply clipped in place.

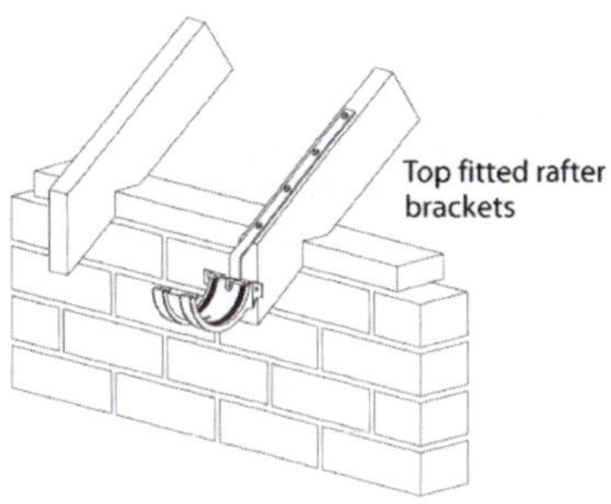

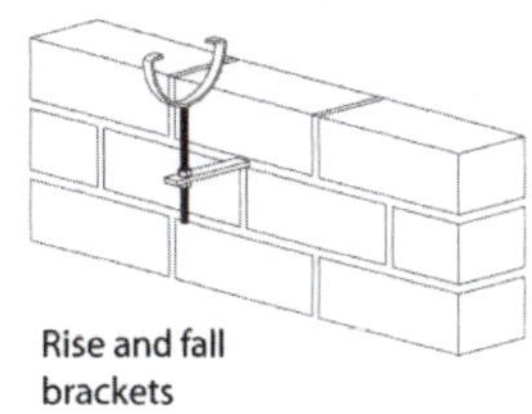

Rafter irons – also called rafter brackets, these are used where the rafters are exposed without the use of facia or soffit boards. They are made from zinc coated steel and screw directly to the top or side of the rafter.

Rise and fall irons – these are used where there are no exposed rafters or facia board. They are installed into the brickwork joints of the building and adjusted to give the correct fall by raising or lowering the gutter bracket by the 10mm screw threaded rod. Mainly used on older and listed/historic buildings.

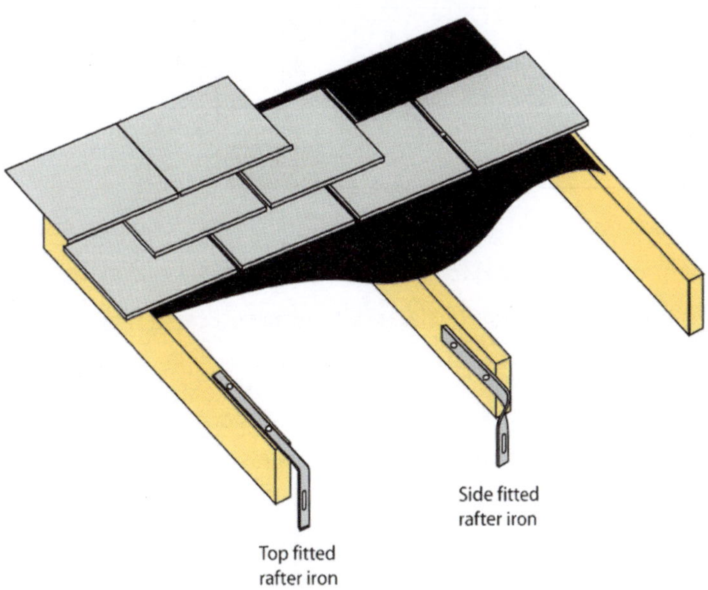

How to use rafter irons

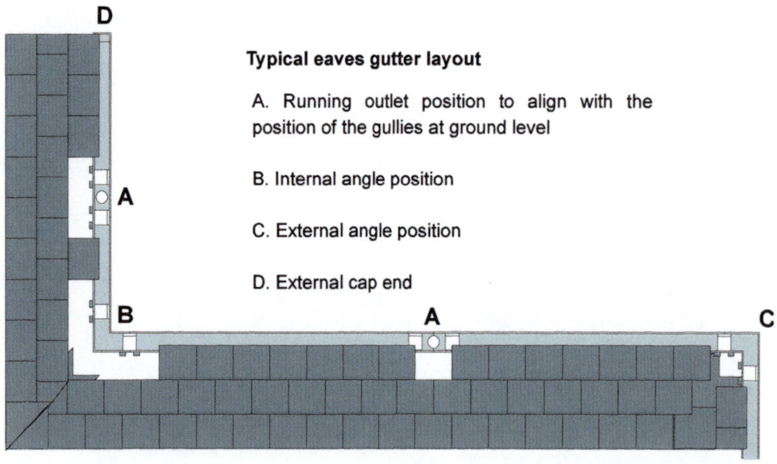

a. **Running outlets** – these should be positioned to align with the gullies at ground level.

b. **Gutter angles** – Positioned at every change of direction.

c. **Gutter unions** – positioned as required by the gutter system design.

d. **Stop ends** – positioned at the end of a gutter run to prevent water flowing out and causing a nuisance.

e. **Specialist unions between different gutter materials** – these are used when converting from Cast iron to PVCu (or PVCu to Cast iron). They are positioned as dictated by the installation.

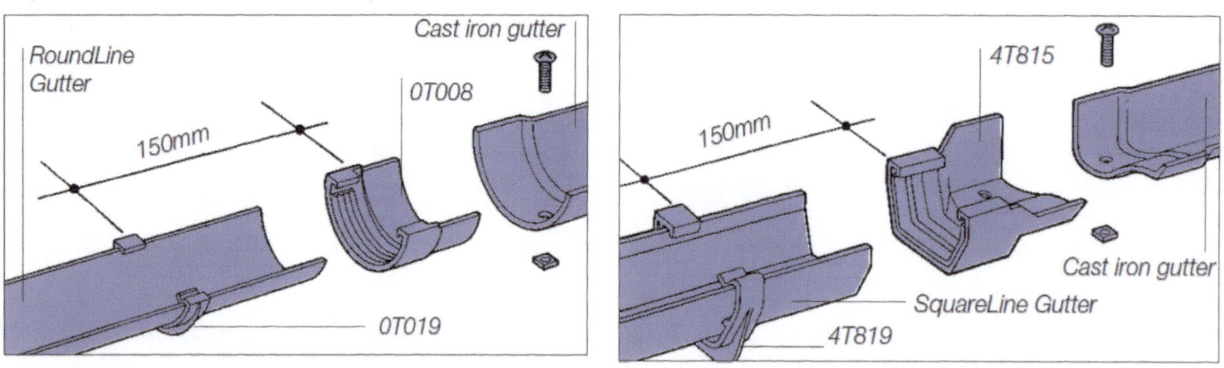

The use of cast iron to PVCu adapters

AC1.7 Identify sources of information required when undertaking work on rainwater systems

The installation of gutters and rainwater systems are subject to various documents and procedures to ensure that the systems we install will not be detrimental to the overall building fabric:

a. **Building Regulations Approved Document H3: 2015 – Rainwater drainage:** Document H3 is directly linked to **BSEN12056-3:2000** and contains information for both the design and installation of rainwater systems. It states that adequate provision must be made for rainwater to be safely carried away from buildings.

b. **BSEN12056-3:2000** – this document contains information regarding the calculation and design of rainwater systems. It covers subjects such as rainfall intensity, the provision of outlets, run off design and layout. Linked to **Document H3** of the Building Regulations.

c. **Manufacturer technical instructions** – every manufacturer of gutter systems and components designs and manufactures their components to comply with the Regulations and British Standards. They also advise on the best way to design and install the systems so that those requirements are adhered to. Manufacturer's technical instructions should be consulted and adhered to.

AC1.8 Explain working principles of rainwater recycling systems

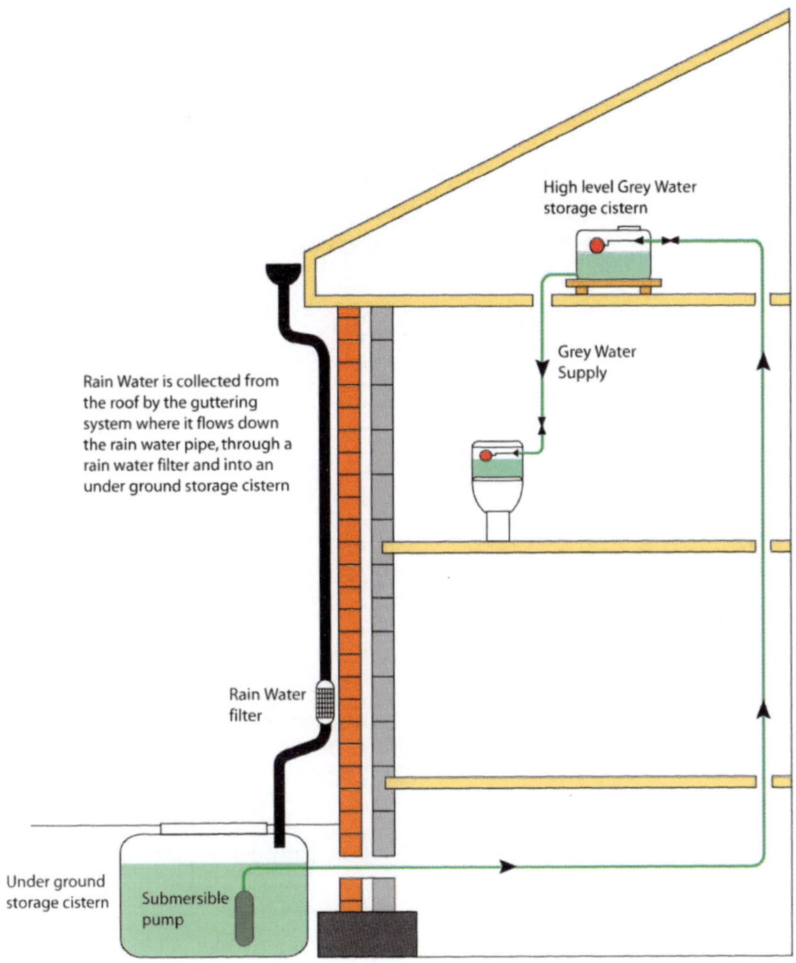

Rainwater recycling systems, also known as 'Rainwater Harvesting', is a system that collects rainwater from a building's roof and uses it in the home to flush WCs, water gardens etc. Since the water has not undergone water treatment, it MUST NOT be used for drinking or any form of human consumption.

The water runs off the roof via the guttering and rainwater pipes and is collected in tanks buried in the ground, where it is filtered to remove any debris. It is then partially sterilised to prevent smells and other bacteria forming, before being pumped to a cistern in the roof space of the building. From here, the water flows by gravity to the WC cistern.

IMPORTANT NOTE: It must be remembered that, because this water has not been treated, it MUST NOT be allowed to come into contact with any appliance, tap or other outlet that is supplied with water from the water undertaker's mains supply. It is an offence under the Water Supply (water fittings) Regulations 1999 to cross-connect any form of rainwater harvesting or grey water recycling system with fluid category 1 water. The water in a rainwater recycling system MUST be treated as fluid category 5. Because of this requirement, the system of pipework both internally and externally must be totally separate from the pipework supplying water for human consumption.

AC1.9 Install rainwater systems

Before gutters are installed, it Is a good idea to conduct a pre-installation survey:

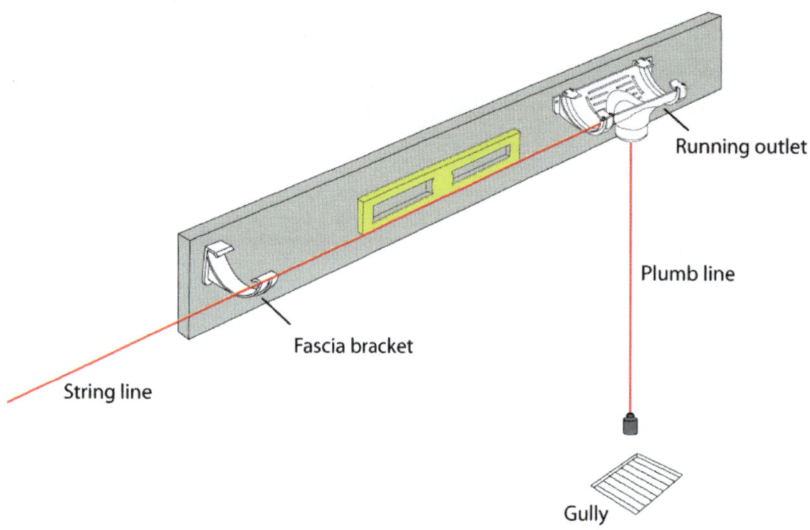

Setting the fall

New installations/new build properties:

- Check that the gutter type and profile chosen will be adequate for the size of the roof and expected rainwater run-off.
- Check that the fascia board is installed, and the gullies are in the correct position.
- Check that the under felt is in place.

Existing installations:

- Establish the type of drainage system being used and whether it is suitable for the rainwater pipework.

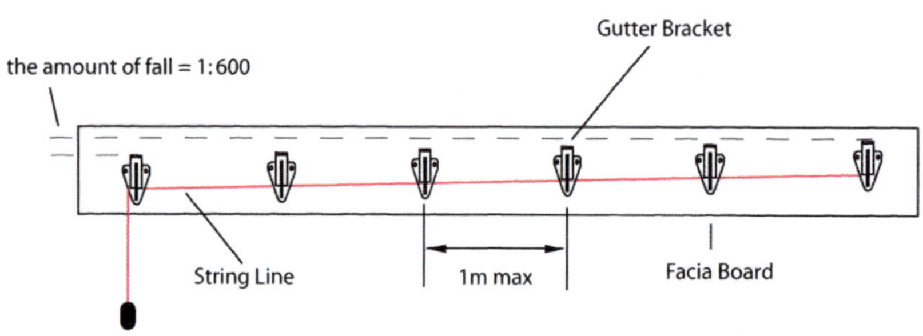

Installing the fascia brackets

- If any soakaways are being used, these should be tested to ensure they are fit for purpose and not backing up or flooding.
- Check with the customer to see whether the existing gutter was functioning correctly and where there may be problem spots on the system.
- Check with the customer to ensure the correct colour and profile are being installed.
- Check that the gutter type and profile chosen will be adequate for the size of the roof and expected rainwater run-off.
- Remove the existing gutters and check the fascia board for any signs of damage or rot.
- Check to ensure that the building structure is sound where the rainwater pipes are being positioned.

Installing gutters

- Gutters should be laid to a fall of 1:600. Measure the length where the gutter is to be installed and calculate the required fall.

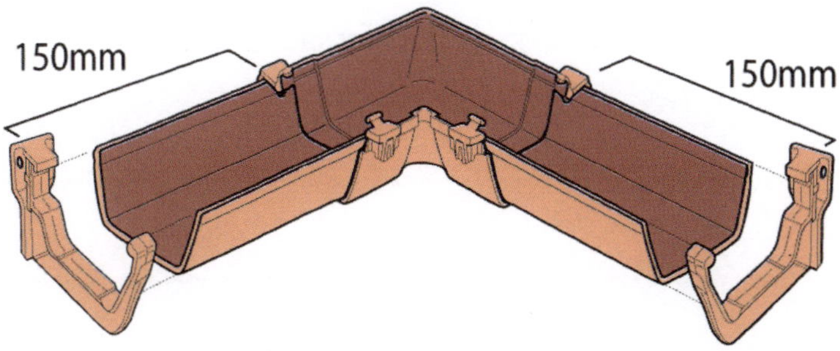

Position of fascia brackets near internal and external corners

- Fix gutter brackets at the high points on the system. If the high points are internal or external corners, then the brackets should be fixed 150mm from where the end of the fitting is. (see drawing right).

- Position the running outlet over the gully using a plumb-line or string-line and fix the outlet so as to give the required fall calculated earlier (see drawing below). Using a level ensures that the gutter will have the required the required fall towards the outlet.

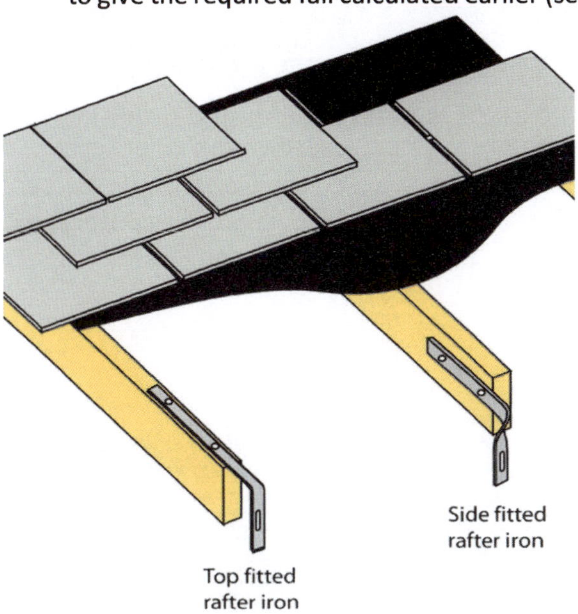

Rafter irons used for fixing fascia brackets

- Now, fasten a string line between the facia bracket at the high point and the running outlet. The line needs to be tight to ensure that there is no sagging. The line will be used to position the rest of the fascia brackets.

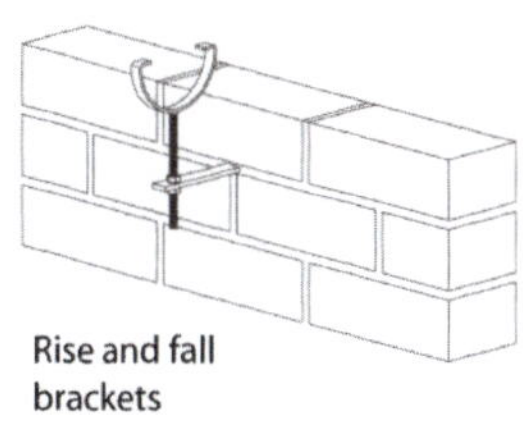

Rise and fall irons

- Fascia brackets should be fixed along the length of the string line at intervals of between 750mm and 1m, depending on whether the area suffers from heavy snow fall. In general, a distance of 800mm is usually used. Always check with the manufacturer's instructions. Use 40mm x 10 zinc plated screws to fix the fascia brackets.

For buildings without fascia boards: gutter brackets can be fixed using either galvanised steel top or side fitted rafter irons on which are fitted the fascia brackets. These should be spaced every rafter or as required by the manufacturer's instructions. They are shown in position in the drawing left.

For buildings without fascia boards or exposed rafters: **Drive in rise and fall irons can be used. These are either built into the wall or driven in at the appropriate point in the brickwork. Manufacturer's gutter fascia bracket spacing requirements should be observed.**

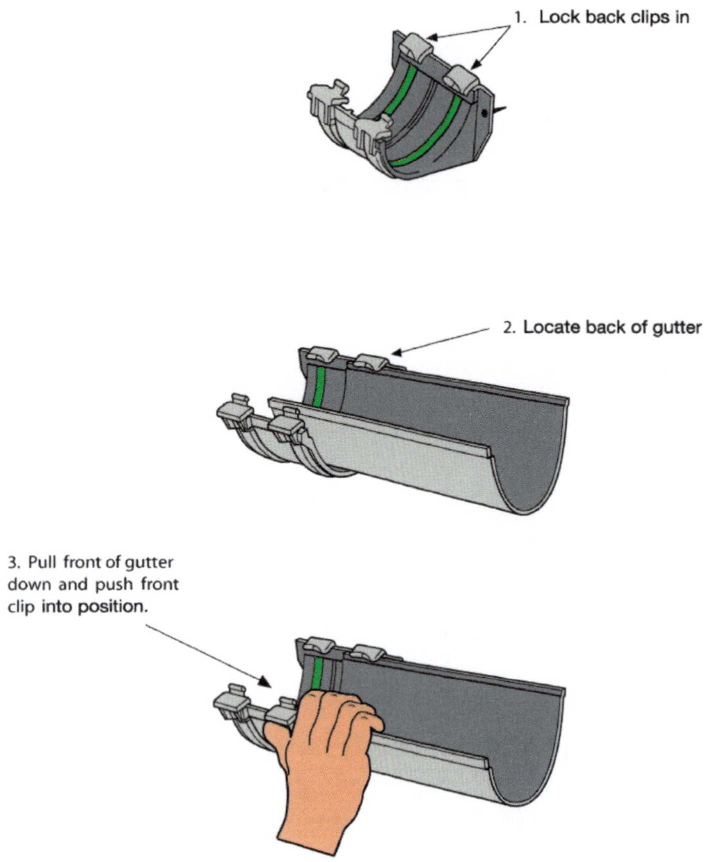

• Once all fascia brackets have been installed, and working away from the running outlet, the gutter can be put into the fascia brackets and clipped into place by pulling the front of the gutter and the bracket downwards together. Make sure that the gutter is clipped into place at the back of the gutter first. The same method is used for making the gutter joints. Ensure the joint retaining clip is clipped properly over the gutter joints. This is shown in the drawing left. Do not insert the gutter beyond the expansion point in the fitting. Start at the joint on the running outlet and work away from that point.

• Ensure that the distance between joints is measured carefully, taking into account the expansion points.

• Remember to support cap ends with a fascia bracket no more than 150mm away from the end of the fitting.

Installing rainwater pipes

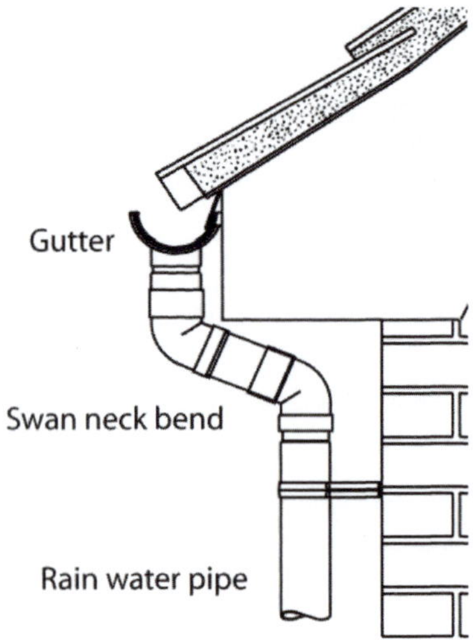

Rainwater clip support centres		
Pipe size (mm)	**Vertical (m)**	**Horizontal (m)**
55 mm	1.2	0.6
62 mm	2.0	1.2
68 mm	2.0	1.2
70 mm	2.0	1.2
82 mm	2.0	1.2
110 mm	2.0	1.2

- Installing rainwater pipes can be a little tricky and it is advisable to fabricate and install the swan neck (offset) bend that connects the running outlet of the gutter to the rainwater pipe first. The swan neck is made from 2 x 112.5° bends that are joined together so that, once the swan neck is installed, the rainwater pipe is the correct pipe clip distance from the wall.
- Measure the distance between the spigot on the swan neck and the connection to drain and carefully cut a length of rainwater pipe to suit.
- Install the pipework and mark the position of the pipe clips, ensuring the correct pipe clip distances are used. These are shown in the table above.
- Choose the correct clip for the installation. There are several different pipe clip styles and these mostly depend on the manufacturer of the pipe. Some are more robust than others. It is possible to

use pipe clips from other manufacturers, although there can be a variation in the colours. The sizes are the same, irrespective of the manufacturer.

- Drill the pipe clip holes using a suitable electric/cordless drill and drill bit.
- Fasten the rainwater pipe back to the wall using suitable plastic wall plugs and 50mm x 10 zinc plated screws.

Connection of Rainwater pipes to drainage systems

a. Rain water pipes can terminate over the top of a gully via a rainwater shoe

b. Direct connection to drainage bend using a suitable drain connector

c. Direct connection to gulley. The pipe must terminate below the grate but above the water line of the gully.

d. Direct connection to a soakaway. A typical soakaway is shown below

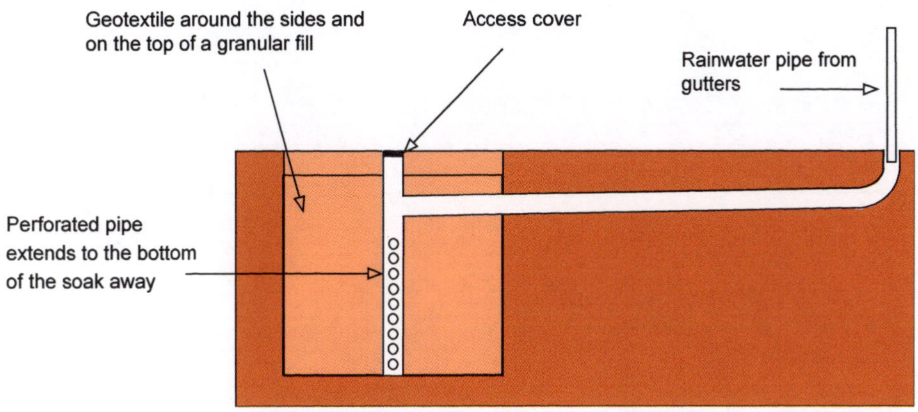

Connection of a rainwater pipe to a soakaway drain

Learning Outcome 2

Decommission Rainwater Systems

Learning Outcome 2: Decommission Rainwater Systems

In Phase 2, you will be investigating the following Assessment Criteria from Learning Outcome 2:

AC2.1 Explain **procedures** for decommissioning systems

AC2.2 Carry out decommissioning

AC2.1 Explain procedures for decommissioning systems

Decommissioning rainwater systems means at some point, you will be working at height. ***You must be aware of the dangers of working at height at all times.***

Decommissioning gutters can be a dirty job, especially if they have not been cleaned out for a while. Moss and debris from the roof collects in the gutters to create a foul-smelling sludge. You should also be aware that they may contain bird faeces, which can be harmful to health. Here are some points for you to consider:

- Inform the householder that you are going to begin the removal of their gutters and rainwater pipes. if the customer has any private property that may be affected, such as a car on the drive, it is a good idea to ask them to remove it until the process is complete.
- Apply warning notices that you are working overhead in areas where the customer or the general public may have access.
- Always wear appropriate PPE, especially when clearing the gutters out. It may be advisable to wear a face mask, especially if the gutter is dry and full of debris.
- Always check for hazardous materials such as asbestos before starting the decommissioning process. This may influence the PPE that you wear.
- Always use the correct type of access equipment. Using a ladder to install and remove gutters is not ideal. Gutters should always be installed and removed from a correctly erected scaffold wherever possible.
- Be aware that some gutter materials are heavy. Cast iron, for instance, is a heavy but brittle material that needs to be taken down with extreme care. It can often snap when being handled roughly.
- Always dispose of any waste materials appropriately. Gutters and rainwater pipes MUST be disposed of at a recognized recycling centre, since metals and most plastics are recyclable.

AC2.2 Carry out decommissioning

As part of your Phase 2 assessment you will be asked to install, test and decommission gutters and rainwater pipes. You may find it helpful to note the points made within this chapter to assist you.

REMEMBER:

1. Be aware of your surroundings and other people working in the vicinity
2. Always use access equipment responsibly
3. Always dispose of waste or unwanted materials safely and responsibly
4. Always leave the work area in a clean and tidy manner.

Learning Outcome 4

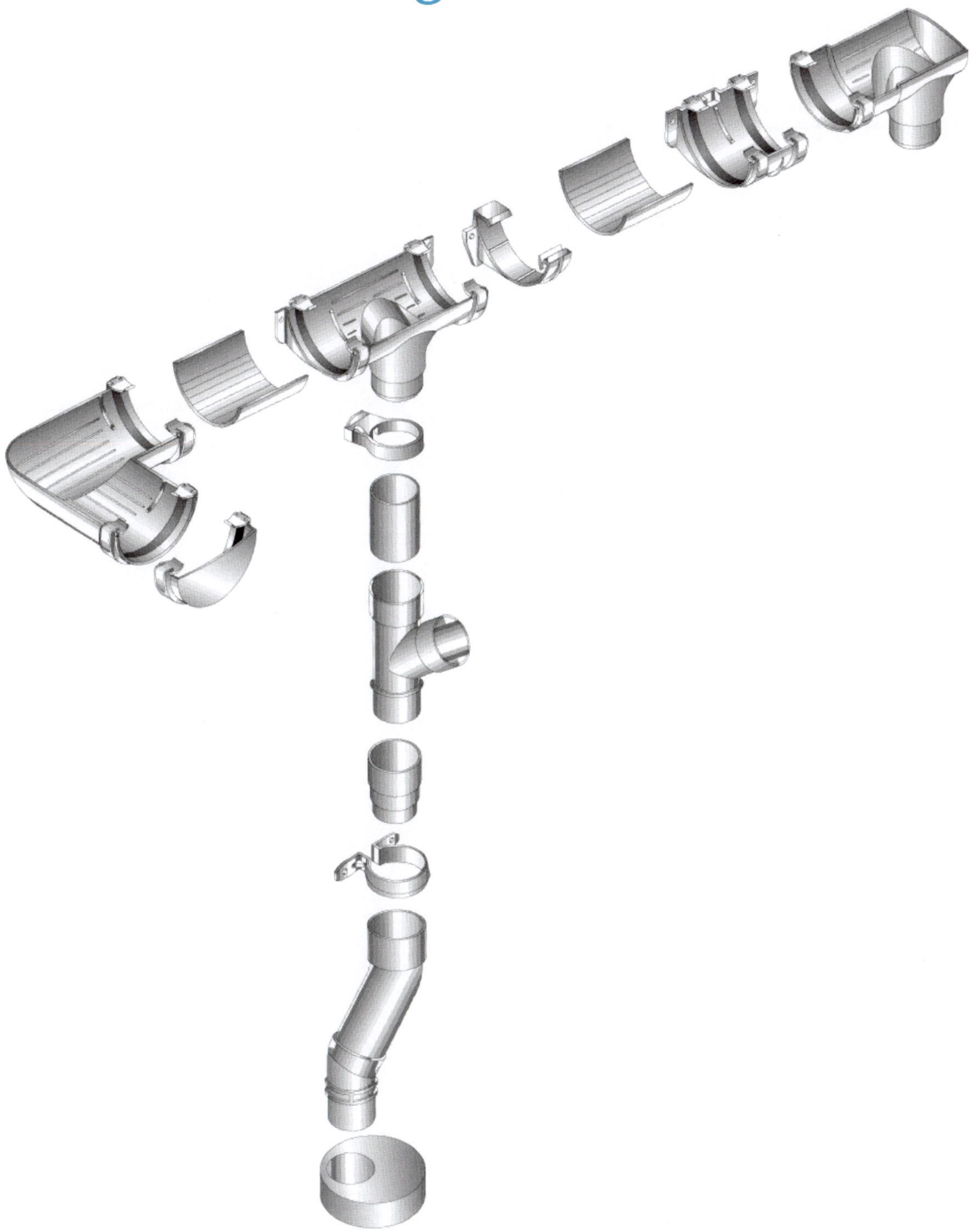

Perform a Soundness Test and Commission Rainwater Systems

Learning Outcome 4 – Perform a Soundness Test and Commission Rainwater Systems

In Phase 2, you will be investigating the following Assessment Criteria from Learning Outcome 4:

AC4.1 Identify information sources required to complete testing and commissioning

AC4.2 Describe a **visual inspection** of a rainwater system to confirm that it is ready to be soundness tested

AC4.3 Describe a **soundness test** to industry requirements on rainwater, gutter systems pipework and components

AC4.4 Describe **operational checks** required during commissioning

AC4.5 Identify actions that must be taken when commissioning reveals defects

AC4.6 Describe the procedure for handing over to the end user

AC4.7 Carry out soundness testing and **commissioning procedures**

AC4.1 Identify information sources required to complete testing and commissioning

Please refer to AC1.7 Identify sources of information required when undertaking work on rainwater systems for a comprehensive list of appropriate documents to be used when testing and commissioning rainwater systems.

AC4.2 Describe a visual inspection of a rainwater system to confirm that it is ready to be soundness tested

Before testing rainwater gutters and rainwater pipes, it is advisable to perform a visual inspection of the system. You should check that:

- The installed gutter is clear of any debris which may block the rainwater pipes or hinder the flow of water
- The gutters and rainwater pipes are installed:
    - To the manufacturer's instructions
    - In line with the building plan (new build properties)
- The gutters have been installed to the correct fall, that rainwater pipes have been installed plumb and that they correctly discharge into the drain or gulley.
- The gutter and rainwater pipes are correctly clipped and that the correct clip spacing have been used
- All joints have been correctly formed.

AC4.3 Describe a soundness test to industry requirements on rainwater, gutter systems pipework and components

Before testing

Before testing takes place, the customer should be notified. If there is any personal customer property that may cause a hindrance to the testing and commissioning procedures, ask them move it to a safe location while the testing is completed.

The testing procedure

Testing should be completed using either a hose pipe or buckets of water. The water should be discharged on to the roof tiles at the gutters highest point(s) and not directly into the gutters as this can often splash over the gutters. Discharging the water on to the roof will cause the water to spread simulating rainfall.

AC4.4 Describe operational checks required during commissioning

With the gutters in operation visually check that the water has a gentle flow (not a fast flow, as this would indicate that the fall of the gutter is incorrect) towards the outlets and that pooling inside the gutter does not occur. Also check for any signs of joint leakage and that the water is discharging from the rainwater pipes in the correct location.

AC4.5 Identify actions that must be taken when commissioning reveals defects

Sometimes, defective installation and components are revealed by the testing process. Any problems that do arise must be dealt with swiftly:

- **Dealing with systems that do not meet correct installation requirements** – poor installation is often the cause of problems with gutters and rainwater pipes. Problems such as too much fall, too little fall or no fall at all can cause major time concerns as the affected work must be taken down and reinstalled correctly. Sagging gutters too can cause aesthetic problems as well as performance related issues.
- **Remedial work associated with defective gutter and pipework bracketing** – as discussed earlier, insufficient fascia brackets and incorrectly placed pipe clips can cause gutter systems to fail in a relatively short space of time. Again, the affected work must be taken down and reinstalled correctly.
- **Remedial work associated with leakage from systems** – most leaks on new installations are caused by three very common issues:
    - Incorrectly jointed gutters
    - Defective components
    - Incorrectly positioned and jointed rainwater pipe fittings.

Of these three, incorrectly jointed gutters accounts for the vast majority of the problems encountered. This is usually due to incorrect measurements being taken and the gutter being cut too short for the gutter union or fitting. It is rare for gutter components to be supplied defective and when they are, most are due to missing sealing rubbers in the joints. Rainwater sockets and fittings are often fitted upside down with the socket facing downwards allowing water to run down the outside of the pipe. In all cases, the rectification is very simple to complete.

AC4.6 Describe the procedure for handing over to the end user

With new build installations of gutters and rainwater pipes, it may be necessary to complete a commissioning report before the installation will be accepted. The report may include the following:

- Name of the company
- Name of the installer
- Name of the commissioning engineer

- The property address
- The type of property i.e. detached, semi-detached etc
- The type of gutter used
- The date of commissioning
- The result of the soundness test and commissioning procedure

AC4.7 Carry out soundness testing and commissioning procedures

During your Phase 2 practical assessment, you will be asked to carry out a soundness test to industry standards.

Unit 9

Sanitation Systems – Phase 2

Unit 9 – Sanitation Systems – Phase 2

In Phase 2, there are 3 Learning Outcomes:

LO1 - Install sanitary appliances and pipework systems:

AC1.1 Identify types and **layout** features of sanitary appliances pipework **systems (completed at Phase 1)**

AC1.2 State advantages and disadvantages of sanitary appliances pipework systems

AC1.3 Identify typical pipe sizes and maximum distances permitted in sanitary appliances pipework systems within dwellings

AC1.4 Describe working principles of sanitary appliances pipework systems (positioning fixing, connection and operation of components

AC1.5 Identify the positioning and fixing of pipework within the **building fabric**

AC1.6 Explain expansion and contraction in sanitary appliances pipework systems and negative effects

AC1.7 Identify sources of information required when undertaking work on sanitary appliances pipework systems

AC1.8 Identify different types of **sanitary appliances** and components used in dwellings

AC1.9 Identify factors that lead to trap seal loss in sanitary pipework systems

AC1.10 Outline the suitability of **below ground drainage systems** to receive **waste water**

AC1.11 Installation requirements of sanitary facilities and equipment in dwellings for the disabled including wet rooms

AC1.12 Identify **jointing methods** used in sanitary appliances pipework systems

AC1.13 Explain working principles of greywater recycling systems

AC1.14 Install sanitary appliances, pipework systems and components

LO2 – Decommission sanitary appliances, pipework systems and components

AC2.1 Explain **procedures** for decommissioning systems

AC2.2 Carry out decommissioning **procedures**

LO4 – Perform a soundness test and commission sanitary appliances, pipework systems and components

AC4.1 Describe a **visual inspection** of sanitary appliances and pipework systems to confirm that it is ready to be soundness tested

AC4.2 Describe a **soundness** test to industry requirements on sanitary appliances, pipework systems and components

AC4.3 Describe **operational checks** required during commissioning

AC4.4 Identify actions that must be taken when commissioning reveals defects

AC4.5 Describe the procedure for handing over to the end user

AC4.6 Carry out soundness testing and **commissioning procedures**

Learning Outcome 1

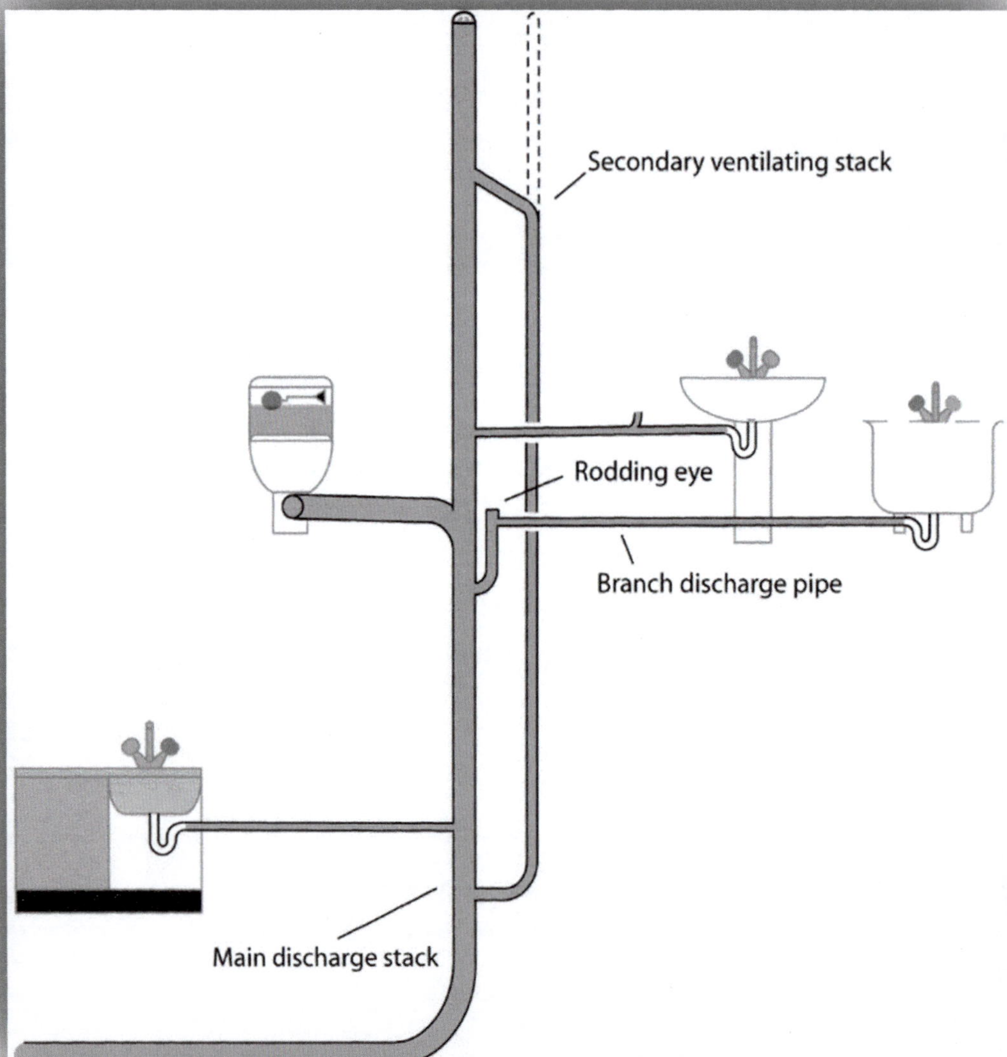

Install Sanitary Appliances and Pipework Systems

Learning Outcome 1: Install Sanitary Appliances and Pipework Systems

In Phase 2, you will be investigating the following Assessment Criteria from Learning Outcome 1:

LO1 - Install sanitary appliances and pipework systems:

AC1.1 Identify types and **layout** features of sanitary appliances pipework **systems (completed at Phase 1)**

AC1.2 State advantages and disadvantages of sanitary appliances pipework systems

AC1.3 Identify typical pipe sizes and maximum distances permitted in sanitary appliances pipework systems within dwellings

AC1.4 Describe working principles of sanitary appliances pipework systems (positioning, fixing, connection and operation of components)

AC1.5 Identify the positioning and fixing of pipework within the **building fabric**

AC1.6 Explain expansion and contraction in sanitary appliances pipework systems and negative effects

AC1.7 Identify sources of information required when undertaking work on sanitary appliances pipework systems

AC1.8 Identify different types of **sanitary appliances** and components used in dwellings

AC1.9 Identify factors that lead to trap seal loss in sanitary pipework systems

AC1.10 Outline the suitability of **below ground drainage systems** to receive **waste water**

AC1.11 Installation requirements of sanitary facilities and equipment in dwellings for the disabled including wet rooms

AC1.12 Identify **jointing methods** used in sanitary appliances pipework systems

AC1.13 Explain working principles of greywater recycling systems

AC1.14 Install sanitary appliances, pipework systems and components

AC1.2 State advantages and disadvantages of sanitary appliances pipework systems

A. The Primary Ventilated Stack

The Primary Ventilated Stack

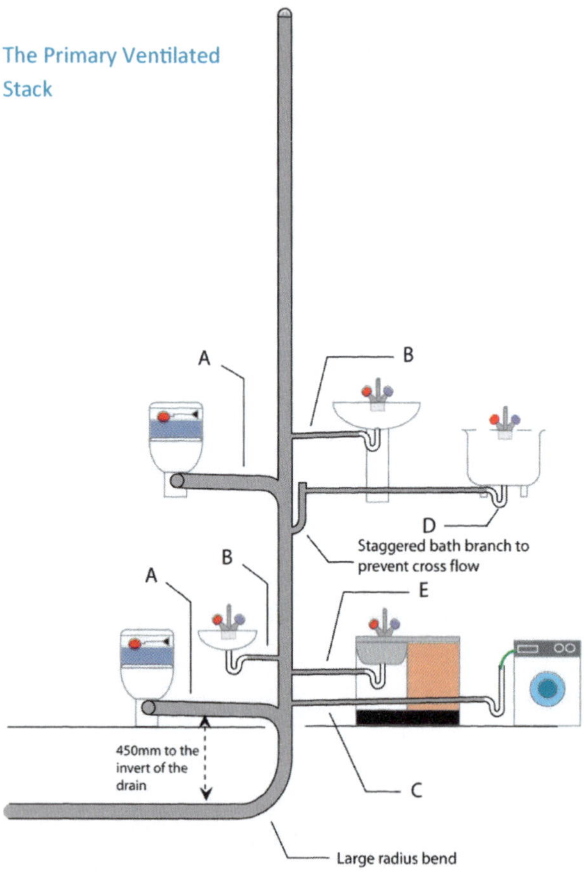

This is the most common soil and vent stack in use in the UK. It is installed in almost every domestic building in the UK. It relies on the fact that most of the appliances are close to or closely grouped around the stack. Because of this, waste pipe lengths are restricted to the lengths and sizes stated in the table below. It is used where the discharge stack is large enough to limit any pressure fluctuations that would otherwise destroy the water seal in the appliance traps. There is no need for extra ventilation via a separate ventilating stack.

The waste pipe sizes, lengths and gradients play a crucial role in determining whether a secondary ventilating stack is required. Waste pipes should be installed with enough fall for the water to reach a ***self-cleansing velocity***.

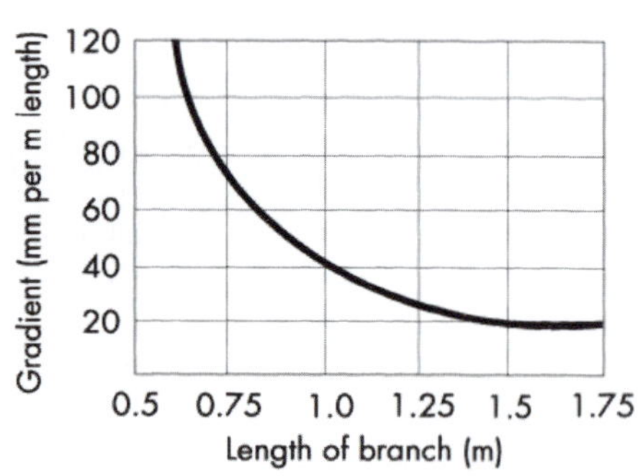

Washbasin waste pipes are slightly different insomuch that if the maximum distance of 1.7m is used, the gradient of the pipe is 18-22mm/m. If the pipe is shorter than 1.7m, the fall can be increased. This is shown in the graph right. If, for instance, the washbasin is 1m away from the stack, then the gradient is 40mm/m.

Appliance		Pipe size (mm)	Max. Length (m)	Gradient (mm/m)	Trap seal depth (mm)
A	WC branch	75 - 100	6	18	50
B	Wash basin and bidet	32	1.7	18 - 22 See graph	75
C	Washing machine/dish washer	40	3	18 -90	75
D	Bath	40	3	18 - 90	50
E	Kitchen/utility sink	40	3	18 - 90	75
Where these lengths are exceeded, then the next pipe size up should be used. 40mm appliances will need to increase to 50mm pipe (see below)					
	Appliances with 50mm waste pipe		4	18 - 90	75

Advantages and disadvantages of the Primary Ventilated Stack

Advantages	Disadvantages
Most common system in domestic properties	Appliances must be closely group around the stack
Uses less pipework than other sanitary systems	Waste pipe lengths and sizes must be adhered to if problems with trap seal loss are to be avoided
Two or more appliances can be connected to one waste pipe	Where more than one appliance is connected to a waste pipe, an increase in waste pipe size is required
Does not require extra ventilation pipework	

B. The Secondary Ventilated Stack

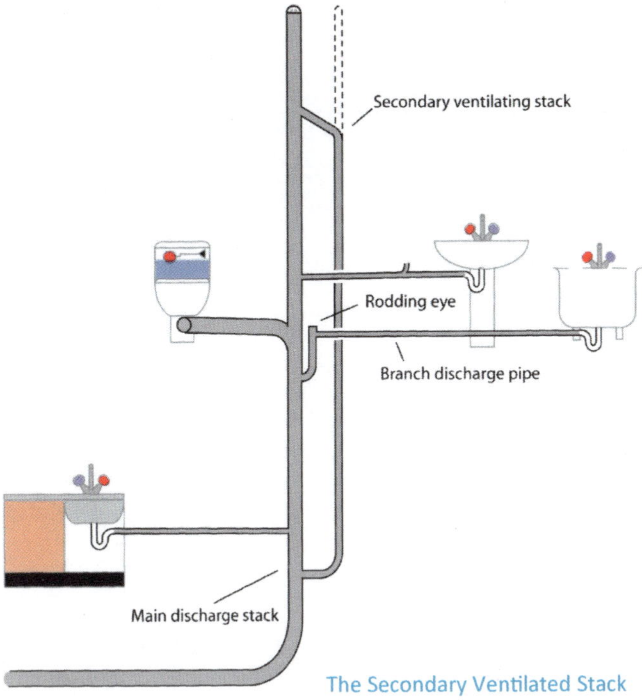

The Secondary Ventilated Stack

With the Secondary Ventilated stack, a smaller secondary ventilating stack is connected to the main discharge stack. This prevents positive and negative pressure fluctuations. The system is shown left.

The secondary ventilated stack is traditionally used in high rise buildings where pressure fluctuations can adversely affect the trap seals of the appliances fitted to the stack. This is due in part to the unsteady flow of water through the pipework. An incorrect air balance can result in blowback of the water into the appliance as well as induced syphonage. Both conditions need to be avoided.

Installing the secondary ventilated stack requires more installation time and materials and takes up much more space than the Primary Ventilated stack. This is an important point if space is an issue. The ventilating stack is usually half the diameter of the main discharge stack.

Advantages and disadvantages of the Secondary Ventilated Stack

Advantages	Disadvantages
Prevents pressure fluctuations within the pipework	Requires extra ventilation pipework
Less risk of trap seal loss	More expensive to install
Can be used where many appliances are connected to the stack	Requires more space for the pipework
Mainly used in multi-storey buildings	

C. The Ventilated Branch Discharge Stack

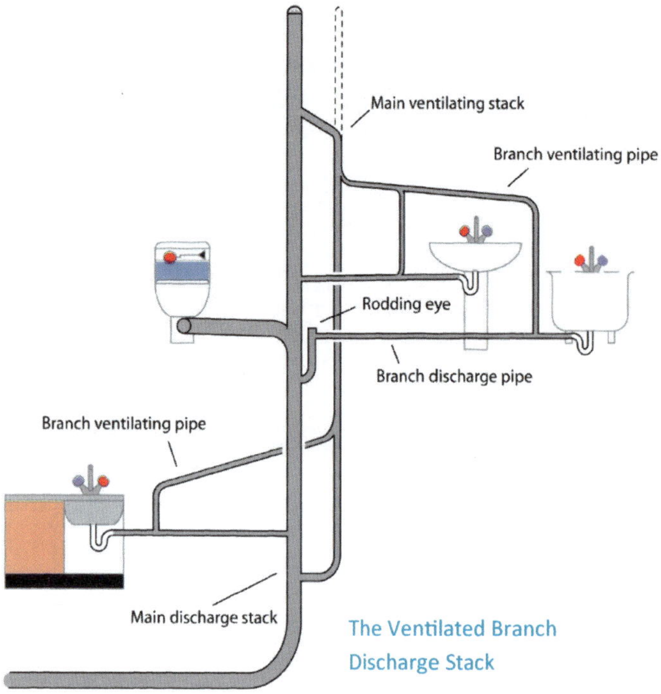

The Ventilated Branch Discharge Stack

The Ventilated Branch Discharge stack is used in situations where the maximum lengths of the waste pipes of the Primary Ventilated Stack may be exceeded. It is also used in larger installations where many appliances may be fitted or where groups of appliances are fitted in ranges. Trap seal loss due to pressure fluctuations caused by excessive waste pipe length is prevented by adding a ventilating pipe that connects from the main ventilating stack to within 750mm of the appliance.

Methods of ventilating branch discharge pipework

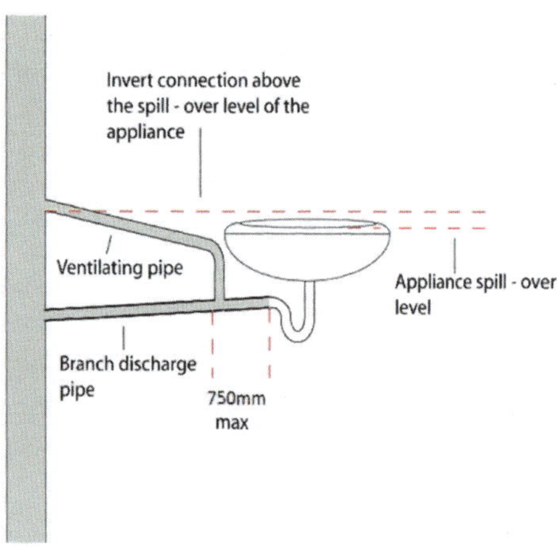

Branch Ventilating pipe rules

Where branch ventilating pipes are required the following rules must be observed:

1. Ventilating pipes must be connected to the discharge stack above the spill over level of the highest appliance installed on the stack.
2. The ventilating pipe must rise away from the appliance and not fall to the stack.
3. The minimum size of ventilating pipe to any single appliance is 25mm. If the ventilating pipe is longer than 15m or serves more than one appliance, the size MUST be increased to at least 32mm diameter (min).
4. The main ventilation stack must be at least 75mm in diameter (this applies to the Primary Ventilating Stack also)

Advantages and disadvantages of the Ventilated Branch Discharge Stack

Advantages	Disadvantages
Prevents pressure fluctuations within the pipework	Requires extra ventilation pipework to every appliance/appliance groups with long waste pipes
Trap seal loss is prevented	
Can be used where many appliances are connected to the stack	More expensive to install
	Requires a lot more space for the pipework
Allows installations where the appliances are further away from the soil stack	Difficult system to install
	Pipework is unsightly

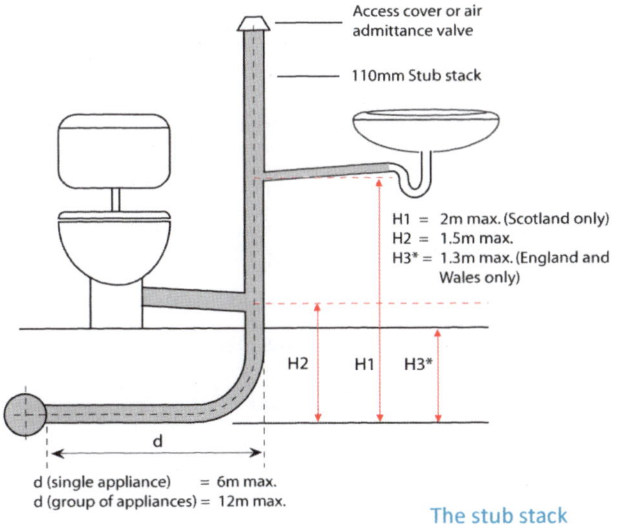

The stub stack

D. The Stub Stack

This is usually used where a small group of appliances, such as a WC and washbasin in a downstairs cloakroom, is to be connected directly to a drain. However, ventilation of the drain must be provided when the connection from the invert of the drain to the connection to the highest appliance exceeds 2m, or the crown connection to the invert exceeds 1.3m. Stub Stacks can be ventilated by an air admittance valve.

Advantages	Disadvantages
Ideal system for small numbers of appliances	Not suited for every soil stack installation as the drains require ventilation direct to air
The system is relatively cheap to install	
Can be used with or without air admittance valves	Should only be used when there are other stacks directly ventilated to air
	Strict rules surrounding stub stacks must be observed (see drawing above)

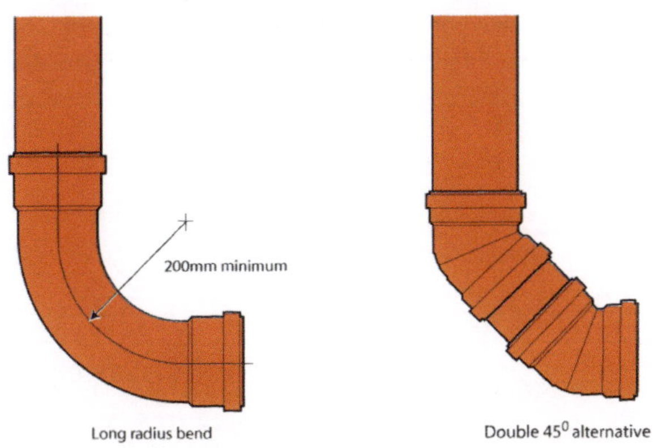

Large radius bend at the foot of the stack

Table 1

Stack size (mm)	Max. Capacity (l/sec)
50*	1.2
65*	2.1
75**	3.4
90	5.3
100	7.2

* No WCs installed

** Not more than 1 WC with an outlet size <80mm

Design Recommendations for Discharge Stacks

Rule 1 – The foot of any discharge stack should only connect to a drain and should have as large a radius bend as possible but not less than 200mm radius. Two 45° bends are acceptable.

Rule 2 – Where possible, there should be no offsets in the wet part of the discharge stack.

Rule 3 – In situations where rule 2 is unavoidable:

a) Buildings over 3 storeys should have a separate ventilation stack connected above and below the offset,

b) Buildings up to 3 storeys should have no branch connection within 750mm of any offset.

Rule 4 – Stacks above 3 storeys high should be placed inside the building to prevent frost damage to the stack and any branch pipes.

Rule 5 – The stack should comply with the minimum diameters recommended in table 3 of section of AD H1 of the Building Regulations (2013). This is reproduced left. In addition to these sizes, the following diameters in table 2 must also be observed:

Table 2	
Appliance	**Diameter of stack (mm)**
Urinals	50mm
WCs (less than 80mm outlets)	75mm
WCs (greater than 80mm outlets)	100mm

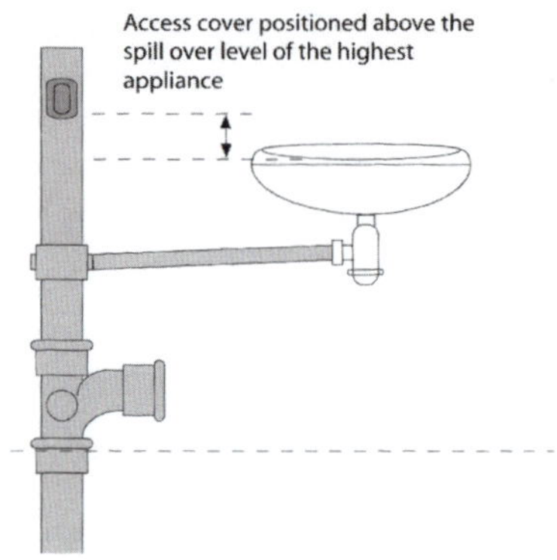

Access positioning

Rule 6 – The diameter of the stack should not reduce in size in the direction of flow.

Rule 7 – The internal diameter of the stack should not be less than that of the largest trap or discharge pipe.

Rule 8 – There should be adequate access points for clearing blockages and all pipework should be accessible for repairs. Any access points should be above the spill over level of the appliance.

Methods of connecting branch discharge pipework into the main stack

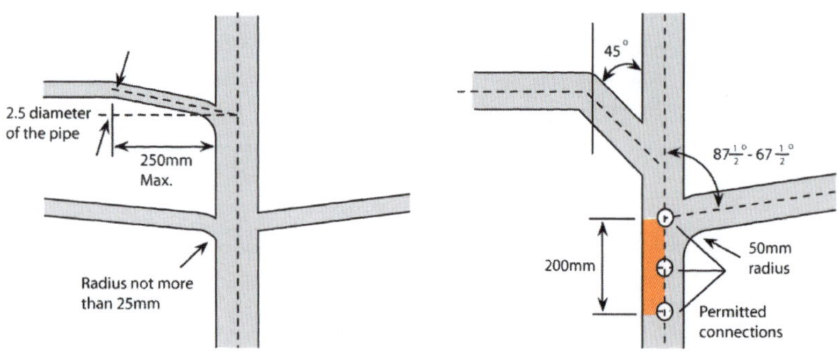

Branch connections to stacks

The drawing below shows that waste pipe connections of less than 75mm diameter should be made at an angle of 45° or with a bend radius of 25mm. The shaded area of 200mm shows that no waste pipe connection can be made within 200mm of a WC connection. Branch connections over 75mm diameter must be made at an angle of 45° or with a bend radius of 50mm.

Branch pipes should not discharge into a stack in a way that causes a cross connection into another branch as this could cause trap seal loss by fluid travelling in the opposite direction up the opposite connection. The following should be observed:

- Stacks up to 82mm in diameter, no connection is allowed for a distance of 90mm
- Stacks up to 110mm, no connections are allowed for a distance of 110mm.
- Stacks up to 160mm, no connections are allowed for a distance of 250mm.

Where it is not possible to meet the requirements of the primary ventilated stack, such as installations with excessive waste pipe lengths, extra ventilation will need to be added to the system to safeguard trap seals:

- The installation of ventilating branch discharge pipework, or
- The installation of secondary ventilating stack pipework.

PREVENTION OF CROSS FLOW

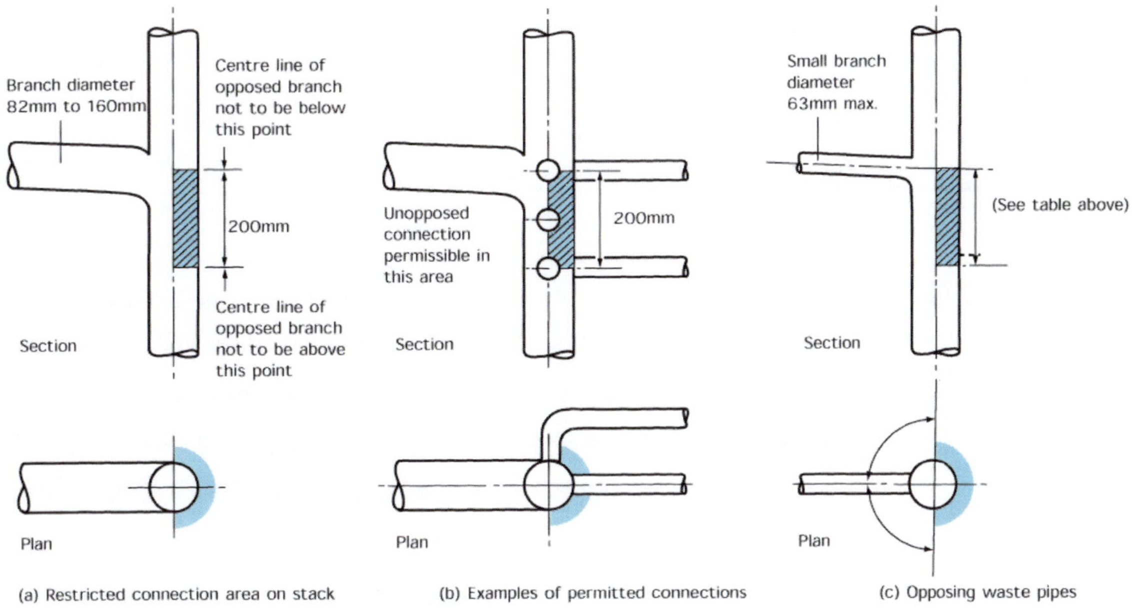

RESTRICTED AREA FOR OPPOSING SMALL DIA. BRANCH PIPES	
Stack dia.	Height of restricted area
82mm	90mm
110mm	110mm
160mm	250mm

- Left hand drawing: where a branch connection into a stack is between 82mm and 160mm in diameter e.g. a WC connection, no other connection is allowed opposite that connection for a distance of 200mm vertically downwards.
- Middle drawing: This drawing shows where connections made at 90o to the branch are allowed
- Right hand drawing: Where the branches are of the same or similar size, then the restricted distance will depend on the stack size:

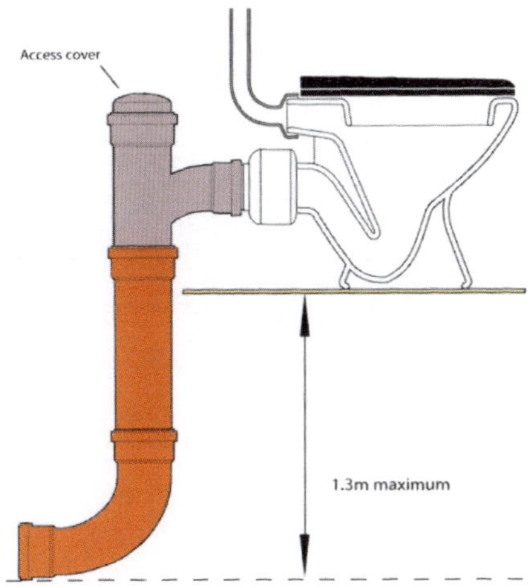

Direct connection of a WC

Low level connections to discharge stacks

For systems up to 5 storeys high, the lowest branch connection should be no lower than 750mm. For low rise dwellings, the distance can be reduced to 450mm.

For systems up to 5 storeys high, the lowest branch connection should be no lower than 750mm. For low rise dwellings, the distance can be reduced to 450mm.

For buildings, more than 5 storeys high, ground floor appliances should be connected to their own stack or drain and for buildings above 20 storeys, the ground and first floor properties should have their own independent stacks.

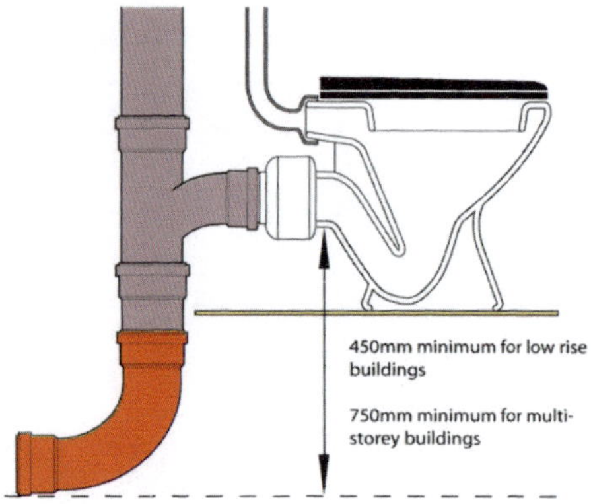

Low level branch connections

WC connections direct to the drain

A WC connection to a ground floor closet should only discharge directly to the drain if the distance from the floor of the closet to the invert of the drain is no more than 1.3m

Waste pipe connections direct to the drain

A direct connection of a waste pipe to a drain is known as a stub waste connection. This unventilated and sealed connection is often used to make remote waste pipe connections in domestic properties such as washing machines and kitchen sinks. The main requirement here, is to ensure that the waste connection is made via a deep seal trap with 75mm of water seal.

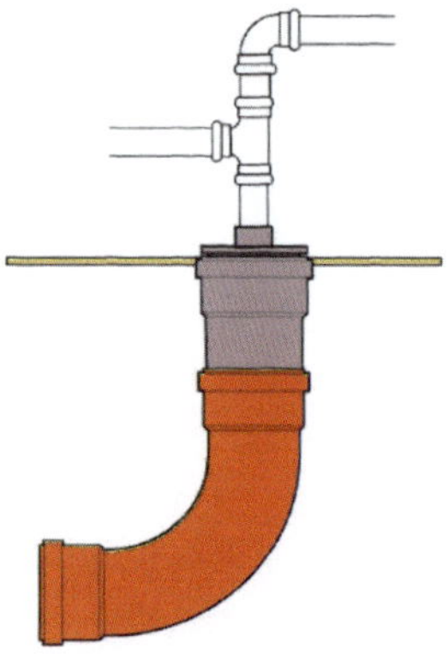

stub waste connections

Waste pipe connections direct to a gully

Kitchen sinks and washing machines are often connected to the drainage system via a gully. The key features of gully connections are:

- A gully capable of accepting a waste pipe connection, such as a back-inlet gully, must be installed.
- The waste pipe must discharge below the grid but above the water level in the gully trap.
- The appliances using a gully connection can use a shallow seal trap with a 38mm water seal. The seal does not have to be 75mm.

Methods of connecting multiple waste appliances to branch discharge pipework

When two or more appliances are to be connected to the same branch discharge pipework, it should be completed in a way that does not cause problems with induced syphonage. The drawing above shows the waste pipe increasing in size as more appliances are connected. The bath is connected in 40mm waste pipe leading to the washbasin where 32mm enters the waste pipe run. The pipe size is then increased to 50mm as it enters the main discharge stack.

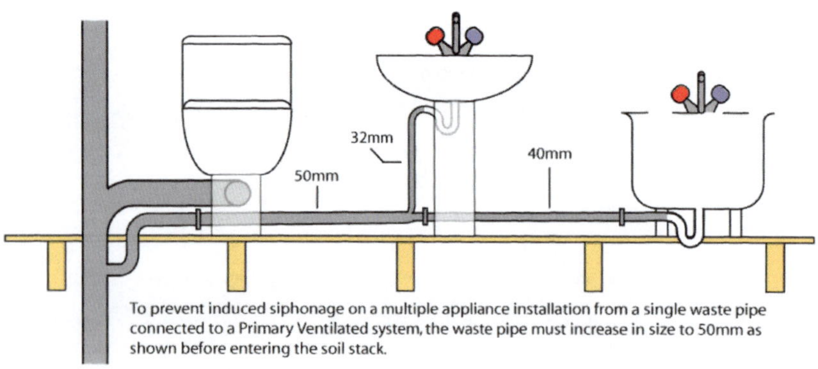

Multiple waste connections

Use of traps and self-sealing valves

An appliance that is connected to the discharge stack or any part of a stack must have a trap or some other, no less effective, method of preventing noxious odours from entering the property, fitted to it. The table on page 266 shows the minimum trap seal depths of sanitary appliances.

As a general rule of thumb, traps connected to a discharge system should be of the deep seal (75mm water seal) variety.

The self-sealing valve, known as the HEP$_V$O valve, can be used as an alternative to normal water filled trap. These can be fitted both in the vertical and horizontal positions.

The self-sealing trap contains a collapsible membrane that opens to allow water to flow through it and then closes again making an air tight seal. Using a waterless self-sealing trap means that problems with induced syphonage, self syphonage and compression are avoided.

AC1.3 Identify typical pipe sizes and maximum distances permitted in sanitary appliances pipework systems within dwellings

Please see the table on page 266.

AC1.4 Describe working principles of sanitary appliances pipework systems (positioning, fixing, connection and operation of components)

In this assessment criteria, we will look at some of the more common components and fittings and how they are used.

Bends

There are a variety of bends available for sanitary pipework and wastepipes. Their primary function, obviously, is to change direction. However, soil and vent stacks often require careful planning because of the size of pipe. 110mm is the most common size and this requires careful handling and route planning. The variety of bend angles helps in this regard. The more common bends are shown following page:

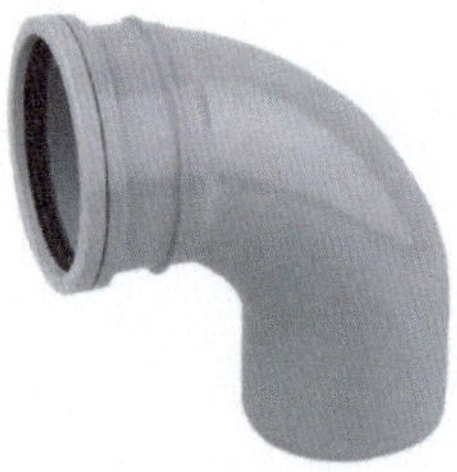

92.5° bend

Designed to be slightly over 90° to give a built in 2.5° fall.

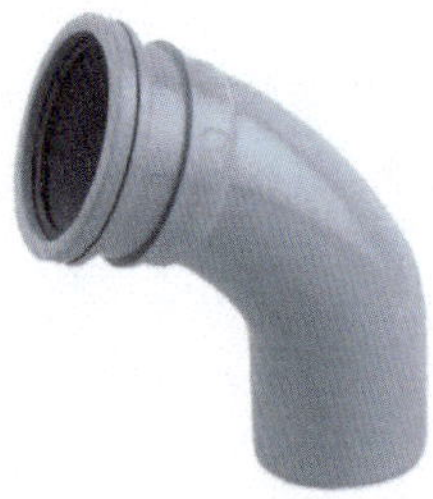

67.5° bend

A multi-purpose bend.

45° bend

Widely used for offsets at the eaves of the property.

30° bend

A multi-purpose bend.

15° bend

A very shallow angle bend. Must be used with caution as it is not aesthetically pleasing when used on exposed pipework.

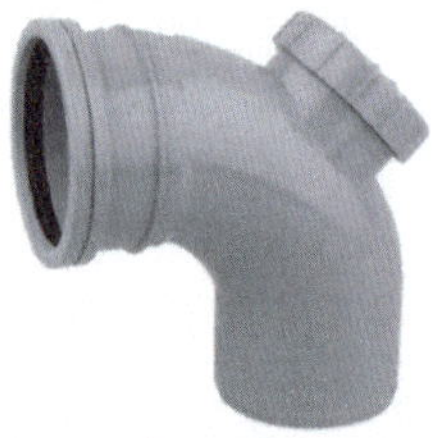

92.5° access bend

Designed to give access to the pipework for maintenance and blockage clearing.

Multi-angle bend

Not aesthetically pleasing when used on exposed pipework. However, this is an extremely useful bend where the angle required cannot be achieved with a fixed angle bend.

Junctions and boss pipes

Junctions

A junction performs the same function as a tee piece. It is used for making branch connections, such as a WC connection on a soil and ventilation stack. Like bends, there are a number of junctions available. Each one differs in the angle of the branch. A selection of Junctions are shown below:

87.5° Branch

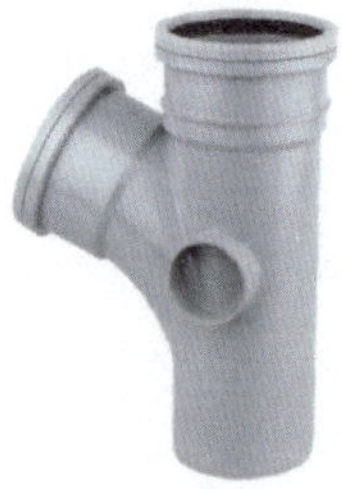

76° Branch

45° Branch

87.5° Double Branch

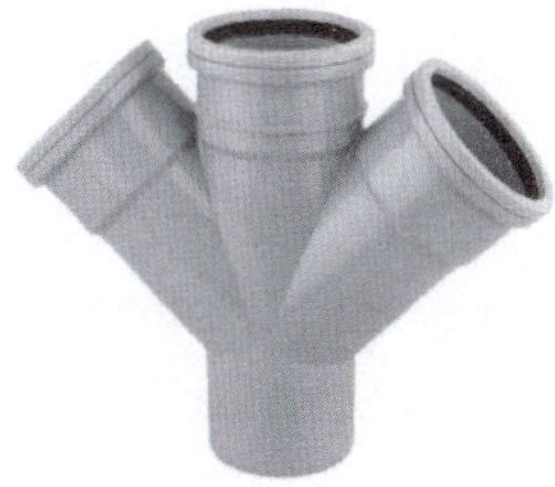

45° Double Branch

87.5° Right angle Double Branch

Boss pipes

A boss pipe allows the connection of a waste pipe to the soil and ventilation stack. The connection to the waste pipe is usually by a rubber insert designed to accept the waste pipe size. The actual boss pipe connection is the same irrespective of waste pipe size. It is the rubber insert that defines the waste pipe size.

Single boss pipe

Boss pipe with access

Triple boss pipe

Single solvent weld boss pipe

Strap boss

Access boss pipe

Vent terminal

Positioned on the top of the vent pipe, the vent terminal helps to prevent wind turbulence and stops birds from nesting on the top of the vent pipe. They are solvent welded on to the end of the pipe.

Waste pipe manifold

Waste pipe manifolds are a neat and simple way of connecting multiple waste pipes into a soil stack. They help to prevent trap seal loss and eliminate the need for multiple boss pipes. They are installed at floor level.

WC Pan Connectors

As the name suggests, these are used to connect WC pans to the soil stack. There are a number of different types, depending on the installation requirements. They simply push into the soil pipe and the WC pan pushes into the pan connector. They use rubber seals to prevent leakage and smells. A variety of pan connectors are shown below:

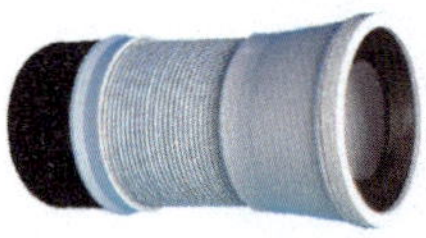

A selection of Waste pipe fittings

Fitting	Waste (solvent weld)	Waste (push fit)	Waste (universal)
Socket			
Knuckle bend		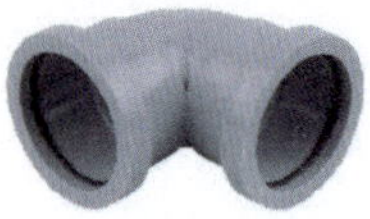	
Large radius bend		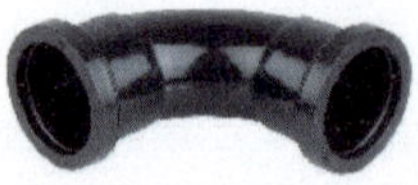	

Tee piece			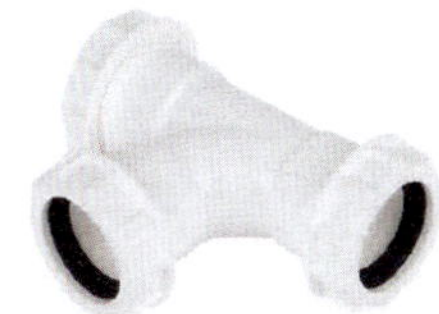
135° bend			
Reducers			

Waste traps

The trap on a sanitary appliance is a key component. It is the trap that stops foul air from entering the building. There are two specific types:

- The 'P' trap
- The 'S' trap

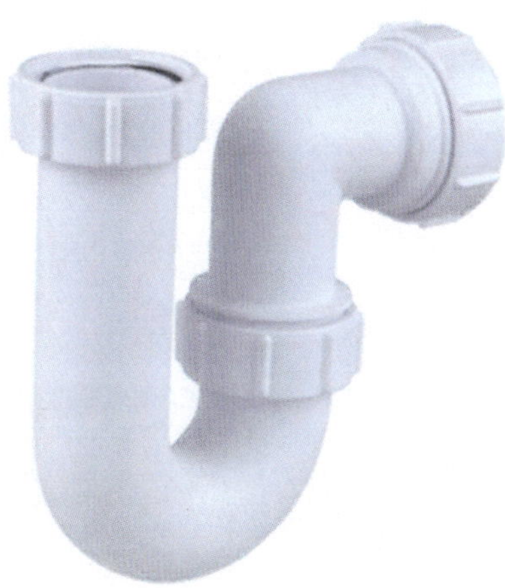

The 'P' Trap

The 'S' Trap

There are a number of trap designs that can be used to suit a variety of installations. Common trap designs are shown in the images below. Most traps are available with compression-type waste pipe connections that are suitable for all types of waste pipe materials.

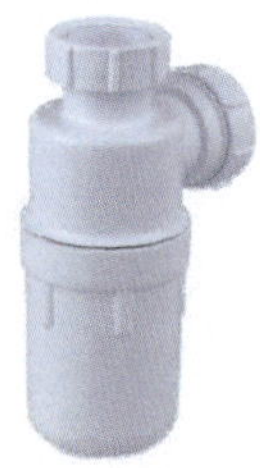

Bottle traps

Made for wash basin installations. They are susceptible to blockage and should not be used on kitchen sinks or baths.

In-line trap

Again, mainly used for wash basin installations to hide the waste pipe behind the wash basin pedestal.

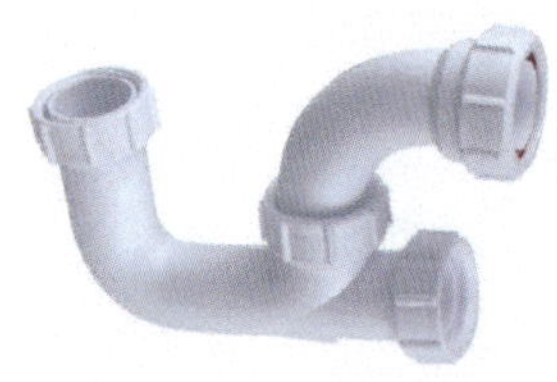

Bath traps

Specifically made for the limited space under baths.

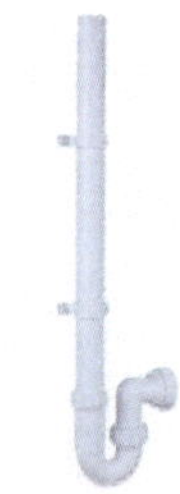

Washing machine traps

These have a long stem that is required when installing washing machines and dish washers.

Standard "P" trap

Can be used for most sanitary appliance applications

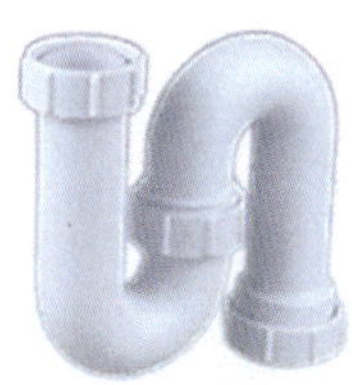

Standard "S" trap

Can be used for most sanitary appliance applications with a vertical discharge.

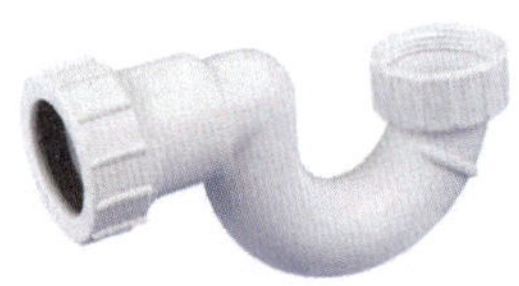

Shallow bath trap

These can only be used when the bath waste pipe discharges to an outside gulley.

Shower tray trap

Installed directly on to the shower tray. It has a low profile for installing in the limited shower tray space.

Traps have three trap depths and it is important to remember which trap depth is used where:

- Traps fitted to appliances discharging in to a soil stack must have a 75mm 'deep seal' trap depth. This is to prevent wind fluctuations and other, multiple appliance installations or installations that are close to the maximum waste pipe length from pulling the trap seal out, allowing foul air into the building.
- Traps fitted to appliances discharging to a gulley can use 50mm 'shallow seal' trap seal depth.
- Shallow seal bath traps discharging to a gulley can use a 38mm trap seal depth.

Appliance	Trap diameter	Depth of seal
Washbasin[1], Bidet	32mm	75mm
Bath, Shower	40mm	50mm
Food waste disposal unit, Sink Washing machine[2], Dish washer[2]	40mm	75mm
WC pan <80mm outlet	75mm	50mm
WC pan >80mm outlet	100mm	50mm

1. Depth of seal may be reduced to 50mm with flush grated wastes without plugs on spray tap washbasins
2. Where the appliance discharges direct to a gully, the seal depth may be reduced to 38mm

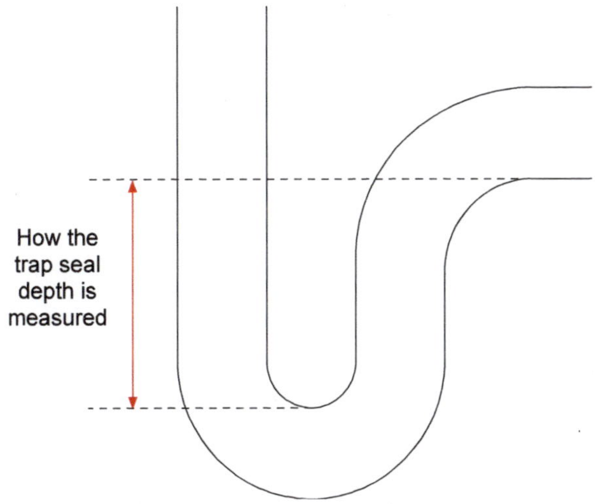

Waterless traps

The self-sealing valve, known as the HEP$_V$O valve, can be used as an alternative to a normal water filled trap. These can be fitted both in the vertical and horizontal positions.

The self-sealing trap contains a collapsible membrane that opens to allow water to flow through it and then closes again making an air tight seal. Using a waterless self-sealing trap means that problems with induced syphonage, self syphonage and compression are avoided.

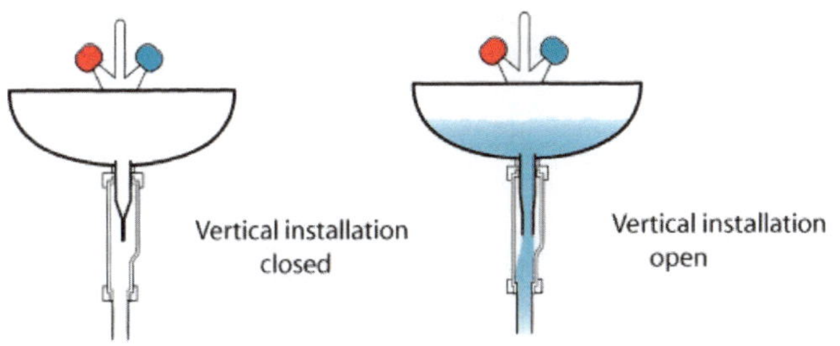

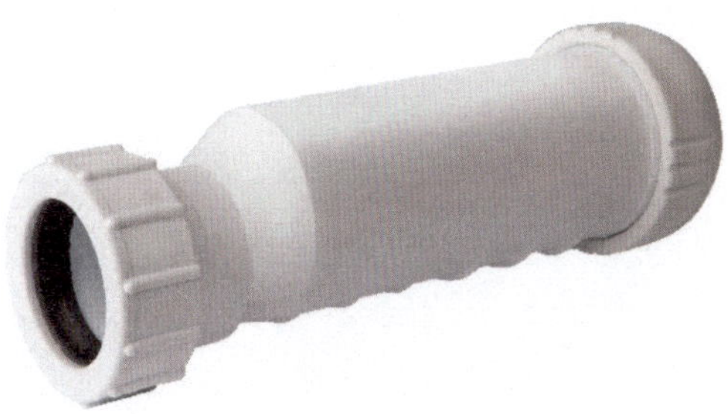

The self-sealing trap

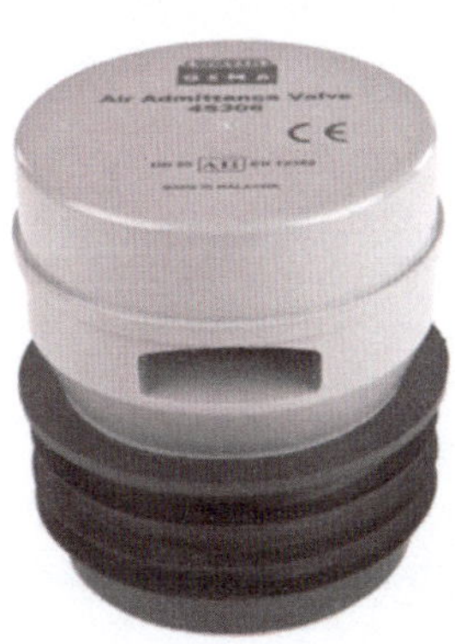

An air admittance valve

Air admittance valves

An air admittance valve allows air into a stub stack to prevent trap seal loss when the appliances fitted to the stack are in operation. When an appliance is used, the waste water running down the stack creates a negative pressure, which opens the air valve to let air in, destabilizing the pressure in the stack. However, it must be remembered that an air admittance valve does not replace the vent pipe and any system may still require a ventilation stack. There are several key points that should be considered:

1. AAVs do not operate correctly if subjected to freezing conditions. Generally, they are not suitable for outside installations.
2. They must be fitted vertically.
3. They should not be fitted in dusty atmospheres.

4. AAVs must be accessible for maintenance purposes. If sited in a boxing, the boxing must include an access panel and have sufficient ventilation in the boxing to allow the AAV to operate correctly.
5. AAVs must be fitted at least 200mm above the flood level of the highest appliance.
6. AAVs must not be used where the below ground drainage system contains an interceptor trap.
7. In the case of multi-storey buildings, the AAV manufacturer will stipulate the maximum number of storeys that the valve may serve in a single building. This is usually between 5 to 10 storeys.
8. Where an underground drain serves more than four properties fitted with an air admittance valve, the following rules apply:
    a) Where five to ten buildings exist, additional conventional ventilation stacks must be installed at the head of the drain run.
    b) Where eleven to twenty buildings exist, additional conventional ventilation stacks must be installed at the head of the drain and at the midpoint in the run of the drain.
    c) All multi-storey domestic properties will require additional conventional ventilation if more than one property is fitted with an air admittance valve and is connected to a common drain, which is not ventilated by a conventional ventilation stack.

Floor gullies

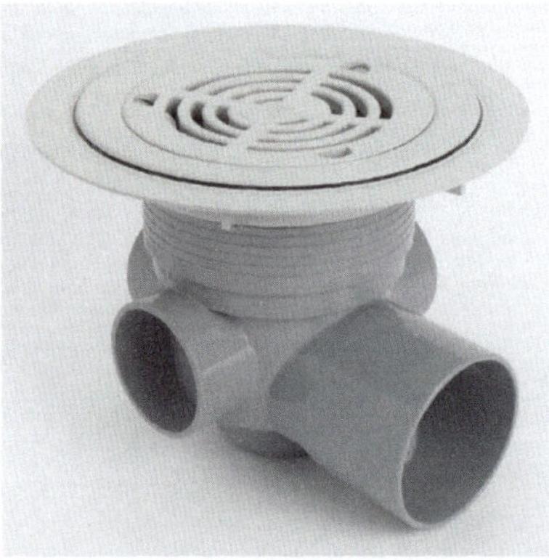

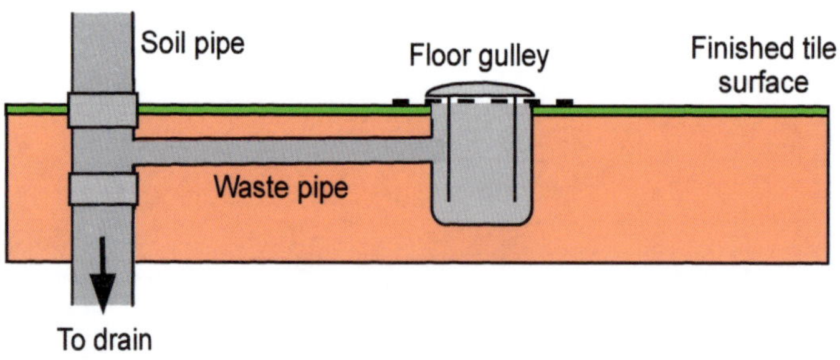

Installation of a floor gulley

Floor gulleys are used in wet rooms to drain the surface water from the finished tile floor. They require special consideration, especially when they are installed where there are living spaces below the wet room. It is important to remember that the water will need to find the gulley, so the finished floor will need to slope from all directions towards the floor gulley in the finished tile floor. Normal wet floor waste systems have a 50mm trap depth. Waterless trap versions are also available and these are often a good alternative, especially where the floor depth is shallow.

Clips/brackets

Like all pipework systems, soil and vent pipes, and waste pipes require clipping. Below are a few clips types that are currently available and their clipping distances. Clips should be screwed to the wall surface using 50mm x 10 zinc plated screws, using red or brown wall plugs. Round head screws are usually the best type to use as these do not cause the clip to split at the screw hole.

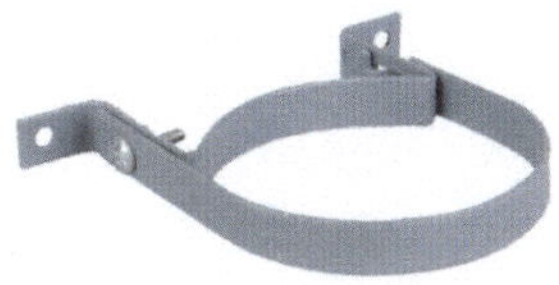

Galvanised steel clip

Mainly used when a more robust clip is required such as in public buildings and commercial premises.

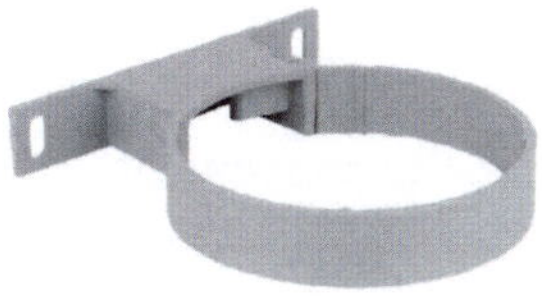

Plastic soil pipe clip

These are the standard clips used externally and internally on most domestic installations.

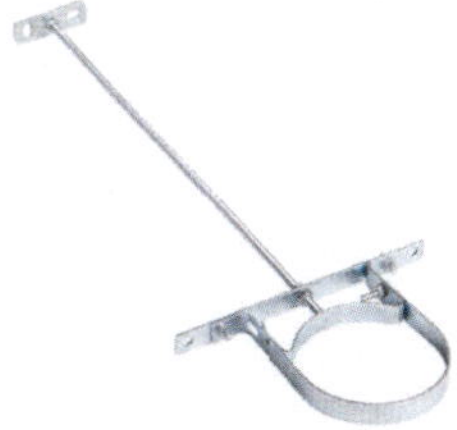

Soil pipe hanging brackets

Used mainly on industrial applications when hanging soil pipes from ceilings at high level.

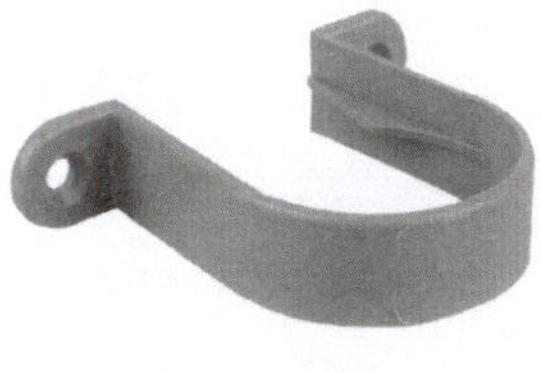

Waste pipe clips

Used for securing wastepipe installations. They are available in sizes 32mm, 40mm and 50mm

Stack Ventilation

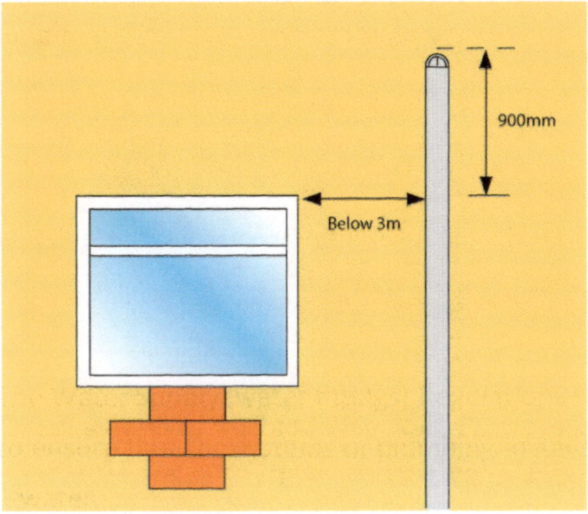

Vent pipe distance above an openable window

Proximity of vent pipe to an openable window – Where a vent pipe (ventilation stack) is installed within 3m of an openable window, the vent pipe should extend above the window by 900mm.

Use of air admittance valves – An air admittance valve allows air into a stub stack to prevent trap seal loss when the appliances fitted to the stack are in operation. When an appliance is used, the waste water running down the stack creates a negative pressure, which opens the air valve to let air in, stabilising the pressure in the stack. However, it must be remembered that an air admittance valve does not replace the vent pipe and any system may still require a ventilation stack. There are several key points that should be considered:

9. AAVs do not operate correctly if subjected to freezing conditions. Generally, they are not suitable for outside installations.
10. They must be fitted vertically.
11. They should not be fitted in dusty atmospheres.
12. AAVs must be accessible for maintenance purposes. If sited in a boxing, the boxing must include an access panel and have sufficient ventilation in the boxing to allow the AAV to operate correctly.
13. AAVs must be fitted at least 200mm above the flood level of the highest appliance.
14. AAVs must not be used where the below ground drainage system contains an interceptor trap.

15. In the case of multi-storey buildings, the AAV manufacturer will stipulate the maximum number of storeys that the valve may serve in a single building. This is usually between 5 to 10 storeys.
16. Where an underground drain serves more than four properties fitted with an air admittance valve, the following rules apply:
    a) Where five to ten buildings exist, additional conventional ventilation stacks must be installed at the head of the drain run.
    b) Where eleven to twenty buildings exist, additional conventional ventilation stacks must be installed at the head of the drain and at the midpoint in the run of the drain.
    c) All multi-storey domestic properties will require additional conventional ventilation if more than one property is fitted with an air admittance valve and is connected to a common drain, which is not ventilated by a conventional ventilation stack.

AC1.5 Identify the positioning and fixing of pipework within the building fabric

This subject is covered in detail in **Unit 5 – Cold water systems** on pages 87 to 91. However, it should be noted that waste pipe systems and soil pipe systems require large pipe sizes that are difficult to position under floors. Notching of joists in such cases, where the pipe is greater in diameter than $^1/_8$ of the joist depth **MUST NOT** take place or the joist strength will be severely compromised. Pipes may be successfully positioned with the joist run but **NEVER** across it.

AC1.6 Explain expansion and contraction in sanitary appliances pipework systems and negative effects

PVCu has a large expansion rate when compared to other pipe materials. Soil pipes and waste pipes expand by 0.06mm/m/°C. This can lead to the pipe movement and joint leakage if the expansion is not accommodated for. For example:

A 6m run of 110mm soil pipe is to be fixed horizontally in a small commercial property. A temperature variation of 20°C is expected in the building. By how much will the pipework expand?

$$6\ x\ 20\ x\ 0.06\ =\ 7.2mm$$

Pipe fittings should be installed so the pipe allows for around 10mm of expansion up to the pipe stop, preventing joint leakage.

AC1.7 Identify sources of information required when undertaking work on sanitary appliances pipework systems

Building Regulations Approved Document H

Approved Document H is set out into six (6) separate documents:

- H1: Foul water drainage
- H2: Waste water treatment systems and cesspools
- H3: Rainwater drainage
- H4: Building over sewers
- H5: Separate systems of drainage
- H6: Solid waste storage

The Building Regulations Approved Document H is a mandatory requirement for anyone that is installing any form of above and/or below ground drainage systems. With each regulation, guidance on how to comply with the requirements is given. They are linked to BS EN 12056-2000: Gravity Drainage Systems Inside Buildings.

A free copy of Approved Document H can be downloaded here:

https://assets.publishing.service.gov.uk/government/uploads/system/uploads/attachment_data/file/442889/BR_PDF_AD_H_2015.pdf

The Water Supply (water fittings) Regulations 1999

The Water Supply (water fittings) Regulations were written under section 74 of the Water Industry Act 1991 to ensure that the systems of plumbing installed in properties prevent the contamination of fluid category 1 water.

Since sanitary appliances are invariably connected to the water supply within the building, it is important to remember that the effluent from an appliance that is running to drain will be classified as not complying to fluid category 1 and, as such, extreme care must be taken to ensure that there is no cross-connection or other potential contamination risk in any part of the installation.

A free copy of the regulations can be downloaded from www.opsi.gov.uk/si/si1999/19991148.htm.

British Standard BS EN 12056

BS EN 12056: Gravity Drainage Systems Inside Buildings is a series of five (5) separate documents, each one covering a different aspect of gravity drainage from buildings:

- **BS EN 12056-1:** 2000 General performance requirements – Part 1 establishes the general and performance requirements of gravity waste water drainage systems only.

- **BS EN 12056-2:** 2000 Sanitary pipework, layout and calculation – Part 2 describes the standards of both layout and calculations required for gravity waste water drainage systems.
- **BS EN 12056-3:** 2000 Roof drainage, layout and calculation – Part 3 deals with roof drainage where the gutter/roof outlets do not limit the flow of water through them. This part of the standard applies to all gutters and material types.
- **BS EN 12056-4:** 2000 Waste water lifting plants, layout and calculation – Part 4 describes the layout, operation and maintenance requirements of lifting plants containing faecal and non-faecal carrying waste water and rainwater from buildings, including pipework and drain connections.
- **BS EN 12056-5:** 2000 Installation and testing, instructions for operation, maintenance and use – Part 5 deals with the installation, fixing requirements, support, thermal movement and testing of gravity waste water drainage systems.

These documents should be consulted before designing and installing drainage systems within dwellings. They are written in such a way that they mirror the mandatory requirements of Approved Document H by giving advice on the design, installation, testing and maintenance of waste water systems so that by following the recommendations within these documents, Approved Document H will be complied with.

Manufacturers' technical instructions

When installing sanitary appliances, the manufacturer's installation instructions, including the connection to the drainage system, are followed. By complying with the manufacturer requirements for the appliance, we ensure that the installation also complies with the App. Doc. H and BS EN 12056.

AC1.8 Identify different types of sanitary appliances and components used in dwellings

Baths

Baths are available in many shapes and sizes and are manufactured using many different materials, including:

- Cast iron
- Fibre glass
- Pressed steel
- Epoxy resin
- Injection moulded polyurethane

Most baths will require assembling when they arrive on site. This assembling procedure – fitting the frame, feet, handles, taps and waste – is known as *'dressing'* the appliance. When installing the bath, care must be taken to ensure that the bath remains level in all directions. This is done using a spirit level on both sides and both ends and adjusting the feet appropriately. The height of the bath panel(s) must also be taken into account, to ensure that the bath is set to the correct height. The bath should be fixed to the wall using the brackets supplied and the feet screwed to the floor to ensure stability and to limit movement. Most lightweight baths will also have a centre fixed foot, to prevent the bath sagging when it is full of water and occupied by a person.

WCs

There are four (4) styles of WC and three (3) different types of WC pan. First, we will take a brief look at the WC styles. WC pans are traditionally made from Vitreous China, but can also be made from stainless-steel and fireclay, depending on the application:

Close coupled

Here the WC pan and the cistern are fixed together into a single appliance. There are many different styles available from traditional to contemporary.

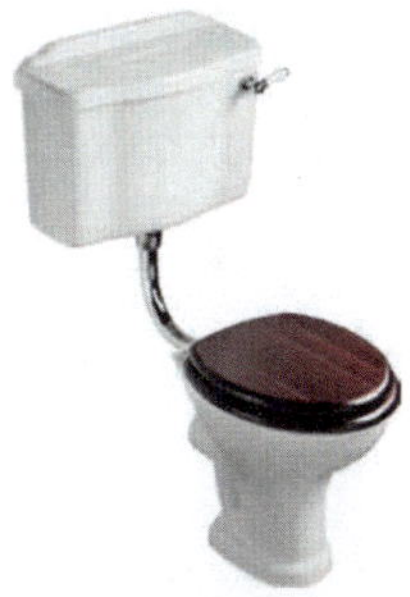

Low level

A very traditional style with a separate pan and cistern connected with a flush pipe.

High level

Very traditional style often used when a 'period' look is required. The height of the cistern gives a very good flush pressure.

Back-to-wall

Designed to be fixed against a wall or cabinet with a hidden WC cistern. These give a clean line to the bathroom, hiding the necessary drain and water connections

WC flushing mechanisms

Traditional dual flush syphon

This uses syphonic action by lifting a column of water, allowing syphonage to take place into the WC pan

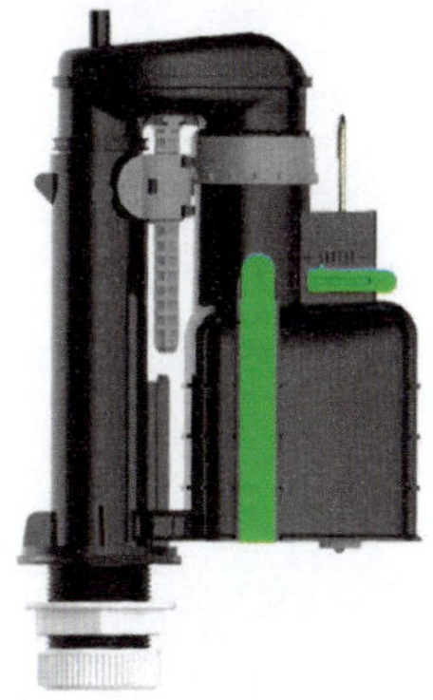

Pneumatic dual-flush valve

This is a simple flap-type valve that is operated by a blast of air from a push button on the WC cistern

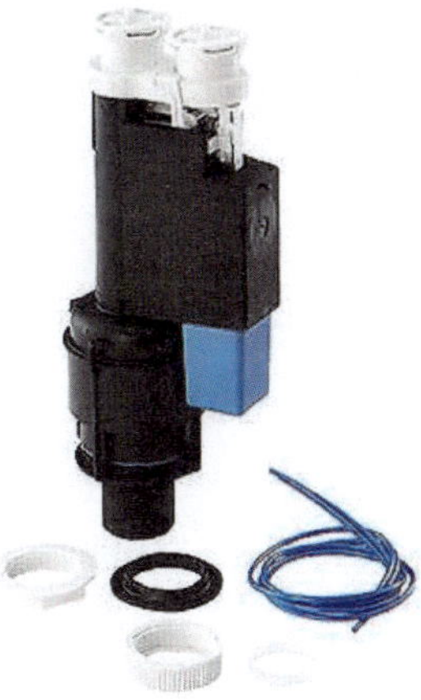

All WC cisterns are dual flush – 4 litre short flush and 6 litre long flush – to comply with Water Supply (water fittings) Regulations 1999. Most cisterns will contain either a syphon or a pneumatic dual-flush valve. These are shown on page 285.

All WC cisterns are provided with a float operated valve to the WC cistern. The float operated valve (FoV) for WC cisterns can either be bottom fed or side fed and all have ½ " BSP male threads to allow connection to the water supply via a ½" x 15mm tap connector. The FoV should be provided with a servicing valve for maintenance, repair and replacement

WC Pan types

There three (3) types of WC pan:

1. The wash down pan
2. The single trap syphonic pan
3. The double trap syphonic pan

Wash down pan

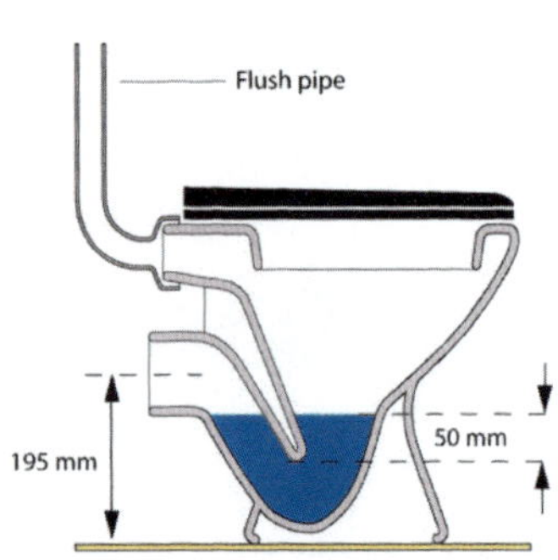

The wash down pan is probably the most common WC pan fitted in the UK. It works by simply using the force of the water flush to wash away the contents of the pan.

Single trap syphonic pan

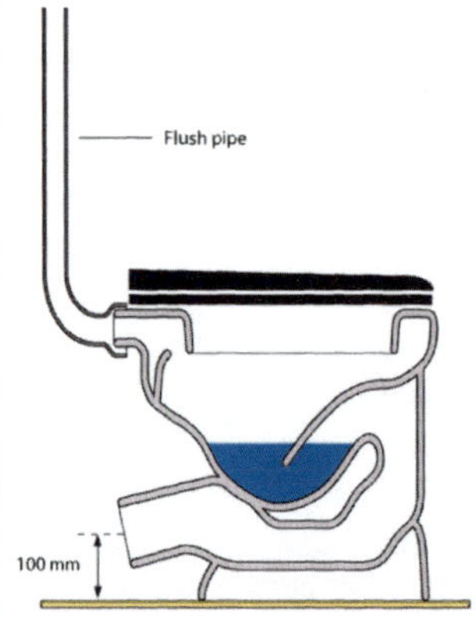

Often called the 'malvern' pan, the single trap syphinic WC pan has a lower outlet than other WC pan designs. Rather than clearing water backwards, this type pushes the water forwards into a narrow chamber. The chamber uses 'Bernoullis principle' to create a rapid increase in water velocity whilst creating a vacuum behind it. It is this vacuum that clears the contents of the pan.

Double trap syphonic pan

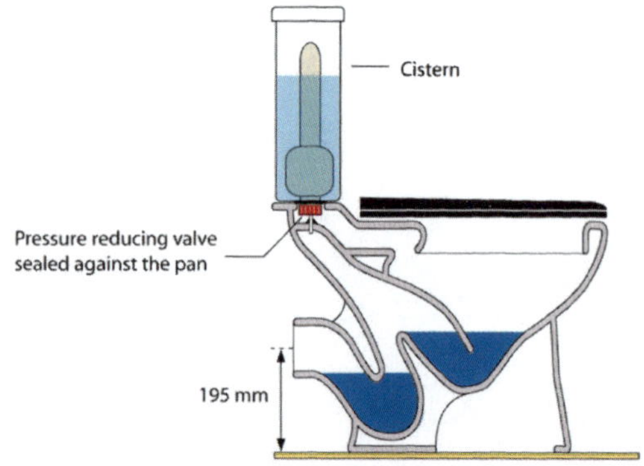

Rarely seen in the UK due to the high cost of manufacture, the double trap syphonic pan has two water traps. When the WC is flushed, a negative pressure is created in the chamber between the two pans by an aspirator (a special pressure reducing device) attached to the syphon in the cistern, which sucks out the air in the centre chamber. This creates a vacuum in the chamber that literally sucks out the contents of the pan through the two traps. The aspirator retains a little water, which is released when the flush cycle is complete, to reseal the traps.

Over rim bidets

Over-rim bidets installed in domestic premises can either have single or combination taps installed. All taps fitted to over-rim bidets must have an AUK2 air gap (air gaps and backflow prevention is covered in AC2.9) provided between the outlet of the tap and the spill-over level of the bidet. The Water Regulations Advisory Scheme (WRAS) Installation Requirement Note IRN R010 also stipulates that the water supply to combination taps shall have reasonably balanced pressures. Where unbalanced water pressures are used, then approved single check valves must be fitted immediately upstream of both hot and cold supplies. Over rim bidets have the same installation requirements as washbasins. The waste size is 1 ¼ " with a wastepipe size of 32mm. Waste pipe gradient and lengths are identical to those of a washbasin. Bidets are manufactured from vitreous china.

Ascending spray bidets

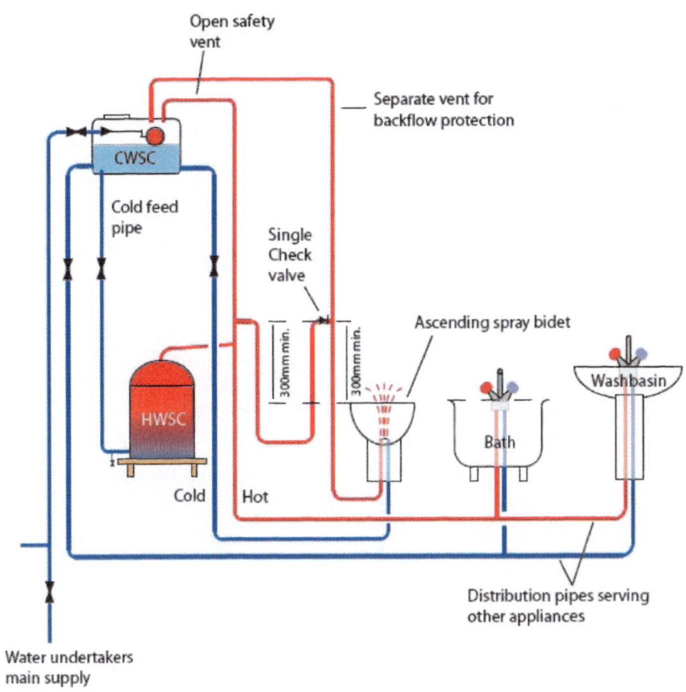

Ascending spray bidet installation

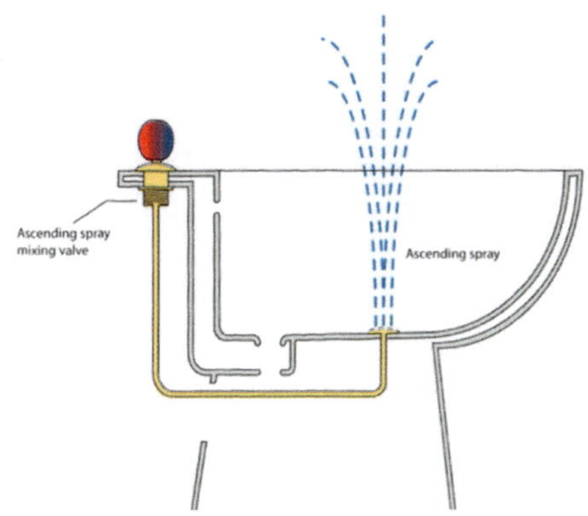

Ascending spray bidet

Ascending spray bidets discharge water from either a mixer tap or from a special outlet in the bottom of the bidet. The special outlet is a spray that sprays water upwards. Because the water in the bidet bowl may be contaminated with human bodily fluids, the water must be classified as ***fluid category 5 – water that carries a significant health risk.*** This means that the hot and cold water supplies to the appliance must be installed in such a way that no cross-connection with any water used for human consumption is possible. The drawing above shows a typical ascending spray installation and the complex pipework system that must be fitted to ensure the safety of the end user.

Ascending spray bidets have the same waste installation requirements as washbasins. The waste size is 1 ¼", with a wastepipe size of 32mm. Waste pipe gradient and lengths are identical to those of a washbasin. Bidets are manufactured from vitreous china.

Wash hand basins

Wash basins come in many shapes and sizes, from small models for cloakrooms to large washbasins for luxury bathroom installations. There are many different types. Four (4) of the more common types are shown below:

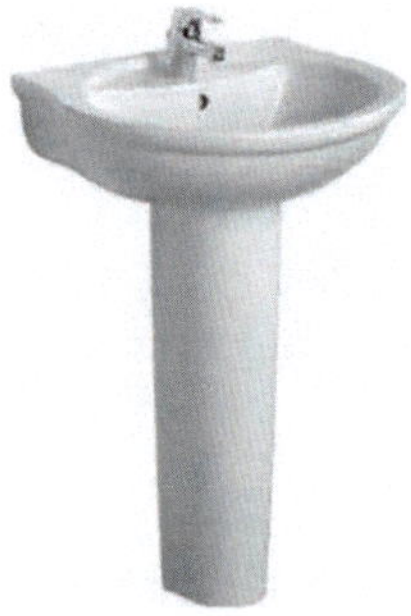

Pedestal washbasin

Wall mounted

Countertop

Semi-countertop

Installing washbasins

Washbasins are usually manufactured from vitreous china but stainless-steel models for commercial and anti-vandal use are also available.

The method of fixing wash basins largely depends on the washbasin type:

- **Pedestal washbasins:** These are often screwed to the wall using brown wallplugs and 2 ½ x 10 brass screws. The pedestal, when screwed to the floor provides extra stability. The pedestal provides enough space for pipework to be hidden behind it, creating a neat, pipe-free appearance.
- **Wall mounted washbasins:** These often use special brackets that are fixed to the wall for the washbasin to hang on. Modern types also have a 'semi-pedestal' to hide the pipework.
- **Countertop/vanity units:** here, the washbasin is fixed using silicone sealant to provide a robust fixing. Some also use small fixing brackets that are hidden within the bottom cabinet.
- **Semi-countertop washbasins:** These are cut into a work surface and hang half-on/half-off the work surface. They are exceptionally neat, with a narrow space saving cabinet to provide storage space. They are fixed using silicone sealant.

Wash hand basins installed in domestic premises can either have single or combination taps installed. Single taps have ½ " male BSP threads while combination taps often have 10mm copper tails. All taps fitted to wash basins must have an AUK2 air gap (air gaps and backflow prevention is covered in AC2.9) provided between the outlet of the tap and the spill-over level of the basin.

Again, Installation Requirement Note IRN R010 also stipulates that the water supply to combination taps shall have reasonably balanced pressures. Where unbalanced water pressures are used, then approved single check valves must be fitted immediately upstream of both hot and cold supplies. The waste size is 1 ¼", with a wastepipe size of 32mm.

Sinks

It would be impossible to state every type of sink, there are simply too many to mention here. Below is a selection of the different types of sink:

London/Belfast sinks[2]

Stainless-steel sink with drainer

Inset sinks

Sinks are available manufactured from a variety of materials, including:

- Fireclay
- Stainless-steel
- Agglomerate marble
- Epoxy resin
- Glass reinforced plastic

There are also a variety of configurations:

- Single bowl
- Single bowl + drainer
- Bowl and ½ + drainer
- Double bowl + drainer

All sinks for domestic purposes, whether manufactured from ceramics or stainless steel must have taps that give an AUK3 air gap. Sink taps can either be single high-neck pillar taps, combination taps or wall mounted bib taps, depending upon the installation requirements. The water in sinks should be regarded as fluid category 5 (see textbook 1). Single taps have ½" male BSP threads while combination taps often have 15mm copper tails. Waste pipe gradients and waste pipe sizes are shown on page 266.

Urinals

Urinals, although not a domestic appliance, require careful consideration. Connecting a urinal to the water supply can be completed in one of either two ways:

- Via an automatic flushing cistern, or;
- A manual or automatic flushing valve

[2] Belfast sinks have an integral overflow. London sinks do not.

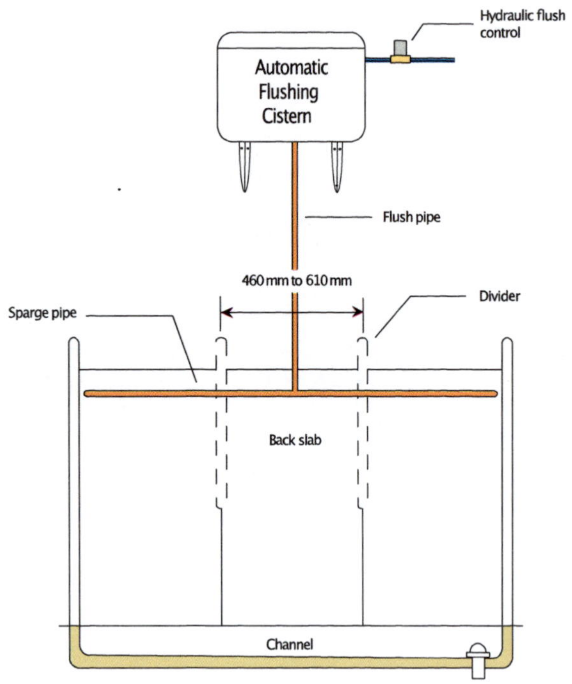

Installation of a slab urinal

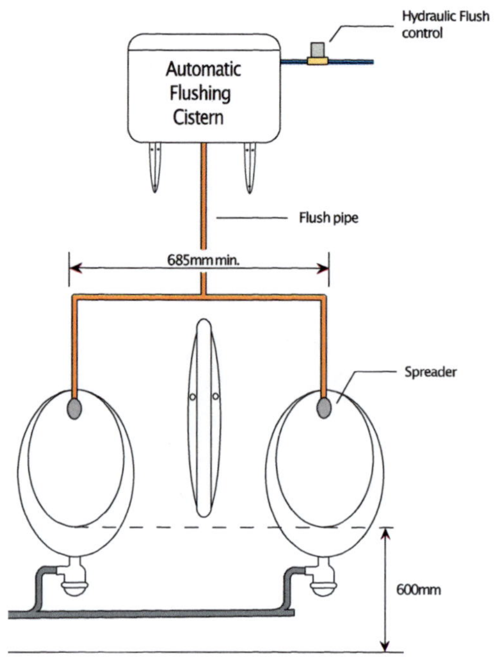

Installation of bowl urinals

Urinals using automatic flushing cisterns must be installed in such a way that does not constitute a wastage of water. WRAS note IRN R075 states that urinals using automatic flushing cisterns should be installed to give a flush rate not exceeding 10 litres per hour for a single urinal bowl and 7.5 litres per hour for each 700mm of a slab urinal or two or more urinal bowls. Urinals using flushing valves must not deliver a flush exceeding 1.5 litres per single flush per bowl each time the valve is operated.

In each case, the water supply must be protected by an air gap or some such no less effective mechanical device in accordance with fluid category 5.

Below are some examples of urinal types commonly installed in the UK.

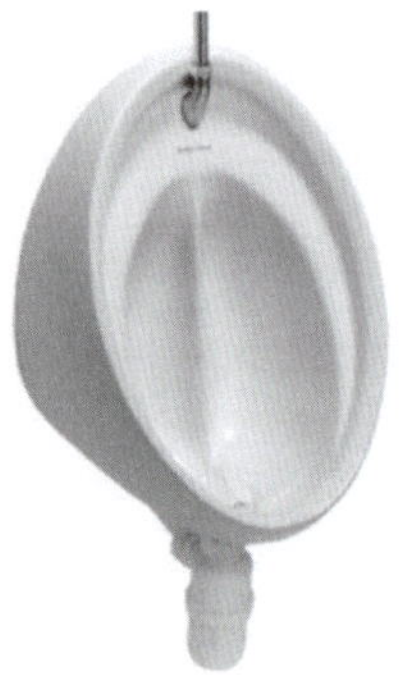

Single bowl urinal

Slab urinal

Trough urinal

Macerators and waste disposal units

Macerators

A macerator is a component that is used on bathrooms and WCs that are installed either too far away from a soil stack or below the level of the drain to remove the waste and foul water, i.e. in a basement or cellar. They contain a series of very sharp stainless-steel blades and a pump. The blades are rotated by a motor that turns any solid human waste and sanitary paper into a thin slurry that is pumped to the soil stack or drain. In some instances, the pipe size can be as little as 22mm. there are certain points to note:

1. If the drain is higher than the installation, the discharge pipe from the macerator must rise vertically first before any horizontal run.
2. The maximum pipe lengths must never be exceeded.
3. 90° Elbows must never be used on the pipework as these have a tendency to block easily. Two 45° bends or machine-made bends are acceptable.
4. Plastic discharge pipe should be solvent welded type and never push-fit. This is to prevent blow-off of the fittings in the event of a blockage.
5. Regular maintenance of macerators is advisable. However, it must be remembered that you are dealing with human waste and the appropriate PPE MUST BE WORN AT ALL TIMES!
6. They should be fitted in accordance with the manufacturer's instructions at all times.

Waste disposal units

A waste disposal unit is fitted to a kitchen sink, and, like macerators, they contain a series of sharp blades that turn solid plant matter into a slurry which is washed down the sink to a drain. On no account must bone waste be placed in the waste disposal as this traps the blades and causes the unit to overheat and cut out.

The end user should be reminded to keep fingers well out of the way of the opening.

AC1.9 Identify factors that lead to trap seal loss in sanitary pipework systems

Trap seal loss in sanitary systems can lead to unpleasant odours entering the property and can often result in people within the building feeling unwell. Trap seal loss is often caused because:

1. the wastepipe is too long

2. the waste pipe is too small
3. the waste pipe has been laid to the wrong gradient
4. incorrect bends installed at the foot of a soil stack
5. too many appliances on a single waste pipe
6. too many direction changes on a single waste pipe.

All of these situations can cause trap seal loss. There are many ways that trap seals are breeched:

Self-siphonage

When water is discharged from an appliance, it forms a plug. As the plug of water travels down the wastepipe, it creates a partial vacuum behind it, because of the water retained by the trap. The faster the water travels or the longer the wastepipe, the more the vacuum created will pull on the water left in the trap. Eventually, in extreme cases, the water in the trap will be pulled out by the travelling plug of water. Induced syphonage occurs on single appliances where:

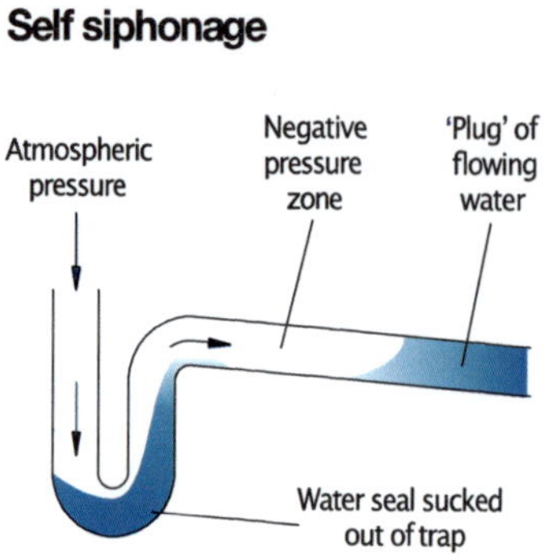

- the waste pipe is too long, or;
- the waste pipe fall is too great causing the water velocity to be excessive.

Induced siphonage

Induced siphonage occurs when water flowing from one appliance pulls the water from the trap of another appliance connected to the same waste pipe. It often occurs where a bath and a washbasin are connected to the same waste pipe or where a group of washbasins are connected to a common waste pipe.

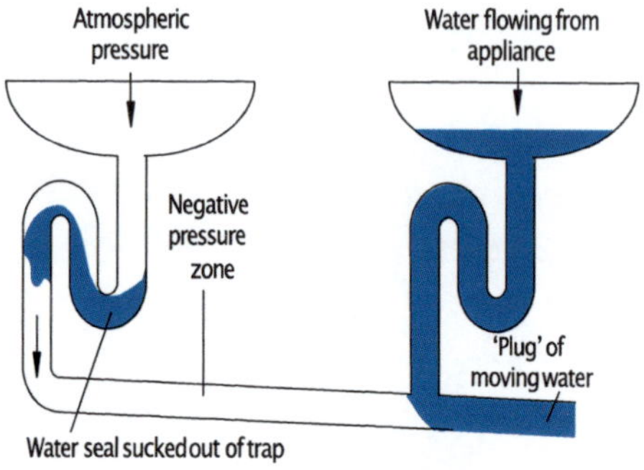

Induced siphonage can be prevented by following some simple rules:

- where a bath and washbasin are to be connected to a single waste pipe, increase the size of waste pipe entering the stack to the next size up from the waste pipe size of the largest appliance. In other words, if the appliance is a bath requiring a 40mm pipe, increase the waste pipe to 50mm.
- install and extra branch ventilation pipe in iline with the recommendations of the ventilated branch discharge system (see page 268)
- where a group of washbasins is being installed, increase the size of the pipe to accommodate all of the washbasins. i.e. increase from 32mm to 40mm with 32mm branches to each washbasin.

Compression

Compression in a soil and vent pipe occurs at the bend at the base of the stack. As water flows from a flushed WC, the water falling down the soil pipe hits the bend at the stack's base and momentarily stops. This sends a shockwave of back pressured air up the stack, which, in extreme cases can be sufficient to blow out the trap from the nearest appliance. It can be eliminated by installing a large radius bend or two x 45° bends at the base of the stack (see left hand drawing).

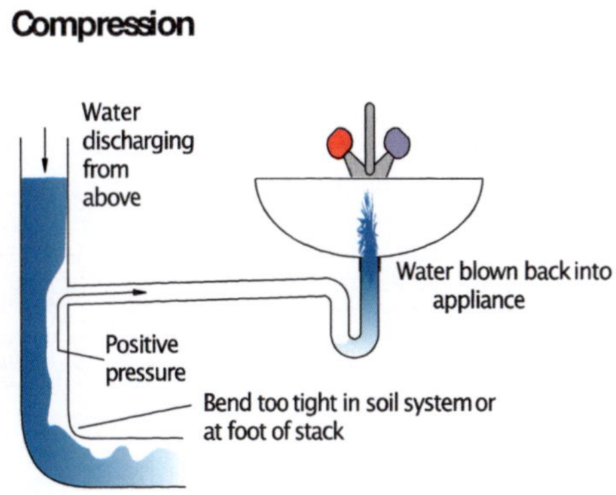

Wavering out

Wavering out is often difficult to cure because it occurs naturally as the wind blows across the top of the soil stack. You will often notice how the water in the WC bounces from side to side when strong winds are blowing. This side to side motion can, in rare occurrences, be strong enough to move water out and over the trap.

An anti-wind cowl, often called a 'mushroom' cowl

The easiest solution to wavering out is to fit a wind cowl on the top of the soil stack so that adverse wind conditions do not affect the soil and vent system.

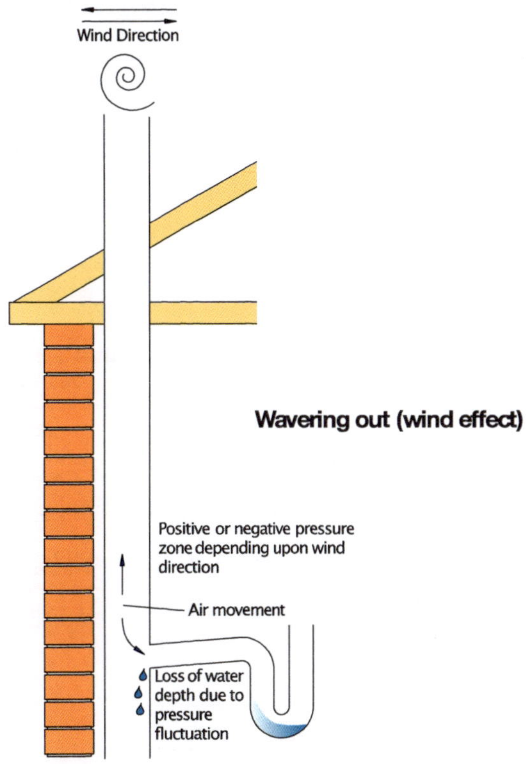

Evaporation

Evaporation is a natural occurrence when appliances go unused for long periods of time. Appliances rely on regular use to keep the traps topped up with water. As soon as the appliance usage drops, then natural evaporation of the water will occur, usually at the rate of 2.5mm per week. This increases during periods of hot, dry weather.

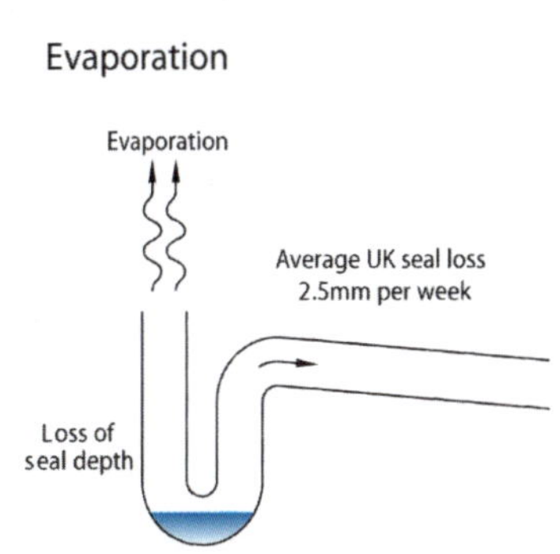

Capillary attraction

This only occurs when there is something in the trap for the capillary action to take place, such as long hair. If the hair gets stuck across the trap seal, then the water will be soaked upwards by the hair and fall downwards through gravity. This will create a weir, slowly releasing all of the water out of the trap.

Momentum

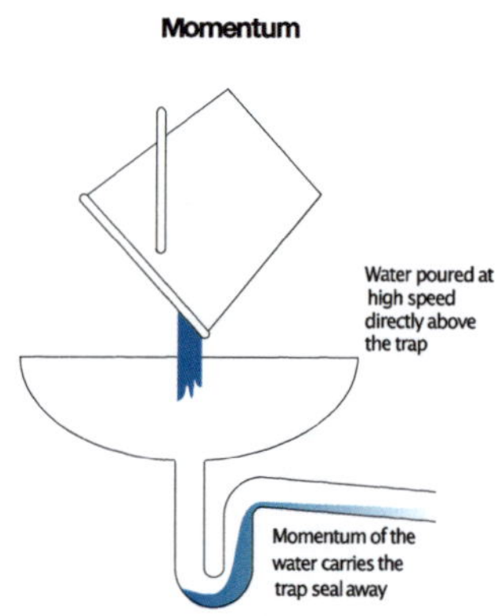

When a large amount of water is discharged down an appliance, such as water being poured from a bucket, then a condition called momentum can take place. It occurs because of the velocity of the water being discharged. The water in the trap simply follows the other water being discharged and the trap seal is lost. It is similar to self-siphonage mentioned earlier.

Foaming

This occurs when too much detergent is used in an appliance. The excessive foam that is created simply backs up the waste pipe soaking away any water in the appliance traps. It can often be detected because of foam emerging from other appliances where the detergent has not been used.

AC1.10 Outline the suitability of below ground drainage systems to receive waste water

So far, we have looked at above ground sanitation systems, components, appliances and fittings. We have investigated how we get the used water out of the appliances and into the pipework, the systems we use and the methods we can employ.

But what happens to the water once it goes below ground? What systems can we use to discharge the water into and what do the pipework arrangements look like?

There are three methods of below ground drainage that can be used to discharge water from sanitary appliances:

- The separate system
- The partially separate system
- The combined system

The separate system

With this system, the water from the building discharges into two very separate drainage systems. The foul and waste water from sanitary appliances discharges into the foul water drain where the effluent terminates at a waste water treatment plant. The rainwater from gutters and rainwater pipes discharge in to the surface water drain, which, in many cases terminates at a water course such as a river.

It is vitally important that we install the appliances on to the correct drainage system. Connection of, say a WC, to a surface water drain will result in the water and all its contents being discharged into a river or other water course with the result that untreated sewage would contaminate that water course. This is an offence under the Building Regulations.

A diagram of the system is shown on the following page:

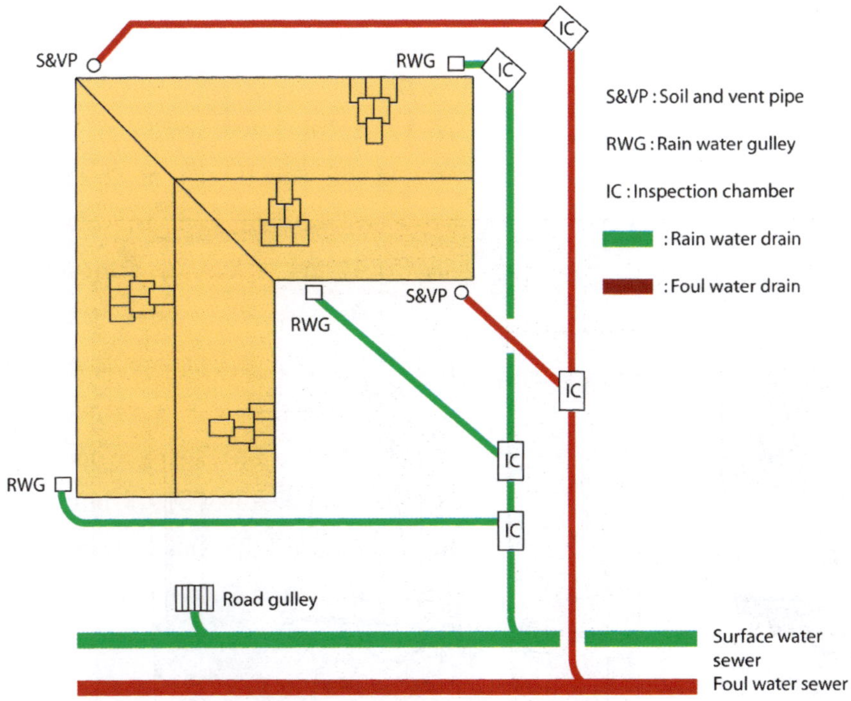

The separate system of below ground drainage

As can be seen from the drawing, the two separate drainage systems lead to two totally separate drains in the road.

The main problem with this system is that it is extremely expensive to install and system planning can be troublesome, especially if the systems need to cross one another. Another potential problem is that the foul water drains are not flushed through during heavy rainfall.

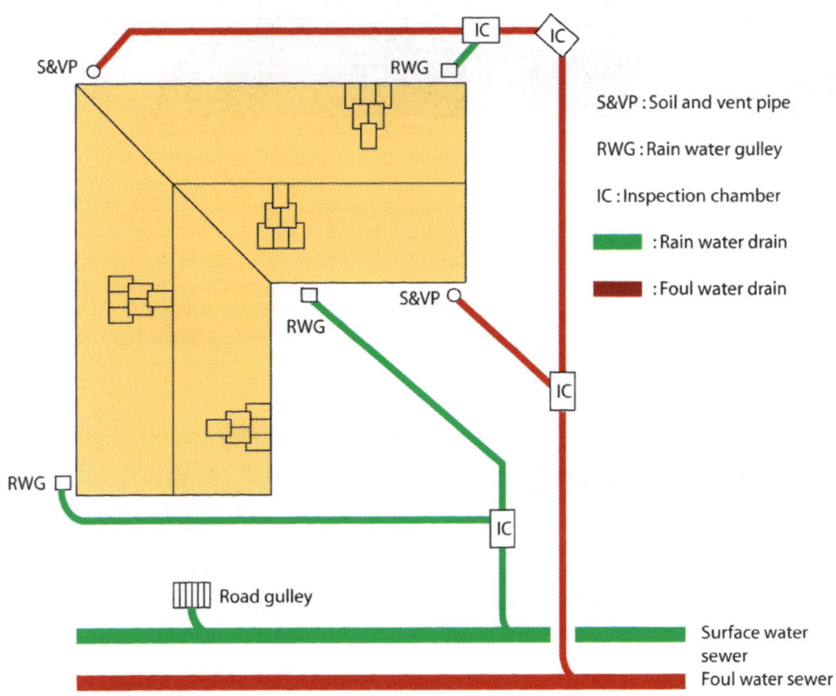

The partially separate system of below ground drainage

The partially separate system

This system is very similar in design to the separate system of below ground drainage. However, here two drains are still used as before but some of the rainwater that cannot be accommodated by the surface drain, is discharged into the foul water drain. It will be noted on the diagram that the rear rainwater discharges into the foul drain. Again, care is required during sanitary appliance installation to ensure that the correct drain is being used.

One of the benefits of this system is that installation costs can be reduced as there is no need to extend the surface water drain to the other rear rainwater outlets and the foul water drain is partially flushed through with rainwater during rainfall.

This system is also known as the partially combined system.

The combined system

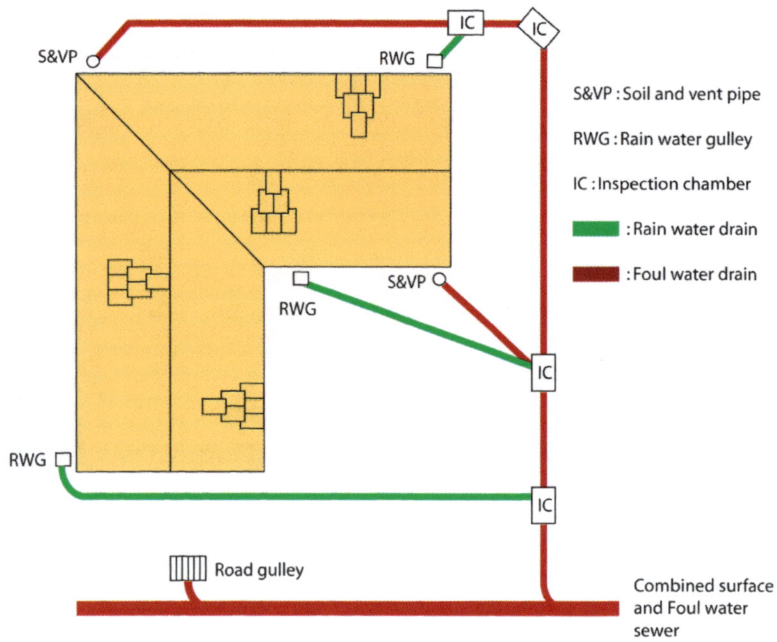

The combined system of below ground drainage

With the combined system of below ground drainage, there is no distinction between foul water and surface water. All waste water – foul and rainwater – discharges into the same drainage system. This is a very economic system to install, since there is only one drain to connect to, saving costs on both materials and installation. It should be noted, however, that this system is not recognized by the Building Regulations as a system that can be used on new installations.

One of the main problems with this system is that, during periods of heavy rain, the drainage system can become flooded and, since all water must pass through the waste water treatment works, treatment is difficult to handle due to the sheer volume of water.

Soakaway drains

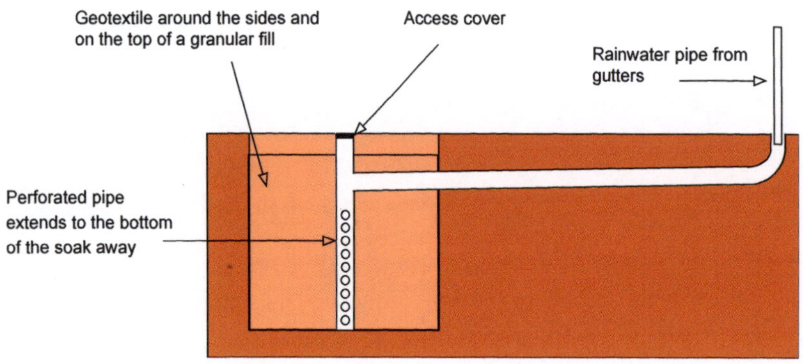

The soakaway is a traditional method of dispersing rainwater run off from roof areas. They are formed from square or circular pits filled with rubble. For small soakaways, serving 25m2 or less of roof area, a design ratio of 10mm rainfall in 5 minutes is quoted in Approved Document H. For soakaways greater than this, reference should be made to BS EN 752 part 4.

Soakaways must not be sited within 5 m of a building or 2.5 m within a boundary wall. They should also not be installed where:

- The ground is unstable
- The water table is high
- Near a drainage field, drainage mound or another soakaway
- Where the rainwater run off could result in contamination of the ground water source.

AC1.11 Installation requirements of sanitary facilities and equipment in dwellings for the disabled including wet rooms

A wet room for a disabled person

A WC for a disabled person

Bathrooms and WC facilities for persons with a physical disability or impaired mobility require special consideration. The requirements for these are found in Building Regulations **Approved Document M Volume 1: Dwellings**, which is split into 3 parts:

- M4(1) Category 1: Visitable dwellings

  Here, reasonable provision in new dwellings should be made for most people, including wheelchair users to enter and access the dwelling's habitable rooms and sanitary facilities at entrance level.

- M4(2) Category 2: Accessible and adaptable dwellings

  Here, reasonable provision should be made in new dwellings for a range of occupants to gain access, including a range of features suitable for those of limited mobility, the elderly and some wheelchair users.

- M4(3) Category 3: Wheelchair user dwellings

  This requirement covers reasonable provision for wheelchair users to live in a dwelling. It also requires that they must be able to use any outdoor space and carparking facilities.

For buildings other than dwellings **Approved Document M Volume 2: Buildings other than dwellings** is applicable, which again, is split into 3 parts:

1. M1 Access and use of buildings other than dwellings

2. M2 Access to extensions to buildings other than dwellings
3. M3 Sanitary conveniences in extensions to buildings other than dwellings

The requirements of M1 to M3 are that reasonable provision must be made to ensure that all buildings are accessible and usable to individuals, regardless of their age, gender or disability with respect to:

- Gain access
- Use their facilities

Both of these requirements are applicable to both workers within the building, those persons that live within the building and any visitors.

The requirement (taken verbatim from the Building Regulations)

Requirement Category 1 – visitable dwelling Access and use M4(1).	
Requirement	*Limits on application*
Reasonable provision should be made for people to— (a) gain access to; and (b) use, the dwelling and its facilities	Requirement M4(1) does not apply to: (a) an extension to a dwelling; or (b) any part of a building that is used solely to enable the building or any service or fitting in the building to be inspected, repaired or maintained.
Requirement Category 2 – accessible and adaptable dwellings M4(2) optional requirement	
Optional requirement	*Limits on application*
(1) Reasonable provision must be made for people to— (a) gain access to; and (b) use, the dwelling and its facilities. (2) The provision made must be sufficient to— (a) meet the needs of occupants with differing needs, including some older or disabled people; and (b) to allow adaptation of the dwelling to meet the changing needs of occupants over time.	Optional requirement M4(2)— (a) may apply only in relation to a dwelling that is erected; (b) will apply in substitution for requirement M4(1); (c) does not apply where optional requirement M4(3) applies; (d) does not apply to any part of a building that is used solely to enable the building or any service or fitting in the building to be inspected, repaired or maintained.
Requirement Category 3 – wheelchair user dwellings M4(3) optional requirement	
Optional requirement	*Limits on application*
(1) Reasonable provision must be made for people to— (a) gain access to, and (b) use, the dwelling and its facilities.	Optional requirement M4(3)— (a) may apply only in relation to ad welling that is erected; (b) will apply in substitution for requirement M4(1);

(2) The provision made must be sufficient to— (a) allow simple adaptation of the dwelling to meet the needs of occupants who use wheelchairs; or (b) meet the needs of occupants who use wheelchairs.	(c) does not apply where optional requirement M4(2) applies; (d) does not apply to any part of a building that is used solely to enable the building or any service or fitting in the building to be inspected, repaired or maintained. Optional requirement M4(3) (2)(b) applies only where the planning permission under which the building work is carried out specifies that it shall be complied with.

Accessible WC facilities

There are two specific types of accessible bathrooms:

Ambulant accessible WCs – these are the most widely used and probably the most straightforward to design and install. Usually, these are installed as part of a larger WC cubicle installation, such as in a place of work or a public facility. They are similar in size to other WC cubicles but there are subtle differences:

a. The WC parts, i.e. WC seat, WC cistern handle, WC pan will be higher than standard installations.
b. Grab rails will be included at strategic positions to aid with moving around the cubicle, sitting and using the washbasin.
c. An outward opening door.

They do not, however, offer wheelchair access.

Dedicated Accessible WCs – These differ greatly from standard WC installations. They are designed to be used by a wider range of users and offer a wider range of specific features:

a. Wheelchair accessibility
b. The WC parts, i.e. WC seat, WC cistern handle, WC pan will be higher than standard installations.
c. Grab rails will be included at strategic positions to aid with moving around the cubicle, sitting and using the washbasin.
d. A washbasin with taps specifically designed for persons with limited hand movement, such as lever operated taps and infra-red operated taps.
e. Emergency alarms (via a red pull cord)
f. Outward opening doors that are at least 900mm wide
g. An overall floor space of 1500mm x 2200mm
h. Sanitary ware should be spaced in accordance with the diagram shown.

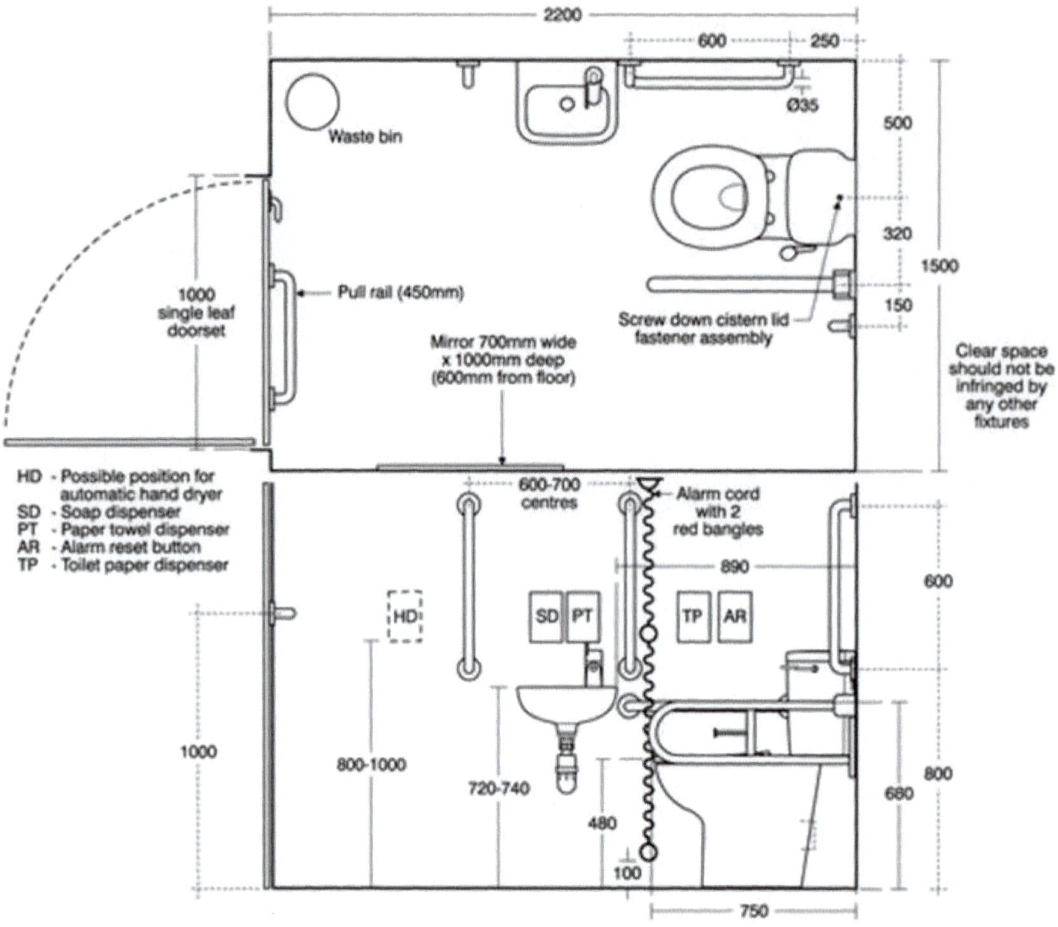

A typical accessible WC floor plan

Accessible bathroom facilities

The main points to observe when designing bathrooms for people with a physical disability is for the need for space, especially if the person is a wheelchair user. For this reason, many people opt for wet rooms and walk-in showers or walk-in baths. Here are some points to remember:

1. The layout of the bathroom will depend upon the needs of the user. There is no 'one solution fits all'.
2. The user may require support, especially in a shower or wet room. This can be provided by the use of strategically positioned grab rails or shower seats.
3. A shower screen would keep the rest of the room dry and possibly prevent slippery floors.
4. A raised height washbasin with lever operated/infra-red operated taps.
5. If a bath is requested, a walk-in bath may be the best option. However, there other considerations:
    a. The material the bath is made from
    b. The length, width and height of the bath
    c. Will a bathboard, bath lift or bath hoist be required?
    d. What alternative bath designs are there? i.e. sits baths, corner baths etc.

AC1.12 Identify jointing methods used in sanitary appliances pipework systems

Jointing plastic soil and waste pipes

There are 3 basic methods of jointing soil and waste pipes. These methods depend on the pipe material. They are:

1. Solvent weld – for PVCu and ABS soil and waste pipes
2. Push-fit:
    a. for PVCu waste pipes
    b. for polypropylene waste pipes
3. Compression – a universal fitting to connect pipework of differing materials.

Solvent weld – for PVCu and ABS soil and waste pipes

Solvent weld fittings use a solvent cement that literally melts the surface of the plastic and fuses the fitting and the pipe together. Care should be taken when using solvent weld adhesive as the fumes given off can be overpowering. It is recommended that its use should only be restricted to outside or a well-ventilated room.

To make a solvent weld joint on PVCu, muPVC and ABS, the following procedure should be followed:

1. Cut the pipe with either a hacksaw or a multi-purpose saw and remove the burr from the inside and outside of the pipe with a file
2. Wipe the dirt and manufacturer's oils off the pipe with a clean, soft cloth
3. Clean the inside of the socket and the outside of the pipe with the correct solvent cleaner
4. Apply the solvent weld cement to the inside of the socket first, and then to the outside of the pipe. This will allow more time to make the joint as the cement cures much quicker when exposed to air.
5. Insert the pipe into the socket and slowly twist it into place. The twisting action allows for an even spread of cement throughout the socket.
6. Wipe off any excess cement with a clean, dry cloth.

After completion, the joint must be left up to 24 hours before testing can begin.

<table>
<tr><th>Solvent weld soil fitting</th><th>Solvent cement</th><th>Solvent cleaner</th></tr>
<tr><td>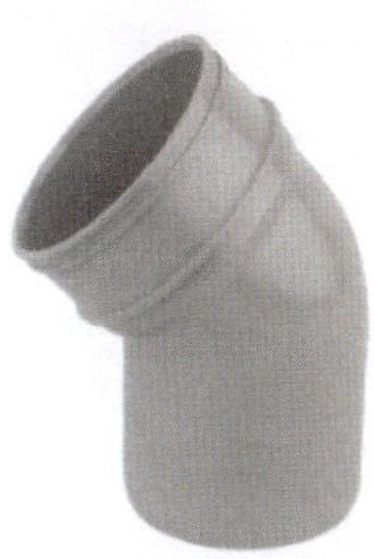</td><td>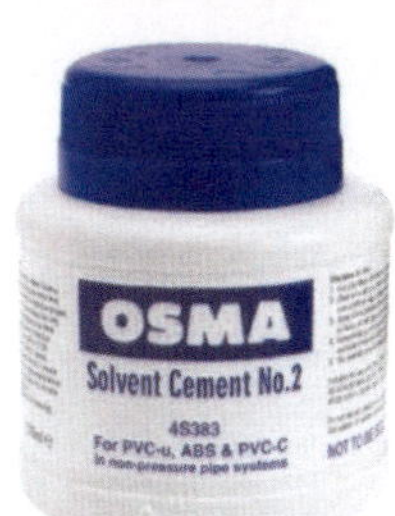
</td><td>
</td></tr>
</table>

Push-fit – for PVCu and ABS soil pipes

Push-fit joints are probably the most common jointing method for soil and vent pipes. The fittings have a single neoprene chamfered rubber sealing ring in them which allows for easy jointing. There are two points to remember when making push-fit joints:

1. Always chamfer the pipe before you try to push the fitting and the pipe together
2. Always use a silicon lubricant to help ease the pipe into the socket. Do not be tempted to use washing up liquid or similar soap because as this dries out, the joint will seize up and movement will be restricted. Also, soap eventually corrodes the rubber seal and so the joint may leak in the future.

To make a push-fit joint on PVCu, muPVC and ABS, the following procedure should be followed:

1. Cut the pipe with either a hacksaw or a multi-purpose saw and remove the burr from the inside and outside of the pipe with a file
2. Chamfer the end of the pipe to about 45° with a rasp or a file
3. Wipe the dirt and manufacturer's oils off the pipe with a clean, soft cloth
4. Apply a little silicon lubricant grease to the end of the pipe and to the rubber seal inside the fitting
5. Push the pipe all the way into the socket
6. Withdraw the pipe 10mm. This will allow for expansion of the pipe. Pipes must be supported by a pipe clip or bracket to ensure that the fitting/pipe does not slip.

Push-fit soil fitting	Silicon spray lubricant	Push-fit fitting for Polypropylene waste pipe systems
		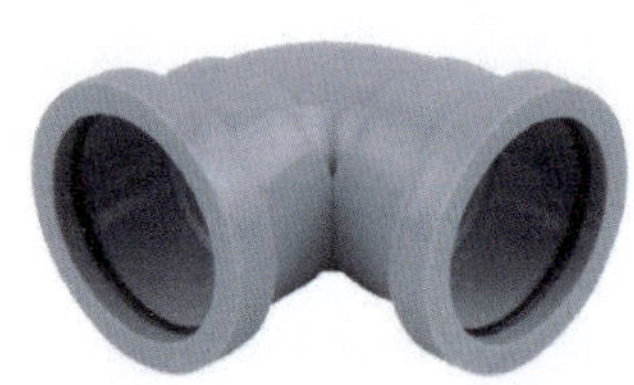

Push-fit – for Polypropylene waste pipes

Polypropylene waste pipe fittings are push-fit containing a single neoprene rubber ring seal. It cannot be solvent welded as polypropylene is unaffected by the solvents used in solvent weld cement.

To make a push-fit joint on polypropylene, the following procedure should be followed:

1. Cut the pipe with either a hacksaw or a multi-purpose saw and remove the burr from the inside and outside of the pipe with a file
2. Chamfer the end of the pipe to about 45° with a rasp or a file

3. Wipe the dirt and manufacturer's oils off the pipe with a clean, soft cloth
4. Apply a little silicon lubricant grease to the end of the pipe and to the rubber seal inside the fitting
5. Push the pipe all the way into the socket

A note about solvent weld and push-fit waste pipes and fittings – push-fit and solvent weld fittings are not compatible with each other. They have different outside diameters and the fittings will only fit their recommended pipe.

AC1.13 Explain working principles of greywater recycling systems

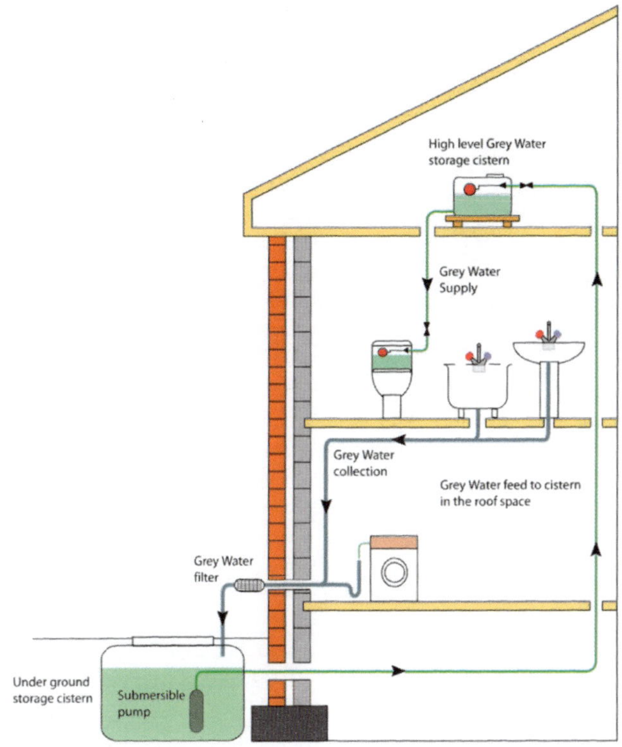

A greywater recycling system

Greywater recycling systems use water from washbasins, showers and baths to perform functions that may not necessarily require water from the water undertaker's cold water supply, such as flushing WCs.

In a grey water recycling system, water is collected from baths, showers, washbasins and washing machines into a cistern installed below ground, where it is filtered and disinfected. The water is then pumped to a holding cistern in the roof space before being distributed to the WCs within the property.

Grey water is usually clean enough for only a light disinfection before use. Problems can sometimes arise where bacteria can multiply, making the water smell. It should also be realised that grey recycled water has only a short life span before it becomes unsafe for use. As with rainwater harvesting, grey water recycling systems MUST be kept totally separate from the mains cold water supply to prevent contamination issues. Cross-connections between grey water recycling and mains cold water are strictly forbidden.

AC1.14 Install sanitary appliances, pipework systems and components

During your Phase 2 practical assessment, you may be required to install various sanitary appliances. Following the procedures in this section will provide you with the knowledge required for a successful Phase 2 completion.

LEARNING FOR LIFE

Learning Outcome 2

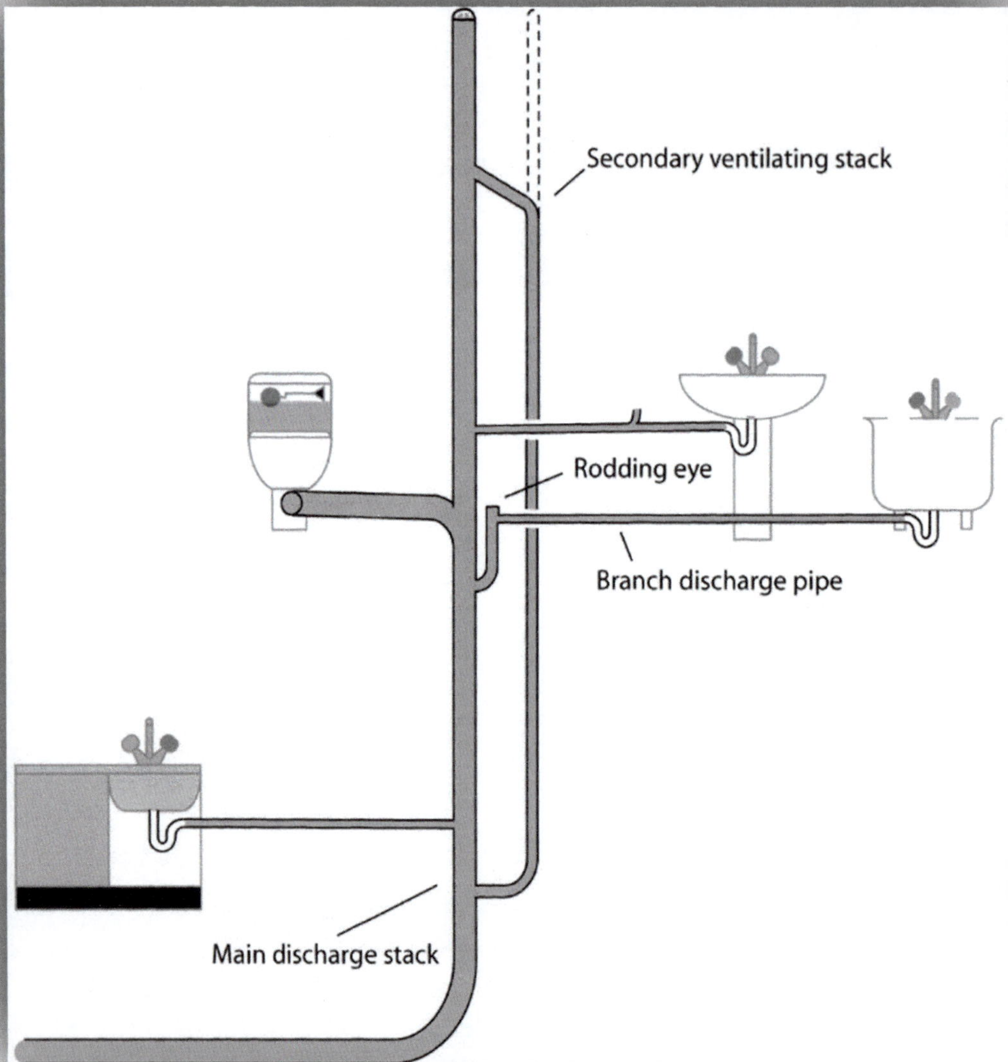

Decommission Sanitary Appliances, Pipework Systems and Components

Learning Outcome 2: Decommission Sanitary Appliances, Pipework Systems and Components

In Phase 2, you will be investigating the following Assessment Criteria from Learning Outcome 2:

LO2 – Decommission rainwater systems

AC2.1 Explain **procedures** for decommissioning systems

AC2.2 Carry out decommissioning **procedures**

AC2.1 Explain procedures for decommissioning systems

Decommissioning sanitary appliances and systems can be a dirty job and carries the risk of contact with human waste. This can be harmful to health. Here are some points for you to consider:

- Inform the householder that you are going to begin the removal of their sanitary appliances and soil and vent pipes. if the customer has any private property that may be affected, such as a car on the drive, it is a good idea to ask them to remove it until the process is complete.
- If the soil and vent pipe are outside the property, then apply warning notices that you are working overhead in areas where the customer or the general public may have access.
- Always wear appropriate PPE, especially when removing WCs and soil and vent pipes. It may be advisable to wear a face mask.
- Always check for hazardous materials such as asbestos before starting the decommissioning process. This may influence the PPE that you wear.
- Always use the correct type of access equipment, if working on external installations. Using a ladder to install and remove soil and vent pipes is not ideal. Installation and removal should be performed from a correctly erected scaffold wherever possible.
- Be aware that some materials are heavy. Cast iron, for instance, is a heavy but brittle material that needs to be taken down with extreme care. It can often snap when being handled roughly.
- Always dispose of any waste materials appropriately. Old appliances and soil/vent pipes MUST be disposed of at a recognised recycling centre, since metals and most plastics are recyclable.
- Always use recognised pipe drain plugs and cap ends when temporarily decommissioning systems and appliances. This will prevent debris from entering the pipework but also prevent unwanted smells from entering the property.

AC2.2 Carry out decommissioning

As part of your Phase 2 assessment you will be asked to install, test and decommission sanitary appliances and waste pipes. You may find it helpful to note the points made within this chapter to assist you.

REMEMBER:

1. Be aware of your surroundings and other people working in the vicinity
2. Always wear the correct PPE
3. Always use access equipment responsibly
4. Always dispose of waste or unwanted materials safely and responsibly
5. Always leave the work area in a clean and tidy manner.

Learning Outcome 4

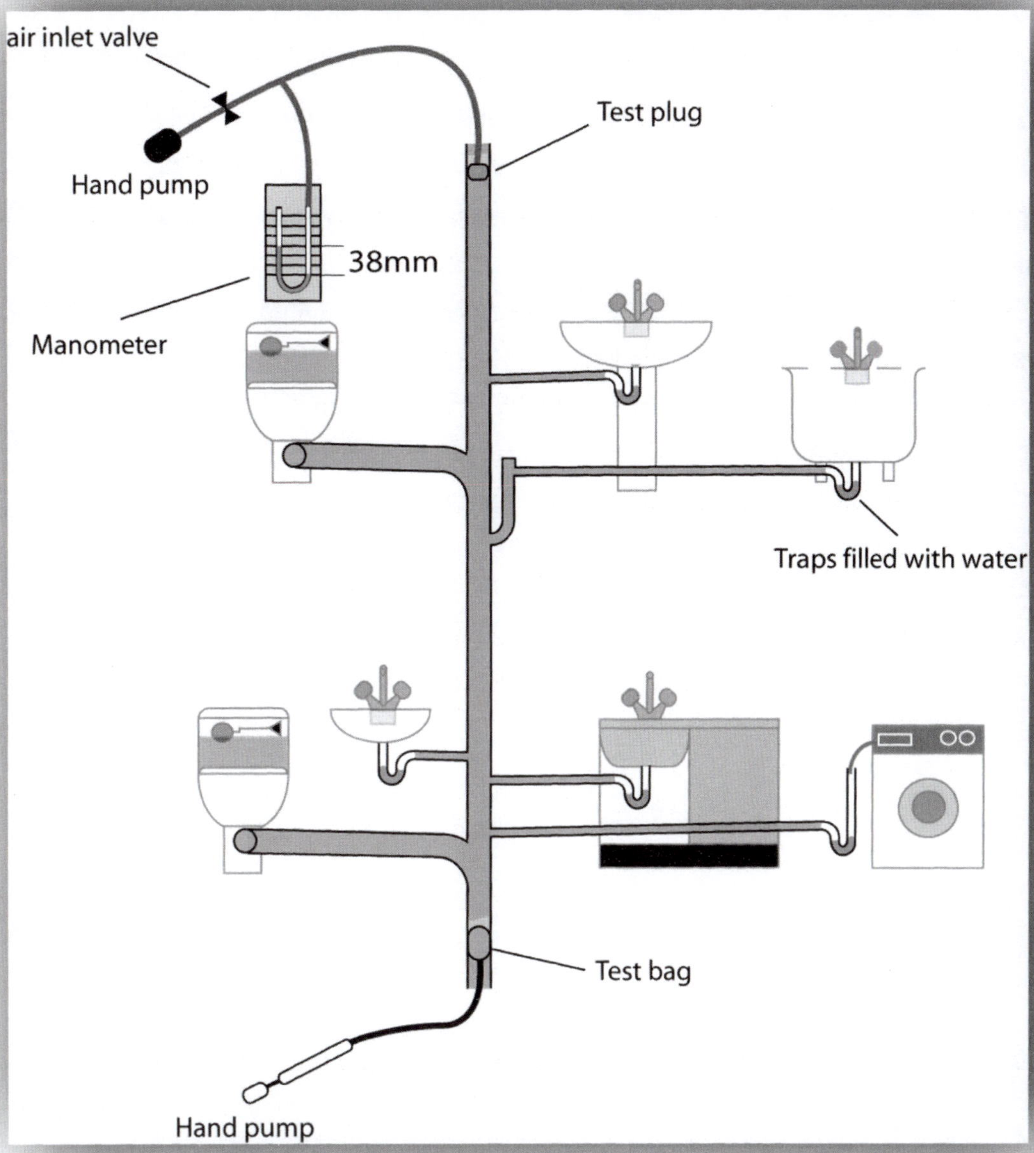

Perform a Soundness Test and Commission Sanitary Appliances, Pipework Systems and Components

Perform a Soundness Test and Commission Sanitary Appliances, Pipework Systems and Components

In Phase 2, you will be investigating the following Assessment Criteria from Learning Outcome 4:

LO4 – Perform a soundness test and commission sanitary appliances, pipework systems and components

- **AC4.1** Describe a **visual inspection** of sanitary appliances and pipework systems to confirm that it is ready to be soundness tested
- **AC4.2** Describe a **soundness** test to industry requirements on sanitary appliances, pipework systems and components
- **AC4.3** Describe **operational checks** required during commissioning
- **AC4.4** Identify actions that must be taken when commissioning reveals defects
- **AC4.5** Describe the procedure for handing over to the end user
- **AC4.6** Carry out soundness testing and **commissioning procedures**

AC4.1 Describe a visual inspection of sanitary appliances and pipework systems to confirm that it is ready to be soundness tested

Before testing the completed sanitary appliance and pipework system, we should first conduct a visual inspection of the installation:

Visual Checks – walk the installation to ensure that you are happy that the system is designed in accordance with **BS EN 12056** and **Approved Document H**, installed correctly and that all joints visibly appear to have been made correctly. Check that all traps have been tightened on to the appliances. it may be a good idea at this stage to put the appliance plugs in the waste outlets so that any accidental discharge of water is prevented from running down the system until you are ready to test. Check that there enough brackets and clips on the pipework and that the pipework is not damaged in any way.

When you are satisfied the installation is correct, then you can perform an initial test for leakage.

Leakage check – fill each of the appliances in turn and release the water down the waste pipe. Check that the trap is sound and that no leakage is present. Check all pipework and joints to ensure that there is no dampness or puddles of water. Flush the WC to ensure that it flushes correctly and that there is no visible leakage.

Once you are satisfied that the installation is water tight, you can begin the testing procedure.

AC4.2 Describe a soundness test to industry requirements on sanitary appliances, pipework systems and components

A soundness test on sanitary pipework uses air pumped into the system by a hand pump. There are several pieces of equipment that are required:

- Drain plugs
- A water gauge
- A hand pump
- Rubber tubing

There are four stages to the soundness test:

Stage 1 At stage 1, the pipework must be sealed top and bottom using drain test plugs. The top plug is sealed at the top of the vent pipe. the bottom plug can be inserted and sealed through the access cover at the base of the soil stack. Run a little water over both of these plugs to ensure a water seal. This can be done by partly flushing the WC with water from a bucket. The

A drain test plug

top plug can be sealed with water from a jug.

Stage 2 Fill all of the appliance traps on the system by running the taps of each appliance for a few seconds. This will, together with the drain plugs, ensure that the system is air tight.

Stage 3 On the top drain plug, attach a rubber tube using the correct attachment fittings. The tube will have a tee junction installed on it. The end that is not attached to the drain plug will be attached to the water gauge; the branch will have a hand pump on it. The water gauge, also known as a manometer, has measurements in mm/watergauge.

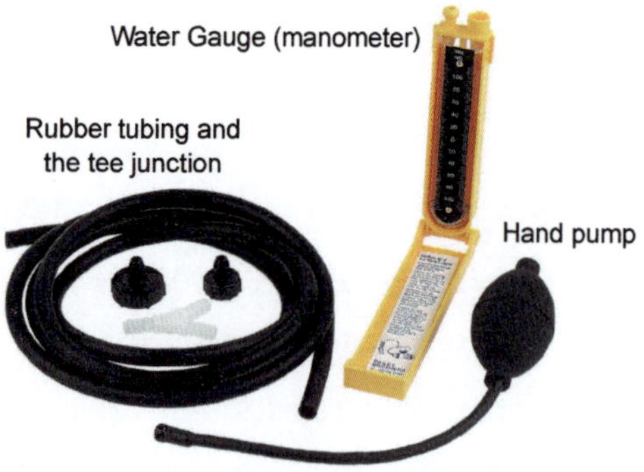

Stage 4 With the water gauge assembled as shown below, pump the hand pump until a reading of 38mm is present on the water gauge. This should hold for a period of 3 minutes, with no visible drop in pressure. Do not exceed 38mm, as the WC only has a water seal of 50mm and any pressure greater than 38mm may cause the trap seal in the WC to break, releasing the test air in the system back to zero (0).

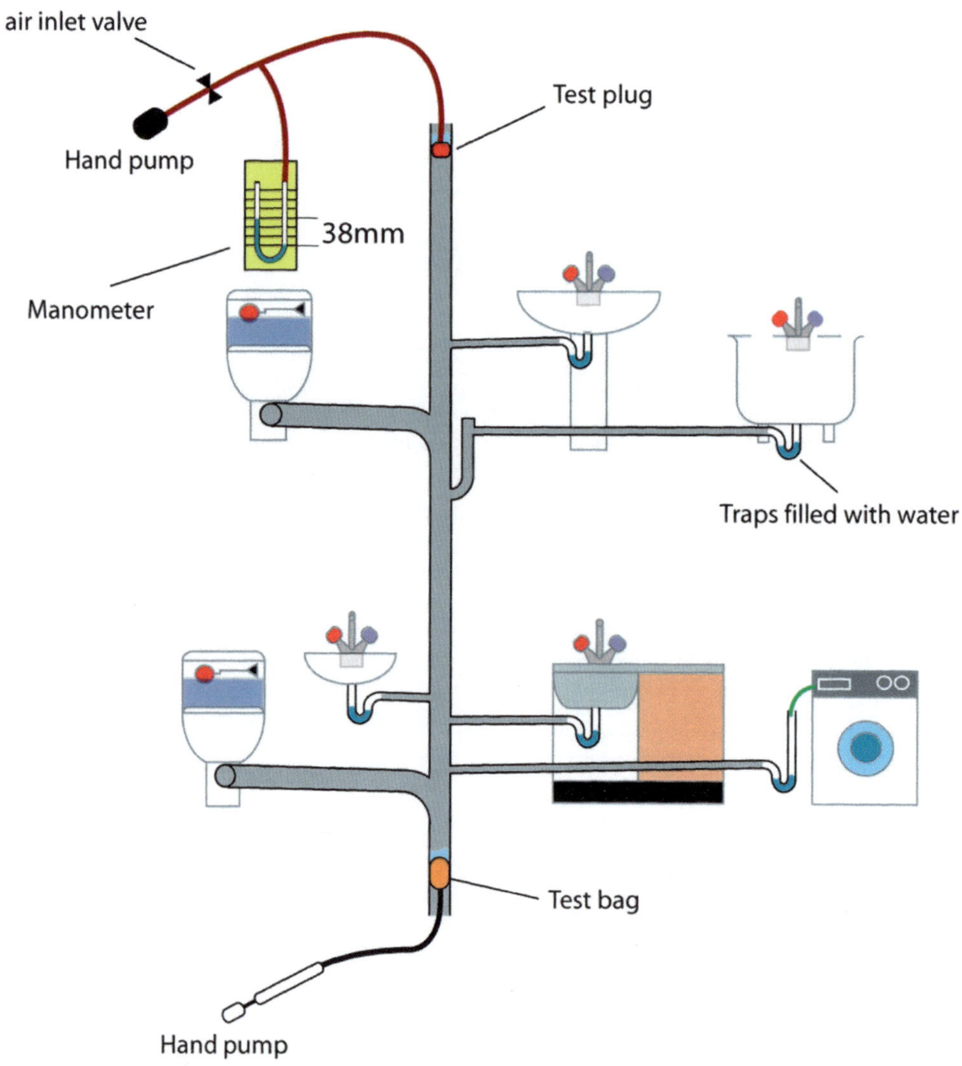

AC4.3 Describe operational checks required during commissioning

Once the soundness test has been completed, then the system can be commissioned. This will involve testing the appliances and checking the depth of the seal in each of the appliances after water has been discharged.

To correctly commission the sanitary pipework, more than one appliance needs to be discharged during the test. The test ensures that the system can perform to at least the minimum recommendations when in use. The test centres on a 'worse case scenario' insomuch that more than one appliance may be discharged at any one time. The number of appliances discharging simultaneously during the test will depend on the size of the stack and the number of appliances installed on it. It works on the principle that not all of the appliances will be in use at once. For domestic properties, 3 appliances are required to discharge simultaneously, with one of those appliances being a WC. The bathroom is the obvious place, since most bathrooms contain at least 3 appliances – 1 bath, 1 WC and 1 Washbasin.

The procedure is as follows:

Stage 1 Insert the waste plugs in the washbasin and the bath and fill. The washbasin can be filled completely to overflow level but the bath only requires about 50mm or so of water.

Stage 2 Ensure that the WC cistern is full and is available to be flushed. It may be an idea to flush the WC first, just to ensure that the WC is flushing correctly. Allow the cistern to fill and the FoV to shut down.

Stage 3 Simultaneously, pull the plug from the bath first (as this contains more water), then pull the plug on the washbasin and flush the WC. Allow the appliances to empty completely and the WC to finish the flush cycle.

Stage 4 Using a dipstick (a slim piece of wood painted matt black) insert the dipstick down the centre of the appliance waste until the bottom of the trap is reached. Withdraw it and measure the 'wetness' that appears on the stick. This is the trap seal depth. Deduct from the mark the size of pipe. For example:

Water mark on the stick 65mm

Size of pipe 32mm

$\mathbf{65 - 32 = 33mm.\ the\ test\ is\ successful}$

For a successful test, it must be no less than 25mm after the appliances have discharged. If the WC water level can be seen visibly, the dipstick need not be used.

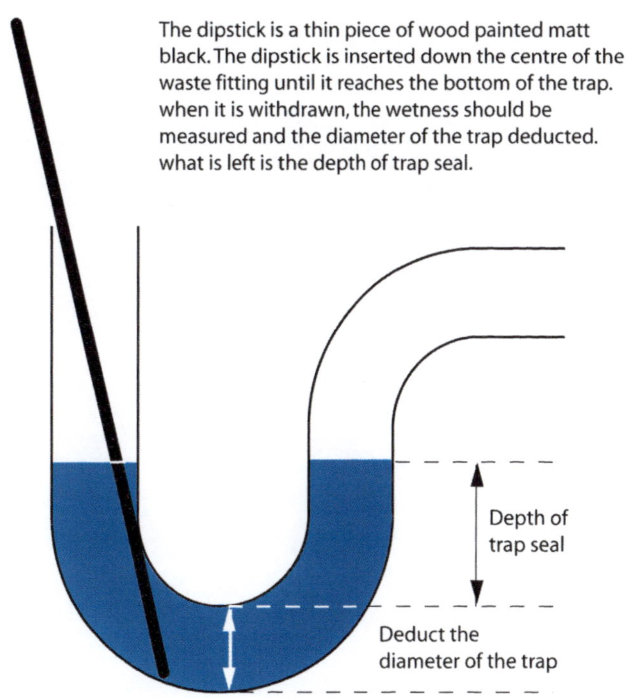

AC4.4 Identify actions that must be taken when commissioning reveals defects

Sometimes, defective installation and components are revealed by the testing process. Any problems that do arise must be dealt with swiftly:

- **Dealing with systems that do not meet correct installation requirements –** poor installation is often the cause of problems with soil/vent and waste pipes. Problems such as too much fall, too little fall or no fall at all can cause major time concerns as the affected work must be taken out and reinstalled correctly. Sagging waste pipes too can cause aesthetic problems as well as performance related issues and blockages.
- **Remedial work associated with defective soil/vent and waste pipework bracketing –** as discussed earlier, insufficient brackets and incorrectly placed pipe clips especially on waste pipes can cause systems to fail in a relatively short space of time. Again, the affected work must be taken out and reinstalled correctly.
- **Remedial work associated with leakage from systems –** most leaks on new installations are caused by three very common issues:
    - Incorrectly jointed solvent weld pipework caused through not following the correct jointing procedures when using solvent cement.
    - Incorrectly jointed push-fit pipework through incorrect preparation of the pipe causing the rubber seal to be pushed out of place.
    - Defective components

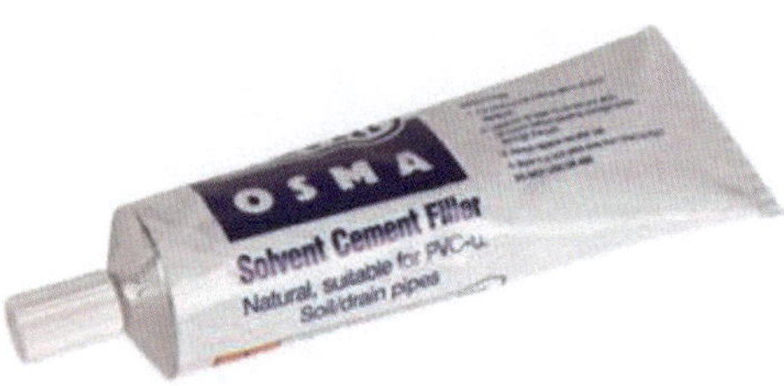

Gap filling cement

Of these three, incorrectly jointed pipework accounts for the vast majority of the problems encountered. This is usually due to incorrect measurements being taken and preparation of the pipe beforehand. It is rare for fittings and components to be supplied defective and when they are, most are due to missing sealing rubbers in the joints. In all cases, the rectification is very simple to complete but may involve the replacement of pipe and fittings, especially when the leakage is from a solvent welded joint. Gap filling cement is available and works in some of the cases described.

AC4.5 Describe the procedure for handing over to the end user

With new build installations of gutters and rainwater pipes, it may be necessary to complete a commissioning report before the installation will be accepted. The report may include the following:

- Name of the company
- Name of the installer
- Name of the commissioning engineer
- The property address
- The type of property i.e. detached, semi-detached etc
- The type of soil pipe and waste pipe used
- The date of commissioning
- The results of the soundness test and commissioning procedure

AC4.6 Carry out soundness testing and commissioning procedures

During your Phase 2 practical assessment, you will be asked to carry out a soundness test to industry standards.

Acknowledgements

BPEC Services Ltd wishes to thank the following companies for allowing the use of their photographs in this publication:

- Rothenberger tools UK
- Aquatech Pressmain Ltd
- Armitage Shanks/Ideal Standard Ltd
- Monument Tools Ltd
- Makita UK
- Pegler Yorkshire Group Ltd
- Wavin UK
- Georg Fischer Ltd
- Screwfix UK
- Waterfit Ltd

Every effort has been made to contact copyright holders. If any have been inadvertently overlooked, we will be pleased to make any necessary arrangements at the first opportunity.

UNIT/LEARNING OUTCOME/ASSESSMENT CRITERIA **PAGE**

bpec LEARNING FOR LIFE

Unit 08 – Rainwater systems – Phase 2

LEARNING FOR LIFE